KW-481-133

01584183

APPLIED ELECTROMAGNETICS AND ELECTROMAGNETIC COMPATIBILITY • *Dipak L. Sengupta and Valdis V. Liepa*

COPLANAR WAVEGUIDE CIRCUITS, COMPONENTS, AND SYSTEMS • *Rainee N. Simons*

ELECTROMAGNETIC FIELDS IN UNCONVENTIONAL MATERIALS AND STRUCTURES • *Onkar N. Singh and Akhlesh Lakhtakia (eds.)*

FUNDAMENTALS OF GLOBAL POSITIONING SYSTEM RECEIVERS: A SOFTWARE APPROACH, Second Edition • *James Bao-yen Tsui*

RF/MICROWAVE INTERACTION WITH BIOLOGICAL TISSUES • *André Vander Vorst, Arye Rosen, and Youji Katsuka*

InP-BASED MATERIALS AND DEVICES: PHYSICS AND TECHNOLOGY • *Osamu Wada and Hideki Hasegawa (eds.)*

COMPACT AND BROADBAND MICROSTRIP ANTENNAS • *Kin-Lu Wong*

DESIGN OF NONPLANAR MICROSTRIP ANTENNAS AND TRANSMISSION LINES • *Kin-Lu Wong*

PLANAR ANTENNAS FOR WIRELESS COMMUNICATIONS • *Kin-Lu Wong*

FREQUENCY SELECTIVE SURFACE AND GRID ARRAY • *T. K. Wu (ed.)*

ACTIVE AND QUASI-OPTICAL ARRAYS FOR SOLID-STATE POWER COMBINING • *Robert A. York and Zoya B. Popović (eds.)*

OPTICAL SIGNAL PROCESSING, COMPUTING AND NEURAL NETWORKS • *Francis T. S. Yu and Suganda Jutamulia*

SiGe, GaAs, AND InP HETEROJUNCTION BIPOLAR TRANSISTORS • *Jiann Yuan*

ELECTRODYNAMICS OF SOLIDS AND MICROWAVE SUPERCONDUCTIVITY • *Shu-Ang Zhou*

RF/Microwave Interaction with Biological Tissues

RF/Microwave Interaction with Biological Tissues

ANDRÉ VANDER VORST
ARYE ROSEN
YOUJI KOTSUKA

IEEE PRESS

A JOHN WILEY & SONS, INC., PUBLICATION

Copyright © 2006 by John Wiley & Sons, Inc. All rights reserved

Published by John Wiley & Sons, Inc., Hoboken, New Jersey
Published simultaneously in Canada

For general information on our other products and services or for technical support, please
contact our Customer Care Department within the United States at 877-762-2974, outside the
United States at 317-572-3993 or fax 317-572-4002.

Wiley also publishes its books in a variety of electronic formats. Some content that appears in
print may not be available in electronic formats. For more information about Wiley products,
visit our web site at www.wiley.com.

Library of Congress Cataloging-in-Publication Data:

Vorst, A. vander (André), 1935–
    RF/microwave interaction with biological tissues / by André Vander Vorst, Arye Rosen,
Youji Kotsuka.
        p. cm.
    Includes bibliographical references and index.
    ISBN-10: 0-471-73277-X
    ISBN-13: 978-0-471-73277-8
        1. Radio waves—Physiological effect. 2. Microwaves—Physiological effect. 3.
Microwave heating—Therapeutic use. I. Rosen, Arye. II. Kotsuka, Y. (Youji), 1941–
III. Title.
    QP82.2.R33V67 2006
    612′.01448—dc22

Printed in the United States of America

10 9 8 7 6 5 4 3 2 1

Contents

Preface

This book addresses the unique needs of today's engineering community with an interest in radio frequency (RF)/microwaves in public health and in medicine as well as those of the medical community. Our decision to embark on this project was made during the time in which the authors served as Members of the IEEE Microwave Theory and Techniques (MTT) Subcommittee on Biological Effects and Medical Applications of Microwaves.

We were even more enthusiastic about writing the book after editing two special issues of *IEEE MTT Transactions* named "Medical Applications and Biological Effects of RF/Microwaves," one in 1996 and the second in 2001. The number of excellent papers accepted for publication in those two special issues required that two volumes be allocated to the subject each year. We then realized that we had the obligation, and the opportunity, to develop a new biomedical course that would encourage further research and produce new researchers in the unique area of RF/microwave interaction with tissue. Thus the book.

The material is divided into six chapters. Chapter 1 summarizes fundamentals in electromagnetics, with the biological mechanisms in mind. Special attention is paid to penetration in biological tissues and skin effect, relaxation effects in materials and the Cole–Cole display, the near field of an antenna, blackbody radiation with the various associated laws, and microwave measurements.

Chapter 2 discusses RF/microwave interaction mechanisms in biological materials. The word *interaction* stresses the fact that end results not only depend on the action of the field but also are influenced by the reaction of the living system. Cells and nerves are described, as well as tissue characterization, in particular dielectric, and measurements in tissues and biological liquids are included. A section is devoted to the fundamentals of thermodynamics, including a discussion on energy and entropy.

Biological effects are the subject of Chapter 3. Dosimetric studies attempt to quantify the interactions of RF fields with biological tissues and bodies. A variety of effects are described and discussed; they include those on the nervous system, the brain and spinal cord, the blood–brain barrier, cells and membranes, effects at the molecular level, influence of drugs, and effects due to extremely low frequency components of signal modulation. Thermal considerations, related to absorption, are the subject of a significant part of this chapter. The possibility of nonthermal effects is also discussed. This leads to a discussion on radiation hazards and exposure standards.

Chapter 4 is devoted to thermal therapy. Thermotherapy has been used as medical treatment in, for example, rheumatism and muscle diseases. In this chapter, the reader will find a description of applicators and an extensive discussion on the foundation of dielectric heating and inductive heating as well as a variety of technological information. Hyperthermia is also discussed as a noninvasive method, and practical thermometry methods are described.

Recently, electromagnetic (EM) environments have become very complex because of the wide and rapid spread of a number of electric or electronic devices, including recent progress and the increase in use in the area of cellular telephony. As a countermeasure, wave absorbers are being used for protecting biological and medical environments, and knowledge about these absorbers has become important. In Chapter 5, we investigate materials for EM wave absorbers, both from a theoretical and an application point of view. Special attention is paid to ferrite absorbers, for which it has long been a challenge to develop an EM wave absorber at the desired matching frequency. The chapter ends with the description of a method for improving the distribution of RF fields in a small room.

Chapter 6 begins with some of the fundamental aspects of major components used in RF/microwave delivery systems for therapeutic applications. The authors have chosen to detail the research done on the subject of cardiac ablation. The chapter also covers new ideas and research done on the use of RF/microwaves in the development of future measurement techniques, such as blood perfusion, for example, and the use of microwaves in therapeutic applications. New inventions would reduce further the need for surgical or invasive procedures, substituting noninvasive and minimally invasive techniques.

The authors acknowledge the enthusiastic help received during the preparation of the manuscript.

Andre Vander Vorst acknowledges the numerous conversations he has had over the years with two past doctoral students, D. Vanhoenacker-Janvier on a variety of microwave issues, more specifically microwave propagation, and Benoît Stockbroeckx about exposure measurements and radiation hazards. The stimulating environment created by some Ph.D. students is also acknowledged, especially Jian Teng, who participated in the investigation of the effect of microwaves on the nervous system of rabbits, and Dirk Adang, who in 2004 started a one-year epidemiological study on rats exposed to microwaves. This author also benefited from conversations with Jacques Vanderstraeten about the effect of microwaves on DNA and with his colleagues of the Belgian Health Council.

Arye Rosen acknowledges the enthusiastic help that he received from then Ph.D. student, now colleague at Drexel University, M. Tofighi, who helped in the preparation of some of the problems included in the book and who, together with Yifei Li, reviewed part of the book. He also wishes to recognize Renee Cohen and Melany Smith, who helped to organize and type part of the material, and Rong Gu, who helped with some of the drawings. He acknowledges that selected, relevant parts, included with permission, have been published previously in *IEEE Transactions* and include contributions by Stuart Edwards, Paul Walinsky, and Arnold Greenspon. He also thanks his son, Harel Rosen, and his wife, Danielle Rosen, both of whose assistance is sincerely appreciated.

Youji Kotsuka acknowledges Shigeyoshi Matsumae, former President of Tokai University, for suggesting valuable topics for thermal therapy. He also expresses his sincere gratitude to Kunihiro Suetake, Professor Emeritus of Tokyo Institute of Technology, for his teaching on pioneering technologies concerning EM-wave absorbers. He thanks Risaburou Satoh and Tasuku Takagi, both Professors Emeritus of Tohoku University, for their encouragement of EMC research, including biological effects.

ANDRÉ VANDER VORST
ARYE ROSEN
YOUJI KOTSUKA

Introduction

Rapid technological advances in electronics, electro-optics, and computer science have profoundly affected our everyday lives. They have also set the stage for an unprecedented drive toward the improvement of existing medical devices and for the development of new ones. In particular, the advances in radio-frequency (RF)/microwave technology and computation techniques, among others, have paved the way for exciting new therapeutic and diagnostic methods. Frequencies from RF as low as 400 kHz through microwave frequencies as high as 10 GHz are presently being investigated for therapeutic applications in areas such as cardiology, urology, surgery, ophthalmology, cancer therapy, and others and for diagnostic applications in areas such as cancer detection, organ imaging, and more.

At the same time, safety concerns regarding the biological effects of electromagnetic (EM) radiation have been raised, including those at a low level of exposure. A variety of waves and signals have to be considered, from pure or almost pure sine waves to digital signals such as those of digital radio, digital television, and digital mobile phone systems.

In this book, we restrict ourselves to a discussion of the applications and effects of RF and microwave fields. These cover a frequency range from about 100 kHz to 10 GHz and above. This choice seems appropriate, although effects at RF and microwaves, respectively, are of a different nature. It excludes low-frequency (LF) and extremely low frequency (ELF) effects. It also excludes ultraviolet (UV) and X-rays, termed "ionizing radiation" because of their capability of disrupting molecular or atom structures. The RF/microwave frequency range covered here may be called "nonionizing."

For many years, hyperthermia and the related radiometry have been a major subject of interest in investigating biological effects of microwaves.

RF/Microwave Interaction with Biological Tissues, By André Vander Vorst, Arye Rosen, and Youji Kotsuka

Numerous publications are available on the subject, with [1–5] as examples. More recently, however, other subjects have received as much attention, in particular EM energy absorption in human subjects, especially in the human head and neck [6], interaction of microwaves with the nervous system [7], influence of the fields of mobilophony on membrane channels [8], and molecular effects [9].

An increasing amount of evidence, derived from in vitro and in vivo studies, indicates that microwaves affect living systems directly. With microwave absorption, however, when experimenting with animals or humans, there are ambiguities concerning the relative contributions to physiological alterations of indirect thermal effects, microwave-specific thermal effects, and possibly direct nonthermal interactions. In spite of these ambiguities, results of in vivo experiments conducted at low specific absorption rates indicate direct microwave effects. Furthermore, unambiguous evidence of direct effects of microwave fields has been provided by the results of in vitro studies. These studies were conducted under conditions of precise and accurate temperature control, revealing direct effects at various frequencies and intensities on a number of cellular endpoints, including calcium binding, proliferation, ligand–receptor-mediated events, and alteration in membrane channels. Interactions occurring at the microscopic level are related to the dielectric properties of biomacromolecules and large molecular units such as enzyme complexes, cell-membrane receptors, or ion channels.

The classical book by Michaelson and Lin [9], which reviews the biological effects of RF radiation, is still recommended to those who want to acquire a solid knowledge of the field. A more recent book by Thuery [10] reviews industrial, scientific, and medical applications of microwaves and lists a number of sources that offer basic information on the interaction between microwave fields and the nervous system. Another handbook by Polk and Postow, published in 1996 [11], reviews the biological effects of EM fields. Specific European research was reviewed in 1993 [12].

In this book, the reader will not find a review in the usual sense of the word, but rather will find an ensemble of aspects, which, in the authors' opinion, are highlights on specific subjects. Although a number of uncertainties about the biological effects of microwaves still persist, there exists today a substantial database on the subject of biological effects of thermal nature, and research is underway on new issues. As a consequence, there are now many applications at RF/microwaves of heating through absorption. Chapter 6 reviews most of those applications in the medical field. It must be emphasized that the application of RF/microwaves in medicine, essentially in cardiology, urology, and surgery, has been made possible due to the enormous amount of research carried out through many years and devoted to the understanding and modeling of microwave absorption in living materials [13].

Investigation of the interaction of EM fields with biological tissues requires a good physical insight and a mathematical understanding of what fields are. A field is associated with a physical phenomenon present in a given region of space. As an example, the temperature in a room is a field of temperature composed of the values of temperature in a number of points in the room. Chapter 1 summarizes fundamentals in electromagnetics, while bearing in mind the biological mechanisms. It emphasizes the fact that, at RF and microwave frequencies, the electric and magnetic fields are simultaneously present: If there is an electric field, then there is a coupled magnetic field and vice versa. If one is known, the other can be calculated.

Special attention is paid to some aspects that are not often handled in a book on electromagnetics but that are of importance as they pertain to biological effects, such as penetration in biological tissues and skin effect, relaxation effects in materials and the Cole–Cole display, the near field of an antenna, blackbody radiation with the various associated laws, and microwave measurements.

Chapter 2 is devoted to RF/microwave interaction mechanisms in biological materials. This is important basic material because the effects of this interaction are the result of the penetration of the EM waves into the living system and their propagation into it, their primary interaction with biological tissues, and the possible secondary effects induced by the primary interaction. The word *interaction* is important. It stresses the fact that end results not only depend on the action of the field but also are influenced by the reaction of the living system. Living systems have a large capacity for compensating for the effects induced by external influences, in particular EM sources. There can, however, be physiological compensation, meaning that the strain imposed by external factors is fully compensated and the organism is able to perform normally; or pathological compensation is possible, meaning that the imposed strain leads to the appearance of disturbances within the functions of the organism, and even structural alterations may result. The borderline between these two types of compensation is obviously not always easy to determine.

Cells and nerves are described as well as tissue characterization, in particular dielectric characterization, and measurements in tissues and biological liquids are included. A section is devoted to the fundamentals of thermodynamics and a discussion on energy and entropy.

Biological effects are treated in Chapter 3. The fact is stressed that only the fields inside a material can influence it, so that only the fields inside tissues and biological bodies can possibly interact with these: The biological effects of RF/microwaves do not depend solely on the external power density; they depend on the dielectric field inside the tissue or the body. Hence, the internal fields have to be determined for any meaningful and general quantification of biological data obtained experimentally. Dosimetric studies that attempt to quantify the interactions of RF fields with biological tissues and bodies are reported.

A variety of effects are described and discussed, such as those affecting the nervous system, the brain, and spinal cord. Special attention is given to effects on cells and membranes, effects at the molecular level, and effects due to ELF components of signal modulation as well as effects on the blood–brain barrier with report on the results of recent studies and the influence of drugs.

Thermal considerations are the subject of a significant part of this chapter. They are related to absorption, and the specific absorption rate is defined and discussed. The possibility of nonthermal effects is also discussed, more specifically as microthermal effects—where microwaves act as a trigger—and isothermal effects—where thermodynamics are a necessary tool. In the latter case, entropy is considered simultaneously with energy. This leads quite naturally to a discussion of radiation hazards and exposure standards, which the reader will find at the end of the chapter.

Chapter 4 is devoted to thermal therapy. The use of heat for medical treatment is not new: Heat has been used to treat lesions for more than 2000 years. Thermotherapy has been used as medical treatment for rheumatism, muscle diseases, and so on. In this chapter the reader will find the description of applicators and an extensive discussion on the foundation of both dielectric heating and inductive heating.

Hyperthermia is discussed in the second part of the chapter. Some reasons for which hyperthermia is effective for treating tumors are described in simple terms, based on biological results on culture cells and tumors of laboratory animals. It is shown that the development of accurate thermometry technology is an important challenge. Temperature measurement methods are generally classified into invasive methods and noninvasive methods, and practical thermometry methods are described.

Recently, EM environments have become very complex because of the rapid spread and wide use of a number of electric and electronic devices, including recent widespread use of cellular telephones. These devices have created an increasing number of EM wave interference problems. Also, there is a growing concern in the population about the possible biological effects induced by such radiation. As a countermeasure, wave absorbers are being used for protecting biological and medical environments, and familiarity with these absorbers has become important. In Chapter 5, materials ideal for use as EM wave absorbers are investigated both from a theoretical and an application point of view.

The absorbers are classified according to constituent material, structural shape, frequency characteristics, and application, and the fundamental theory is established for single- as well as multilayer wave absorbers. Special attention is paid to the ferrite absorbers, for which it has long been a challenge to develop an EM wave absorber at the desired matching frequency. A new method for the effective use of ferrite for microwave absorbers has therefore been proposed where small holes are punched through a rubber ferrite [14]. It is described in detail in this chapter. Such a method is termed an equivalent transformation method of material constant. The chapter ends with a method for improving the distribution of RF fields in a small room.

We have chosen to start Chapter 6 with some of the fundamental aspects of major components used in RF/microwave delivery systems for therapeutic applications. The components treated in detail are transmission lines, such as coaxial cable terminated in a simple antenna, and circular waveguide—both currently used in cardiology, urology, endocrinology, and obstetrics, to name just a few.

Although some of the applications are discussed in this chapter, the authors have chosen to detail the research done on the subject of cardiac ablation. The use of RF/microwaves has become the method of choice in the treatment of supraventricular tachycardia and benign prostatic hyperplasia worldwide. The engineering community should be proud of this accomplishment.

The chapter also covers new ideas and research done on the use of RF/microwaves in the development of future measurement techniques, such as blood perfusion.

In addition, the authors discuss a number of subjects for possible research in the use of microwaves in therapeutic medicine. Hopefully, with the understanding of the fundamentals of RF/microwave interaction with tissue and familiarity with the techniques already in use, these new ideas will encourage the student to become an inventor. New inventions would then reduce further the need for surgical or invasive procedures, substituting noninvasive and minimally invasive techniques.

The authors have organized the material as a textbook, and the reader will find problems at the end of each chapter.

REFERENCES

[1] L. Dubois, J. P. Sozanski, V. Tessier, et al., "Temperature control and thermal dosimetry by microwave radiometry in hyperthermia," in A. Rosen and A. Vander Vorst (Eds.), Special Issue on Medical Application and Biological Effects of RF/Microwaves, *IEEE Trans. Microwave Theory Tech.*, Vol. 44, No. 10, pp. 1755–1761, Oct. 1996.

[2] D. Sullivan, "Three-dimensional computer simulation in deep regional hyperthermia using the finite-difference time-domain method," *IEEE Trans. Microwave Theory Tech.*, Vol. 38, No. 2, pp. 204–211, Feb. 1990.

[3] C. Rappaport, F. Morgenthaler, "Optimal source distribution for hyperthermia at the center of a sphere of muscle tissue," *IEEE Trans. Microwave Theory Tech.*, Vol. 35, No. 12, pp. 1322–1328, Dec. 1987.

[4] C. Rappaport, J. Pereira, "Optimal microwave source distributions for heating off-centers tumors in spheres of high water content tissue," *IEEE Trans. Microwave Theory Tech.*, Vol. 40, No. 10, pp. 1979–1982, Oct. 1992.

[5] D. Dunn, C. Rappaport, A. Terzuali, "FDTD verification of deep-set brain tumor hyperthermia using a spherical microwave source distribution," in A. Rosen and A. Vander Vorst (Eds.), Special Issue on Medical Applications and Biological Effects of RF/Microwaves, *IEEE Trans. Microwave Theory Tech.*, Vol. 44, No. 10, pp. 1769–1777, Oct. 1996.

[6] M. Okoniewski, M. A. Stuchly, "A study of the handset antenna and human body interaction," in A. Rosen and A. Vander Vorst (Eds.), Special Issue on Medical

Application and Biological Effects of RF/Microwaves, *IEEE Trans. Microwave Theory Tech.*, Vol. 44, No. 10, pp. 1855–1864, Oct. 1996.

[7] A. Vander Vorst, F. Duhamel, "990–1995 advances in investigating the interaction of microwave fields with the nervous system," in A. Rosen and A. Vander Vorst (Eds.), Special Issue on Medical Application and Biological Effects of RF/Microwaves, *IEEE Trans. Microwave Theory Tech.*, Vol. 44, No. 10, pp. 1898–1909, Oct. 1996.

[8] F. Apollonio, G. D'Inzeo, L. Tarricone, "Theoretical analysis of voltage-gated membrane channels under GSM and DECT exposure," in *IEEE MTT-S Microwave Int. Symp. Dig.*, Denver, 1997, pp. 103–106.

[9] S. Michaelson, J. C. Lin, *Biological Effects and Health Implications of Radiofrequency Radiation*, New York: Plenum, 1987.

[10] J. Thuery, *Microwaves—Industrial, Scientific and Medical Applications*, Boston: Artech House, 1992.

[11] C. Polk, E. Postow, *Handbook of Biological Effects of Electromagnetic Fields*, New York: CRC, 1996.

[12] A. Vander Vorst, "Microwave bioelectromagnetics in Europe," in *IEEE MTT-S Microwave Int. Symp. Dig.*, Atlanta, 1993, pp. 1137–1140.

[13] A. Rosen, H. D. Rosen, *New Frontiers in Medical Device Technology*, New York: Wiley, 1995.

[14] M. Amano, Y. Kotsuka, "A method of effective use of ferrite for microwave absorber," *IEEE Trans. Microwave Theory Tech.*, Vol. 51, No. 1, pp. 238–245, Jan. 2003.

CHAPTER ONE

Fundamentals of Electromagnetics

1.1 RF AND MICROWAVE FREQUENCY RANGES

The rapid technological advances in electronics, electro-optics, and computer science have profoundly affected our everyday lives. They have also set the stage for an unprecedented drive toward the improvement of existing medical devices and the development of new ones. In particular, the advances in radio-frequency (RF)/microwave technology and computation techniques, among others, have paved the way for exciting new therapeutic and diagnostic methods. Frequencies, from RF as low as 400 kHz through microwave frequencies as high as 10 GHz, are presently being investigated for therapeutic applications in areas such as cardiology, urology, surgery, ophthalmology, cancer therapy, and others and for diagnostic applications in cancer detection, organ imaging, and more.

At the same time, safety concerns regarding the biological effects of electromagnetic (EM) radiation have been raised, in particular at a low level of exposure. A variety of waves and signals have to be considered, from pure or almost pure sine waves to digital signals, such as in digital radio, digital television, and digital mobile phone systems. The field has become rather sophisticated, and establishing safety recommendations or rules and making adequate measurements require quite an expertise.

In this book, we limit ourselves to the effects and applications of RF and microwave fields. This covers a frequency range from about 100 kHz to 10 GHz and above. This choice is appropriate, although effects at RF and microwaves,

RF/Microwave Interaction with Biological Tissues, By André Vander Vorst, Arye Rosen, and Youji Kotsuka

respectively, are of a different nature. It excludes low-frequency (LF) and extremely low frequency (ELF) effects, which do not involve any radiation. It also excludes ultraviolet (UV) and X-rays, called *ionizing* because they can disrupt molecular or atom structures. The RF/microwave frequency range covered here may be called *nonionizing*.

Radiation is a phenomenon characterizing the RF/microwave range. It is well known that structures radiate poorly when they are small with respect to the wavelength. For example, the wavelengths at the power distribution frequencies of 50 and 60 Hz are 6.000 and 5.000 km, respectively, which are enormous with respect to the objects we use in our day-to-day life. In fact, to radiate efficiently, a structure has to be large enough with respect to the wavelength λ. The concepts of radiation, antennas, far field, and near field have to be investigated.

On the other hand, at RF and microwave frequencies, the electric (E) and magnetic (H) fields are simultaneously present: if there is an electric field, then there is a coupled magnetic field and vice versa. If one is known, the other can be calculated: They are linked together by the well-known Maxwell's equations. Later in this book, we shall be able to separate some biological effects due to one field from some due to the other field. We need, however, to remember that we are considering the general case, which is that of the complete field, called the EM field. Hence, we are not considering direct-current (DC) and LF electric or magnetic fields into tissue.

Because we limit ourselves to the RF/microwave range, we may refer to our subject of interaction of electric and magnetic fields with organic matter as *biological effects of nonionizing radiation*. It should be well noticed that, by specifically considering a frequency range, we decide to describe the phenomena in what is called the *frequency domain*, that is, when the materials and systems of interest are submitted to a source of sinusoidal fields. To investigate properties over a frequency range, wide or narrow, we need to change the frequency of the source. The frequency domain is not "physical" because a sinusoidal source is not physical: It started to exist an infinite amount of time ago and it lasts forever. Furthermore, the general description in the frequency domain implies complex quantities, with a real and an imaginary part, respectively, which are not physical either. The frequency-domain description is, however, extremely useful because many sources are (almost) monochromatic.

To investigate the actual effect of physical sources, however, one has to operate in what is called the *time domain*, where the phenomena are described as a function of time and hence they are real and physically measurable. Operating in the time domain may be rather difficult with respect to the frequency domain. The interaction of RF/microwave fields with biological tissues is investigated mostly in the frequency domain, with sources considered as sinusoidal. Today numerical signals, such as for telephony, television, and frequency-modulated (FM) radio, may, however, necessitate time-domain analyses and measurements.

There is an interesting feature to note about microwaves: They cover, indeed, the frequency range where the wavelength is of the order of the size of objects of common use, that is, meter, decimeter, centimeter, and millimeter, depending of course on the material in which it is measured. One may, hence, wonder whether such wavelengths can excite resonance in biological tissues and systems. We shall come back later to this question.

1.2   FIELDS

Investigating the interaction of EM fields with biological tissues requires a good physical insight and mathematical understanding of what are *fields*. A field is associated with a physical phenomenon present in a given region of space. As an example, the temperature in a room is a field of temperature, composed of the values of temperature in a number of points of the room. One may say the same about the temperature *distribution* inside a human body, for instance. We do not see the field, but it exists, and we can for instance visualize constant-temperature or isothermal surfaces.

There are fields of different nature. First, fields may be either *static* or *time dependent*. Considering, for instance, the temperature field just described, the room may indeed be heated or cooled, which makes the temperature field time dependent. The human body may also be submitted to a variety of external sources or internal reasons which affect the temperature distribution inside the body. In this case, the isothermal surfaces will change their shapes as a function of time.

Second, the nature of the field may be such that one parameter only, such as magnitude, is associated with it. Then, the field is defined as *scalar*. The temperature field, for instance, inside a room or a human body, is a scalar field. One realizes that plotting a field may require skill, and also memory space, if the structure is described in detail or if the observer requires a detailed description of the field in space. This is true even in the simplest cases, when the field is scalar and static.

On the other hand, in a *vector* field, a vector represents both the magnitude and the direction of the physical quantity of interest at points in space, and this vector field may also be static or time dependent. When plotting a static scalar field, that is, one quantity, in points of space already requires some visualization effort. On the other hand, plotting a time-dependent vector field, that is, three time-varying quantities, in points of space obviously requires much more attention. A vector field is described by a set of *direction lines*, also known as *stream lines* or *flux lines*. The direction line is a curve constructed so that the field is tangential to the curve in all points of the curve.

1.3 ELECTROMAGNETICS

1.3.1 Electric Field and Flux Density

The electric field E is derived from Coulomb's law, which expresses the interaction between two electric point charges. Experimentally, it has been shown that

1. Two charges of opposite polarity attract each other, while they repel when they have the same polarity, and hence a charge creates a field of force.
2. The force is proportional to the product of charges.
3. The force acts along the line joining the charges and hence the force field is vectorial.
4. The force is higher when the charges are closer.
5. The force depends upon the electric properties of the medium in which the charges are placed.

The first observations showed that the force is about proportional to the square of the distance between them. In 1936, the difference between the measured value and the value 2 for the exponent was of the order of 2×10^{-9} [1]. It is admitted as a postulate that the exponent of the distance in the law expressing the force between the two charges is exactly equal to 2. It has been demonstrated that this postulate is necessary for deriving Maxwell's equations from a relativistic transformation of Coulomb's law under the assumption that the speed of light is a constant with respect to the observer [2, 3]. Hence, Coulomb's law is

$$\bar{f} = \frac{q_1 q_2}{4\pi\varepsilon_0 r^2} \bar{a}_r \qquad \text{N} \tag{1.1}$$

where $\bar{f}$ is the force; q_1 and q_2 the value of the charges, expressed in *coulombs* (C), including their polarity; the factor 4π is due to the use of the rationalized meter-kilogram-second (MKS) system, exhibiting a factor 4π when the symmetry is spherical; and ε_0 measures the influence of the medium containing the charges, equal to approximately $10^{-9}/36\pi$ *farads per meter* (F m^{-1}) in vacuum.

If a test charge Δq is placed in the field of force created by a charge q, it undergoes a force

$$\bar{f} = \frac{q_1 (\Delta q)}{4\pi\varepsilon_0 r^2} \bar{a}_r \qquad \text{N} \tag{1.2}$$

The test charge Δq is small enough to avoid any perturbation of the field of force created by q. The intensity of the *electric field*, in *volts per meter* (V m^{-1}), is then defined as the ratio of the force exerted onto q by the charge Δq, which for the electric field created by a charge q in vacuum yields

$$\overline{E} = \frac{q}{4\pi\varepsilon_0 r^2}\overline{a}_r \qquad \text{V m}^{-1} \tag{1.3}$$

Ideally, the electric field is defined in the limit that Δq tends to zero. It is a vector field, radial in the case of a point charge. It comes out of a positive charge and points toward a negative charge. The lines of electric field are tangential to the electric field in every point. Equation (1.3) is linear with respect to the charge. Hence, when several charges are present, one may vectorially add up the electric fields due to each charge, which yields what is often called the generalized Coulomb's law.

The electric charge may appear in four different forms:

1. It can be punctual, as in Eqn. (1.2). It is then usually denoted q and measured in *coulombs*.
2. It can be distributed in space along a line (material of not). It is then usually denoted ρ_l and measured in *coulombs per meter* (C m^{-1}).
3. It can be distributed in space over a surface (material of not). It is then usually denoted ρ_s and measured in *coulombs per square meter* (C m^{-2}).
4. It can also be distributed in a volume. It is then usually denoted ρ and measured in *coulombs per cubic meter* (C m^{-3}).

When a material is submitted to an applied electric field, it becomes polarized, the amount of which is called the *polarization vector* $\overline{P}$. This is due to the fact that, in many circumstances, electric dipoles are created or transformed into the material, which corresponds to what is called the dielectric properties of the material. Hence, the polarization is the *electric dipole moment per unit volume*, in *coulombs per square meter*.

The total electric field in a dielectric material is the sum of the applied electric field and of an induced electric field, resulting from the polarization of the material. As a simple example, a *perfect electric conductor* is defined as an equipotential material. If the points in the material are at the same electric potential, then the electric field must be zero and there can be no electric charges in the material. When a perfect electric conductor is submitted to an applied field, this applied field exists in all points of the material. To have a vanishing *total electric field*, the material must develop an induced electric field such that the sum of the applied field and the induced field vanishes in all points of the material. The induced field is calculated by taking into account the geometry of the problem and the boundary conditions, which can of course be complicated. As another example, a human body placed in an applied electric field develops an induced electric field such that the sum of the applied field and the induced field satisfies the boundary conditions at the surface of the body. The total field in the body is the sum of the applied field and of the induced field.

A new vector field $\overline{D}$ is then defined, known as the *displacement flux density* or the *electric flux density*, in *coulombs per square meter* similarly to the polarization, defined as

$$\overline{D} = \varepsilon_0 \overline{E} + \overline{P} \qquad \text{C m}^{-2} \tag{1.4}$$

This definition is totally general, applying to all materials, in particular to all biological materials. It indeed holds for materials in which [3]:

1. The polarization vector has not the same direction as the vector electric field, in which case the material is *anisotropic*.
2. The polarization can be delayed with respect to the variation of electric field, as is the case in *lossy materials*. All physical materials are lossy, so this is a universal property. It is neglected, however, when the losses are reasonably small, which is not always the case in biological tissues.
3. The polarization is not proportional to the electric field, in which case the material is *nonlinear*.

In all other cases, that is, when the material is *isotropic*, *lossless*, and *linear*, the definition (1.4) can be written

$$\overline{D} = \varepsilon_0 \varepsilon_r \overline{E} = \varepsilon \overline{E} \tag{1.5}$$

which combines the applied and induced fields, hence the external source field and the induced polarization, into the definition of ε (F m^{-1}), permittivity of the material, product of the permittivity of vacuum ε_0 (F m^{-1}) and the relative permittivity ε_r (dimensionless) of the material. The *electric susceptibility* χ_e is related to the relative permittivity by the expression

$$\varepsilon_r = 1 + \chi_e \tag{1.6}$$

It should be stressed that the use of permittivity, relative permittivity, and susceptibility is limited to isotropy, losslessness, and linearity, which is far from being always the case, in particular in biological tissues.

Dielectric polarization is a rather complicated phenomenon [4]. It may be due to a variety of mechanisms, which can be summarized here only briefly. The simplest materials are gases, especially when they are rarefied. The simplest variety is formed of *nonpolar gases*, in which the molecules have no electric dipole at rest. When an electric field is applied, an electric dipole is induced. This is a simple case for which a simple model can be used for correctly calculating the polarization. The next category is that of *polar gases*, in which an electric dipole does exist at rest. When an external electric field is applied, the dipole orientation is modified; it essentially rotates. For such a polar rarefied gas, which is still a very simple case, the relationship between polarization and applied field is already found to be nonlinear. When the density increases, modeling becomes much more difficult, and classical physics yields wrong models for compact gases, liquids, and of course solids, in particular conductors, semiconductors, and superconductors. Classical physics almost completely fails when trying to establish quantitative models. It can however yield some very illuminating insight on the phenomena involved with the dielectric character of materials, in particular about the influence of frequency, as will be shown now.

The *dipolar polarization*, resulting from the alignment of the molecule dipolar moment due to an applied field, is a rather slow phenomenon. It is correctly described by a first-order equation, called after Debye [5]: The dipolar polarization reaches its saturation value only after some time, measured by a time constant called *relaxation time τ*. The ability to polarize, called the *polarizability*, is measured by the parameter

$$\alpha_d = \frac{\alpha_0}{1 + j\omega\tau} + C \qquad (1.7)$$

where constant C takes into account the nonzero value of the polarizability at infinite frequency. The relative permittivity related to this phenomenon is

$$\varepsilon_r = \varepsilon_r' - j\varepsilon_r'' \qquad (1.8)$$

where N is the number of dipoles per unit volume. It should be observed that the permittivity is a complex quantity with real and imaginary parts. If ε_{r0} and $\varepsilon_{r\infty}$ are the values of the real part of the relative permittivity at frequencies zero and infinity, respectively, one can easily verify that the equations can be written as

$$\varepsilon_r' = \frac{\varepsilon_{r0} - \varepsilon_{r\infty}}{1 + \omega^2\tau^2} + \varepsilon_{r\infty} \qquad \varepsilon_r'' = \frac{(\varepsilon_{r0} - \varepsilon_{r\infty})\omega\tau}{1 + \omega^2\tau^2} \qquad (1.9)$$

The parameter $\varepsilon_{r\infty}$ is in most cases the value at optical frequencies. It is often called the *optical dielectric constant.*

Dipolar polarization is dominant in the case of water, much present on earth and an essential element of living systems. The relative permittivity of water at 0°C is

$$\varepsilon_r = 5 + \frac{83}{1 + j0.113f\,(\text{GHz})} \qquad (1.10)$$

with $1/\tau = 8.84\,\text{GHz}$. The real part of the relative permittivity is usually called the *dielectric constant*, while the imaginary part is a measure of the dielectric losses. These are often expressed also as the tangent of the *loss angle*:

$$\tan\,\delta_e = \frac{\varepsilon''}{\varepsilon'} \qquad (1.11)$$

Table 1.1 shows values of relaxation times for several materials. A high value of the relaxation time is indicative of a good insulator, while small values are typical of good conductors.

TABLE 1.1   Relaxation Time of Some Materials

Material	Relaxation Time
Copper	1.51^{-19} s
Silver	1.31^{-19} s
Sea water	2.01^{-10} s
Distilled water	10^{-6} days
Quartz	10 days

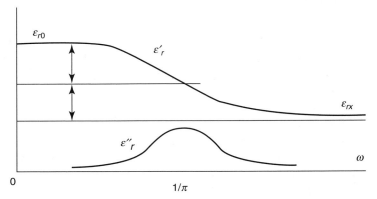

FIGURE 1.1   Relaxation effect.

Figure 1.1 represents the typical evolution of the real and imaginary parts of a relative permittivity satisfying Debye's law, where ε_{r0} and $\varepsilon_{r\infty}$ are the values at frequencies zero and infinity, respectively. It shows the general behavior of the real and imaginary parts of permittivity: The imaginary part is nonzero only when the real part varies as a function of frequency. Furthermore, each part can be calculated from the variation of the other part over the whole frequency range, as indicated by the Kramer and Kronig formulas [6]:

$$\varepsilon'(\omega) = \varepsilon_0 + \frac{2}{\pi}\int_0^\infty \frac{x\varepsilon''(x)}{x^2 - \omega^2}dx$$

$$\varepsilon''(\omega) = -\frac{2\omega}{\pi}\int_0^\infty \frac{\varepsilon'(x) - \varepsilon_\infty}{x^2 - \omega^2}dx \qquad (1.12)$$

It can easily be seen that $\varepsilon'' = 0$ if ε' is frequency independent. The variable of integration x is real. The principal parts of the integrals are to be taken in the event of singularities of the integrands. The second equation implies that $\varepsilon''(\infty) = 0$. The evaluation of Eqn. (1.12) is laborious if the complex $\varepsilon(\omega)$ is not a convenient analytical function. It is interesting to observe that the formulas are similar to those relating the real and imaginary parts of impedance in general circuit theory [7].

The structure of Maxwell's equations shows that permittivity and conductivity are related parameters. To keep it simple, one may say that they express the link between current density and electric field: When both parameters are real, the permittivity is the imaginary part while the conductivity is the real part of this relationship. This can be written as

$$\overline{J} = (j\omega\varepsilon + \sigma)\overline{E} \qquad \text{A m}^{-2} \tag{1.13}$$

When the permittivity is written as complex, there is an ambiguity. There are, however, too many parameters, as can be seen in the expression

$$\begin{aligned}\overline{J} &= j\omega(\varepsilon' - \varepsilon'')\overline{E} + \sigma\overline{E} \\ &= j\omega\varepsilon'\overline{E} + (\sigma + \omega\varepsilon'')\overline{E}\end{aligned} \tag{1.14}$$

from which it appears that the real part of the relation between the current and the electric field can be written either as an effective conductivity equal to

$$\sigma_{\text{eff}} = \sigma' + \omega\varepsilon'' \qquad \text{S m}^{-1} \tag{1.15}$$

or as an effective imaginary part of permittivity equal to

$$\sigma''_{\text{eff}} = \frac{\sigma}{\omega} \qquad \text{F m}^{-1} \tag{1.16}$$

It should be observed that these two expressions are for the conductivity and permittivity, respectively, and not just the relative ones. Both expressions are correct and in use. Generally, however, the effective conductivity is used when characterizing a lossy conductor, while the effective imaginary part of the permittivity is used when characterizing a lossy dielectric. At some frequency, the two terms are equal, in particular in biological media. As an example, the frequency at which the two terms σ' and $\omega\varepsilon''$ are equal is in the optical range for copper, about 1 GHz for sea water, 100 MHz for silicon, and 1 MHz for a humid soil. Although both expressions are correct, one needs to be careful in such a case when interpreting the results of an investigation.

One more comment, however, is necessary. It has just been said, and it is proven by Kramer and Kronig's formulas (1.12), that, if the permittivity varies as a function of frequency, it must be a complex function. In fact, this is true of all three electromagnetic parameters: permittivity, conductivity, and permeability. Hence, by writing the conductivity as a real parameter in Eqns. (1.13)–(1.16), we have assumed that it was independent of frequency, which is about the case of the steady conductivity. If it varies with frequency, then it has to be written as a complex parameter $\sigma' = \sigma' + j\sigma''$ and Eqns. (1.13)–(1.16) have to be modified consequently.

Ionic polarization and *electronic polarization* are due to the displacement of the electronic orbits with respect to the protons when an electric field is applied. This phenomenon is much faster than dipolar polarization. It is a movement and is described adequately by a second-order equation, characterized by possible resonance [3, 5]. Dielectric losses induce damping, which makes the permittivity complex in the frequency domain, as for the first-order effect described by Eqn. (1.6). The real and imaginary parts of the relative permittivity are classical expressions for a second-order equation:

$$\varepsilon_r' = 1 + \frac{q^2 N}{\varepsilon_0 m} \frac{\omega'^2 - \omega^2}{\left(\omega'^2 - \omega^2\right)^2 + \gamma^2 \omega^2}$$

$$\varepsilon_r'' = \frac{q^2 N}{\varepsilon_0 m} \frac{\gamma \omega}{\left(\omega'^2 - \omega^2\right)^2 + \gamma^2 \omega^2} \tag{1.17}$$

with

$$\omega_0'^2 = \omega_0^2 - \frac{q^2 N}{\varepsilon_0 m} \tag{1.18}$$

where m is the moving mass, γ the damping term, $-q$ the moving charge, and N the number of active elements per unit volume. When the damping term vanishes, the permittivity is purely real:

$$\varepsilon_r = 1 + \frac{q^2 N}{\varepsilon_0 m} \frac{1}{\omega'^2 - \omega^2} \tag{1.19}$$

In fact, the oscillation frequencies $\omega_0'^2$ are not identical for all the electrons in a given volume, which requires a generalization by adequately summing all the contributions. Figure 1.2 represents the typical evolution of the real and

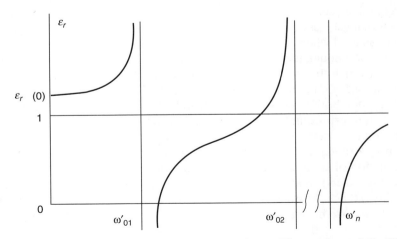

FIGURE 1.2   Ionic and electronic polarization (from [3], courtesy of De Boeck, Brussels).

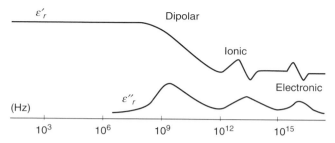

FIGURE 1.3   Permittivity as a function of frequency (from [3], courtesy of De Boeck, Brussels).

imaginary parts of a relative permittivity satisfying the second-order equation of movement. When the damping term decreases, the variation of the real part of the permittivity becomes steeper while the maximum of the imaginary part increases and tends to infinity at the oscillation frequency.

When submitted to electric fields, an actual material exhibits a variety of relaxation and resonance phenomena. Figure 1.3 represents a typical variation of the real and imaginary parts of the relative permittivity as a function of frequency.

Some materials exhibit unusual properties when submitted to electric fields. For instance, barium titanate and some other materials are *ferroelectric*: The electric polarization does not vary linearly with the amplitude of the applied field, exhibits hysteresis, and varies significantly with temperature. A critical temperature, above which ferroelectricity disappears and termed the *Curie temperature*, is associated with the phenomenon. Some other materials are called *antiferroelectric*: Their electric dipoles are very much organized as well; they are antiparallel, however.

1.3.2   Magnetic Field and Flux Density

The magnetic field $\overline{H}$ is derived from Ampere's law, which expresses the interaction between two loops of wires carrying currents. The total force experienced by a loop is the vector sum of forces experienced by the infinitesimal current elements comprising the second loop. Experimentally, it has been shown that [3, 4–7]

1. Two currents of equal directions attract each other, while they repel when they have opposite directions.
2. The magnitude of the force is proportional to the product of the two currents and to the product of the lengths of the two current elements.
3. The magnitude of the force is inversely proportional to the square of the distance between the current elements.

4. The force acts along the cross product of one current element by the cross product of the other current element by the unit vector along the line joining the two current elements.
5. The force depends upon the magnetic properties of the medium in which these are placed.

It is usual to calculate directly the contribution to the magnetic field evaluated at an observation point P of a current element I', expressed in *amperes* (A), directed along

$$\overline{dH}(\bar{r}) = \frac{I'(\bar{r}')\overline{dl'}x(\bar{r} - \bar{r}')}{4\pi|\bar{r} - \bar{r}'|^3} \qquad \text{A m}^{-1} \qquad (1.20)$$

where the factor 4π is due to the use of the rationalized MKS system. The magnetic field is a vector field, circumferential in the case of a linear current. The lines of magnetic field are tangential to the magnetic field in every point. Equation (1.20) is linear with respect to the current.

This law shows similarities with Coulomb's law,

$$\overline{E} = \frac{q}{4\pi\varepsilon_0 r^2}\bar{a}_r \qquad \text{[Eq. (1.3)]}$$

since, for a current element located at the origin of the coordinates ($\bar{r}' = 0$), the modulus of the magnetic field equals

$$|\overline{dH}| = \frac{I\,dl'}{4\pi r^2}\sin(\overline{dl'},\,\bar{r}) \qquad (1.21)$$

The total value of the magnetic field at point P is obtained by integrating Eqn. (1.14) over all the possible currents:

$$\overline{H}(\bar{r}) = \int_{\text{source}} \frac{I'(\bar{r}')\overline{dl'}x(\bar{r} - \bar{r}')}{4\pi|\bar{r} - \bar{r}'|^3} \qquad (1.22)$$

The electric current may appear in three forms:

1. It can be a current element as in Eqn. (1.20). It is then usually denoted $\overline{I}$ and expressed in *amperes*.
2. It can be a surface current density. It is then usually denoted $\overline{K}$ and expressed in *amperes per meter* (A m^{-1}).
3. It can be a current density. It is then usually denoted $\overline{J}$ and expressed in *amperes per square meter* (A m^{-2}).

When a material is placed in an applied magnetic field, it becomes magnetized, the amount of which is called the *magnetization vector* $\overline{M}$. This corresponds to what is called the magnetic properties of the material, which can be due to a variety of effects. In particular, as for the electric polarization, it may be due to the fact that magnetic dipoles are created or transformed into the material. In this case, the magnetization is the *magnetic dipole moment per unit volume*, in *amperes per square meter*.

The total magnetic field in a material is the sum of the applied magnetic field and an induced magnetic field, resulting from the magnetization of the material. Earlier in this section, the perfect electric conductor has been defined as an equipotential material in which all the points are at the same electric potential. In such a material, the total electric field must be zero, which implies that the material must develop an induced electric field such that the sum of the applied field and the induced field vanishes in all points of the material. On the other hand, one may define the *perfect magnetic conductor* as the dual of the perfect electric conductor, in which all the points are at the same magnetic potential. In such material, the total magnetic field must be zero, which implies that the material must develop an induced magnetic field such that the sum of the applied field and the induced field vanishes in all points of the material. The definition of the perfect electric conductor is very useful, largely because there are in nature materials whose properties are very close to those of the perfect electric conductor, for example, gold, silver, and copper. There is, however, no physical material with properties very close to those of the perfect magnetic conductor.

The induced magnetic field is calculated by taking into account the geometry of the problem and the boundary conditions, which can of course be complicated. The human body is essentially nonmagnetic, hence transparent to magnetic fields, except however for some localized magnetic properties.

A new vector field $\overline{B}$ is then defined, known as the *magnetic flux density*, in *webers per square meter*, similarly to the magnetization, or *tesla*, defined as

$$\overline{B} = \mu_0(\overline{H} + \overline{M}) \qquad \text{Wb m}^{-2}, \text{ T} \qquad (1.23)$$

This definition is totally general, applying to all materials. It indeed holds for materials in which [3]:

1. The magnetization vector has not the same direction as the vector magnetic field, in which case the material is *anisotropic*.
2. The magnetization can be delayed with respect to the variation of magnetic field. This is the case in all *lossy materials*, and all materials are lossy, so this is a universal property. It is neglected, however, when the losses are reasonably small.
3. The magnetization is not proportional to the magnetic field, in which case the material is *nonlinear*.

In all other cases, that is, when the material is *isotropic, lossless*, and *linear*, the definition (1.23) can be written

$$\overline{B} = \mu\overline{H} \qquad (1.24)$$

which combines the applied and induced fields, hence the external source field and the induced magnetization. The *permeability* of the material is defined by

$$\mu = \mu_0(1 + \chi_d) = \mu_0\mu_r \qquad \text{H m}^{-1} \tag{1.25}$$

It is the product of the permeability of vacuum μ_0 (in henrys per meter) by the relative permeability μ_r (dimensionless) of the material. Equation (1.25) defines at the same time the *magnetic susceptibility* χ_d. The use of permeability, relative permeability, and magnetic susceptibility is limited to isotropy, losslessness, and linearity, which is far from being always the case. It should be stressed again, however, that biological tissues are essentially nonmagnetic. Magnetization is a very complicated phenomenon [3, 4–7]. It may be due to a variety of mechanisms, which can be summarized here only briefly.

All materials are diamagnetic: *Diamagnetism* is a general property of matter. It is due to the fact that electrons placed in a magnetic field have their rotation speed increased, which results in an induced magnetic moment opposed to the applied magnetic field, hence in a decrease of magnetization. The diamagnetic susceptibility is negative and very small, normally of the order of -10^{-8} to -10^{-5}. It does not depend explicitly upon the temperature; it depends however upon the density of the material. A number of metals used in engineering are only diamagnetic, for instance, copper, zinc, gold, silver, cadmium, mercury, and lead. As an example, the relative permeability of copper is 0.999991.

When the atoms of a material have a permanent magnetic moment, this moment tends to align itself with the applied magnetic field to minimize the magnetic energy. This phenomenon is similar to electric polarization. It is a rather slow phenomenon, correctly described by a first-order law. It characterizes *paramagnetism*, in which the magnetization is positive: It reinforces the effect of the applied field. It should be noted that paramagnetic materials are also diamagnetic. Diamagnetism, however, is much smaller than paramagnetism and is not observable in this case. Paramagnetic magnetization is modeled similarly to electric polarization. It depends upon temperature and density. The paramagnetic susceptibility is positive and, in general, much smaller than 1. Some metals are paramagnetic, for example, aluminum, platinum, manganese, magnesium, and chromium. As for electric polarization, when the material density increases, modeling becomes much more difficult, and classical physics yields wrong models. Classical physics almost completely fails when trying to establish quantitative models. It can, however, yield some very illuminating insight on the phenomena involved with the dielectric character of materials, in particular about the influence of frequency.

The relative permeability related to this phenomenon is

$$\mu_r = \mu_r' - j\mu_r'' \tag{1.26}$$

It is a complex quantity, with real and imaginary parts. The imaginary part is a measure of the magnetic losses. These are often expressed also as the tangent of the *loss angle*:

$$\tan \delta_m = \frac{\mu''}{\mu'} \tag{1.27}$$

As for the electric polarization, the imaginary part of the permeability is nonzero only when the real part varies as a function of frequency. Furthermore, each part can be calculated from the variation of the other part over the whole frequency range.

In certain materials, the application of a magnetic field may induce an extremely strong magnetization, of the order of 10^8 A m^{-1}, which is 10 times more than the highest fields which can be produced in laboratory. This characterizes the phenomena that are called *ferromagnetism, ferrimagnetim, antiferromagnetism*, and *antiferrimagnetism*. An even simple explanation of the underlying physics is far beyond the scope of this book. Classical physics indeed fails in trying to model these phenomena. In fact, quantitative modeling of ferromagnetism was the reason Dirac developed quantum mechanics. The phenomenon is based on an exchange probability for some electrons between neighboring atoms in the crystalline structure, which corresponds to a spin coupling. The interaction from neighbor to neighbor is positive if the spins are parallel, which characterizes ferromagnetism; it is negative if the spins are antiparallel, which characterizes antiferromagnetism. The conditions for the exchange to take place are the following:

1. The distance between the electron and nucleus may be neither too large nor too small: it must be of the order of 0.27–0.30 nm; the materials in which this distance is smaller than 0.26–0.27 nm are practically antiferromagnetic.
2. The atoms must have nonsaturated layers to be able to accept electrons.
3. The temperature has to be lower than a critical temperature, called the *Curie temperature*, to avoid too much thermal disorder; above this temperature, the materials become paramagnetic.

Conditions 1 and 2 imply the material has a nonsaturated layer at the right distance from the nucleus: It is the third layer, and it is nonsaturated in the case of iron, nickel, and cobalt, which are the ferromagnetic materials.

Ferrites are ferrimagnetic materials. They are ceramics, generally composed of oxides of iron and other metals. Their chemical formula is very often $ZO \cdot Fe_2O_3$, where Z represents one or more divalent metals (cobalt, copper, manganese, nickel, and zinc). In most common ferrites, Z is for the combinations Mn + Zn or Ni + Zn. The size of the grains in ferrites is of the order of 1–20 μm. The oxygen anions play an important role in ferrimagnetics: Their presence results in a magnetization much smaller than that of ferromagnetic materials, because the magnetic cations are distributed in a nonmagnetic oxygen network. Furthermore, because of the oxygen ions, the alignment of the cation moments is antiparallel in some places of the network. When the antiparallel subnetworks have magnetic moments equal per unit volume, the material is antiferromagnetic.

Most ferrites are polycrystalline. They usually have a high permeability, relatively small losses, and a high electrical resistivity. The saturation magnetic flux density is lower than in ferromagnetic materials: It is of the order of 0.5 T in a MnZn ferrite instead of 2 T in iron.

Several types of energy have to be taken into account in investigating ferro- and ferrimagnetic materials:

1. Exchange energy, corresponding to the energy difference between the elements with parallel spins and those with antiparallel spins.
2. Magnetostatic energy, associated with free magnetic poles in the material.
3. Magnetostriction energy, related to the fact that a magnetic material undergoes an elastic deformation in the direction of magnetization, which may be positive or negative, corresponding to an elongation or a contraction, respectively—this deformation is of the order of a few millionths.
4. Anisotropic energy of the crystal, associated with the difference between the direction of magnetization and those of easy magnetization.

1.3.3   Electromagnetic Field

In DC situations, the electric field is calculated correctly from the laws of electrostatics. The same is true at extremely low and low frequencies, although the approximation is less applicable when the frequency increases. Similarly, the magnetic field is calculated from the laws of magnetostatics in DC situations. The same is true at extremely low and low frequencies, with however a quality of approximation decreasing when the frequency increases. Values of the frequency at which the approximation is not valid anymore depend on the geometric, electrical, and magnetic properties of the problem.

When the frequency increases, the electric and magnetic fields cannot be separated from each other: If one of the fields exists, so does the other. One cannot consider one field as the source of the other. They are linked to each other in every situation, and this is described by *Maxwell's equations*. Maxwell, a British physicist, formalized the laws of electromagnetics around 1880, without writing the equations however as we know them today. He made at this occasion two essential contributions:

1. One was to say that all the former laws, essentially based on experimental measurements made by Gauss, Ampere, Lenz, and others, were valid but that they had to be considered as a system of equations.
2. The other was to point out that, in the ensemble of these laws, one term was missing: The displacement current did not appear in the former law and it was to be added.

A consequence of the first contribution is that the electric and magnetic fields are linked together; hence there is a coupling between electricity and magnetism. As a consequence of the second contribution, the electric current can take one more form than previously known, so that one could have the following (the electric current passing through a given area is obtained by surface integrating the corresponding current density):

1. A *convection current density*, due for instance to a density of electric charges moving in vacuum, is described by

$$\bar{J}_{conv} = \rho \bar{v} \tag{1.28}$$

2. A *conduction current density*, due to the conductivity of some materials, is described by a relationship based on the electric current in the conductor:

$$\bar{J}_{cond} = \sigma \bar{E} \tag{1.29}$$

3. A *displacement current density*, due to the time variation of the electric field, is equal to the time derivative of the displacement field (the electric current density):

$$\bar{J}_d = \frac{\partial \bar{D}}{\partial t} \tag{1.30}$$

If we consider an alternate-current source feeding a capacitor formed of two parallel plates in vacuum through an electric wire, the displacement current ensures the continuity between the conduction current circulating in the wire and the electric phenomenon involved between the two capacitor plates: The displacement current density, integrated over the area of the plate and time derived, that is, multiplied by $j\omega$, is equal to the conduction current in the wire [3, 7].

4. A *source current density*.

Maxwell's equations form a system of first-order equations, vector and scalar. They are usually considered as the generalization of the former laws describing electricity and magnetism. This is a historical point of view, usually accompanied by the comment that up to now no EM phenomenon has been pointed out which does not satisfy Maxwell's equations. It has been demonstrated, however, that Maxwell's equations can be derived from a relativistic transformation of Coulomb law, under the constraint of the speed of light constant with respect to the observer. The demonstration is not too difficult in the case of linear motion, because then special relativity may be used [1–3]. It involves, however, a significant amount of vector calculus. In the case of rotation, however, general relativity must be used, which makes the calculation much more complicated [8]. As a result, it can be concluded that if an EM phenomenon is found which does not satisfy Maxwell's equations, Coulomb's law, relativity, or the constancy of the speed of light has to be revisited.

Maxwell's equations are not valid at boundaries, where *boundary conditions* have to be used. They may be written in either the time or frequency domain. They may also be written in either differential or integral form. As a consequence, they can take four different forms [3]. The *differential* forms, in the time or frequency domain, are used to describe the EM phenomena in continuous media. They are punctual and valid in every point of a continuous medium submitted to EM fields. The *integral* forms, on the other hand, link average quantities; they are applicable to volumes, surfaces, and contour lines. They are generally used to investigate circuits, mobile or not. *Time-domain* descriptions evaluate real, physically measurable quantities. They must be used when the physical consequence is not in synchronism with the cause, for instance, when investigating some optoelectronic phenomena. They must also be used when investigating wide-band phenomena. They are valid at any frequency. *Frequency-domain* descriptions are extremely useful when investigating monochromatic or narrow-band phenomena. The quantities they evaluate are complex numbers or vectors: They are not real, physically measurable quantities, although they may have a real physical content. They are valid at any time.

Solving Maxwell's equations in general cases may be quite a difficult task. Decomposing the vector equations into scalar ones, the system to be solved has a total of 16 equations. There are also 16 unknowns: the 5 vector quantities of electric field, electric flux density, magnetic field, magnetic flux density, and current density, that is, 15 scalar quantities, and the scalar charge density. Fortunately, in many cases, the system is much simpler, for instance because of symmetry, simplicity of the materials involved, slowly varying phenomena, and low frequency. A fair amount of symmetry, in particular, may greatly simplify calculations.

In living tissues, electromagnetic phenomena are usually slow, when compared to the extremely broad variety of phenomena to be evaluated in physics and engineering. The shortest biological response time indeed is of the order of 10^{-4} s, while most biological reactions are much slower. Hence, Maxwell's equations are most generally not used for evaluating biological effects in living tissues and systems. Furthermore, in this book we are interested in RF and microwave stimulation. At RFs, the wavelength in vacuum is large with respect to living tissues and systems, including human beings as a whole: At 1 MHz, the wavelength in vacuum is equal to 300 m. On the other hand, at microwaves, the period of oscillation is small, equal to 10^{-9} s at 1 GHz, which is much smaller than the fastest biological responses. This implies that the evaluation of biological effects in living tissues and systems due to EM stimulation, at RF as well at microwaves, in most cases does not necessitate the use of Maxwell's equations. Hence, in practice, quasi-static approaches are quite satisfactory in biological material, and the electric and magnetic fields are very often considered separately, even at RF/microwaves.

1.3.4   Electromagnetic Wave

Investigating the structure of Maxwell's equations, Hertz found, around 1888, both theoretically and experimentally, that it included the notion of propagation of EM waves because of the specific coupling between the electric and magnetic fields due to the particular form of the vector equations: The coupling between space variations and time variations of the electric and magnetic fields results in wave motion. The waves were propagating at a finite speed, the "speed of light." A few years later, in 1895, Marconi starting experimenting this at larger and larger distances, establishing what was called *wireless communications*.

Propagation has a precise mathematical definition: All the components of fields and associated physical quantities, such as current and charge densities, have a z dependence expressed as the factor e^{-jkz} in a cylindrical coordinate system or an r dependence expressed as the factor e^{-jkr} in a spherical coordinate system. Such an ensemble of fields is called an EM wave. Hence, the words *propagation* and *wave* are closely related [9].

Maxwell's equations are first-order equations. Eliminating one of the fields in these equations yields a second-order equation for the other field, which is called the *wave equation* or *Helmholtz equation*. The general solution of the wave equation is a propagating EM wave. Even in a medium as simple as vacuum, there are a variety of waves satisfying the wave equation, with transverse and/or longitudinal components of the fields. The field structure of the wave also depends on the fact that the propagation medium is either infinite or bounded: A coaxial cable and a waveguide have lateral bounds. When the medium is unbounded, it is often called *free space*. When the medium is bounded, *reflection* occurs on the interfaces. When edges or corners are present, *refraction* may occur.

On the other hand, the orientation of the fields may vary in space when the wave is propagating, depending on the type of wave and the medium. What characterizes this variation is called the *polarization*. It is related most often to the electric field, although it may also be related to the magnetic field. In this case, however, it should be mentioned explicitly.

The simplest EM wave structure is well known from elementary physics. It has a magnetic field perpendicular in space to the electric field, both being perpendicular in space to the direction of propagation. The fields are stationary neither in space nor in time: The wave displaces itself in space as a function of time at constant speed and the field amplitudes vary as a function of time, sinusoidal in a monochromatic case. During this variation, when the fields remain parallel to themselves, the polarization is called *linear*. When, on the contrary, the fields rotate at constant amplitude in a plane perpendicular to the direction of propagation, the polarization is called *circular*. Circular polarization may be right handed of left handed. If the amplitude and the direction of the fields both vary when propagating, the polarization is termed *elliptic*. In practice, many RF and microwave sources operate at linear polarization. It should not be assumed, however, that this is always the case: Circular polarization is also used in technical applications.

One way of classifying waves is based on geometric considerations, characterizing the geometry of the surface of constant phase of the wave. Depending on whether this surface is a plane, a cylinder, or a sphere, the wave is *planar*, *cylindrical*, or *spherical*. Furthermore, when the amplitude of the field is the same at every point of the surface of constant phase, which makes the surface of constant phase also a surface of constant amplitude, the wave is *uniform*; otherwise it is *nonuniform*. A *uniform plane wave* is a wave whose surfaces of constant phase are planes on which the amplitude of the field is a constant. As an example, the simplest wave structure described above is a uniform plane wave.

Another way of classifying waves is based on the number of field components. When a wave has only transverse field components for both the electric and magnetic field and no longitudinal components, the wave is *transverse electromagnetic* (TEM). A TEM wave has only four components. When, in addition, the wave has one longitudinal component only, it is called *transverse electric* (TE) if the longitudinal component is magnetic and *transverse magnetic* (TM) if the longitudinal component is electric. Both TE and TM waves have only five components. The most general wave has six components. It is a linear combination of TE and TM waves, possibly together with a TEM wave, depending upon the boundary conditions.

Transverse electromagnetic waves are very much appreciated in practice because they have only four components, with no longitudinal components: $E_z = 0$ and $H_z = 0$. On the other hand, uniform plane waves also characterize a very simple structure. It should, however, be well noted that a TEM wave is not necessarily a uniform plane wave. As an example, the very well known coaxial cable has a TEM wave as the main propagating mode. This TEM wave, however, is not a uniform planar wave. The surface of constant phase is indeed a plane, perpendicular to the direction of propagation, hence the wave is planar. The field, however, is not constant over the plane because it varies between the two conductors according to a $1/r$ law. Hence, the TEM wave in a coaxial cable is a nonuniform plane wave.

Specific parameters describe EM waves. The product of the *wavelength* λ (in meters) and the frequency is a constant, called the *speed of light* in vacuum, more correctly called the *phase velocity* in the medium where the propagation takes place, for instance inside a human body:

$$f\lambda = c(\text{vacuum}) = v_{\text{ph}}(\text{material}) = \sqrt{\frac{\varepsilon}{\mu}} \qquad \text{m s}^{-1} \qquad (1.31)$$

The speed of light is about $300{,}000\,\text{km}\,\text{s}^{-1}$ in vacuum. The last equality is strictly valid for TEM waves only. Being interested essentially in RF and microwave excitation, we shall find useful to note that the equation can be written as

$$f(\text{MHz}) \times \lambda(m) = 300 \text{ in vacuum} \qquad (1.32)$$

Equation (1.31) shows that the wavelength is inversely proportional to the frequency. In vacuum, the wavelength is 6000 km at 50 Hz, while it is 300 m at 1 MHz, 0.3 m at 1 GHz, and 1 mm at 100 GHz. It also shows that the phase velocity decreases when the relative permittivity and/or permeability increases. It is important to observe that the phase velocity may vary significantly from one material to the other, in particular because of the presence of water. Water is a dielectric material with a very high dielectric constant, of the order of 80 at low frequencies, as will be shown later. Most living tissue contains a significant amount of water. As a consequence, the phase velocity at 1 GHz in a human body is almost 9 times smaller than in vacuum because the wavelength is almost 9 times smaller than in vacuum. At higher frequency, however, the permittivity decreases and the values of wavelength and phase velocity are closer to their values in vacuum.

Some parameters are typical of propagation. The *wave number* measures the number of wavelengths per unit length. In physicochemistry, for instance in spectroscopy, it is still rather usual to characterize a frequency by the corresponding wave number. This always looks surprising to a physicist or an engineer, especially when the unit length is a familiar unit rather than the meter. As an example, a wave number of $1000 \, \text{cm}^{-1}$ characterizes a frequency at which there are 1000 wavelengths over a length of 1 cm, hence with a wavelength of $1 \, \mu\text{m}$, that is, in the infrared (IR).

On the other hand, the ratio of electric to magnetic field amplitudes is the *intrinsic impedance* of the propagation medium. In vacuum, it is about 377 (or 120π) ohms. It is also called the *wave impedance*. In the nineteenth century, the impedance of the propagation medium, in particular a vacuum, has long been considered as puzzling. It even led some to imagine the presence of a specific media, the *ether*, with no other characteristics than the permittivity and permeability of a vacuum, offering the wave an adequate medium to propagate. The wave impedance does not correspond to power absorption because the wave can propagate in a lossless medium. This is due to the fact that, even in the simple TEM structure, the electric field is not in phase with the magnetic field: Both are in quadrature, as has been said before. The wave impedance expresses the ability of a propagating wave to transport power density from one point to another in space and have it possibly absorbed there, in total or in part. This will be explained in Section 1.4.1.

Power absorption is a very important concept when investigating biological effects, as will be illustrated in Chapters 1–3. It is also important when designing materials for protecting biological systems in an EM environment, including the medical environment. Recently, EM environments have become very complex because of the wide and rapid spread of many kinds of electric or electronic devices, as exemplified by recent cellular telephone progress. As a result, EM wave interference problems due to these devices have increased in frequency. Also, biological effects based on these kinds of EM wave radiation have been feared. As a method of countermeasure for protecting biological and medical environment as well as the measurement involved, knowledge of EM wave absorbers has become significant. This will be analyzed in detail in Chapter 5.

Phase velocity is the velocity at which the phase of a wave propagates. It is also the velocity at which energy propagates, when the medium properties do not vary with frequency. It may not totally characterize the propagation of a wave, however. When the EM properties of the propagating medium vary with frequency, the energy does not propagate at the phase velocity; it propagates at the *group velocity*, which is the frequency derivative of the phase velocity. When the EM properties of the medium vary very significantly with frequency, it may even happen that the *energy velocity* is not the group velocity any more. This is not to be considered in biological tissues and systems, however.

1.3.5   Antennas and Near Field

The evaluation of biological effects, including hazards, and medical applications is related to situations where biological tissues and living systems are placed in specific EM environments. In this book, we limit ourselves to RF and microwave environments. Antennas transmit the fields. For instance, they may be placed in free space and have other purposes than illuminating human beings, such as transmitting television, FM radio, or mobile telephony signals. They may also be placed in specific locations, within a part of a human body, for instance, to exert a specific medical effect. In this case, the antennas are often called applicators. On the other hand, fields may have to be measured, for instance in radiometric applications. The antenna being most generally a *reciprocal device*, transmitting and receiving properties are similar. The fields radiated by antennas of finite dimensions are spherical waves. Some applicators are cylindrical and transmit cylindrical waves over the cylindrical portion of the antenna. Because of the importance of the antenna in the exposure situations, some elements about antenna theory have to be reviewed [10].

The space surrounding an antenna in a transmitting or a receiving mode is divided into different regions, usually three: (a) reactive near-field, (b) radiating near-field (Fresnel), and (c) far-field (Fraunhofer) regions (Fig. 1.4). The field structure is different in each region, without abrupt changes, however, from one region to the neighboring one.

The *reactive near-field region* is the region immediately surrounding the antenna, where the reactive field predominates. For most antennas, this region is commonly taken as interior to a distance $0.62(D^3/\lambda)^{1/2}$ from the antenna surface, where λ is the wavelength and D the largest dimension of the antenna. The region is called reactive because the reactive power density predominates in this region.

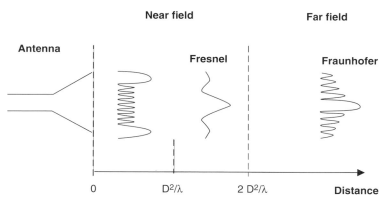

FIGURE 1.4 Angular distribution of fields around an antenna in the zone of near fields (Fresnel zone) and far fields (Fraunhofer zone).

The *radiating near-field (Fresnel) region* extends from the reactive near-field limit to a distance $2D^2/\lambda$, where D is the largest dimension of the antenna. For this expression to be valid, D must also be large compared to the wavelength. If the antenna has a maximum overall dimension which is very small compared to the wavelength, this field region may not exist. For an antenna focused at infinity, the radiating near field is sometimes referred to as the Fresnel region on the basis of analogy to optical terminology. In this region the field pattern is, in general, a function of the radial distance and the radial field component may be appreciable.

The *far-field (Fraunhofer) region* is commonly taken to exist at distances greater than $2D^2/\lambda$ from the antenna. This criterion is based on a maximum phase error of $\pi/8$. The reference to Fraunhofer is due to analogy to optical terminology, when the antenna is focused at infinity. In this region the fields are essentially transverse and the angular distribution is independent of the radial distance where the measurements are made.

In RF and microwave communications, far-field situations are usually the ones of most practical concern. This is very convenient because approximations can then be made to obtain closed-form solutions for the fields. This is not always the case when evaluating biological effects, and it is very important to clearly distinguish between near-field and far-field exposure. The evaluation of *hazards* due to RF/microwave exposure on human beings or animals is usually made in far-field conditions. Transmitting stations are normally far enough from living and working situations on the one hand while the antenna of a mobile telephone is so small with respect to the wavelength that the head of an end user is in the far field of the antenna on the other hand. The evaluation of *specific biological effects*, however, in particular in medical applications, is usually made in near-field conditions, as will be obvious later in this book. In such cases, near-field calculations have to be done, which may require much more attention.

As has been said before, antennas are reciprocal, except for very special devices. Hence, near-field situations have to be taken into account not only for transmitting antennas such as TV and FM radio transmitters but also for receiving antennas. Antennas can indeed be implanted in living tissues and organisms for medical applications.

When the antenna is used to deliver microwave power to heat tissue, the size and location of the microwave field have to be carefully located to control the affected tissue. Hence, the type and shape of the antenna are very much dependent upon the specific application, and there are a variety of applicators. A main problem is of course that of matching the applicator to the tissue. However, EM energy transfer depends, to a great extent, on the absorption properties of the tissue. It also depends on the frequency. As an example, sources at millimeter waves yield results similar to IR frequencies. These aspects will be investigated in detail in Section 1.5.

1.4   RF AND MICROWAVE ENERGY

1.4.1   Power and Energy

Adequately combining Maxwell's equations yields what is called *Poynting's theorem*. In the time domain, it expresses *equality between the spatial variation of EM power and the time variation of EM energy*, the sum of the electric and magnetic energies [3, 6, 7, 11]. The cross product of the electric field and the magnetic field is called the Poynting vector, with units of volt-amperes per square meter (watts per square meter). Being a cross product, it is perpendicular to the plane of the two vectors. In Section 1.3.4, we have seen that the TEM wave is the simplest wave structure, with the electric field perpendicular to the magnetic field, both fields being perpendicular to the direction of propagation. Hence, the Poynting vector of a TEM wave is in the direction of propagation.

Power and energy have not to be confused. Electromagnetic power is represented by the Poynting vector, just described. The integration of this vector over an open surface yields the power flow through the surface, in *watts*. The integration over a closed surface, with the normal to the surface considered as positive when extending outside the surface, also yields the power flow through the surface, that is, the total power coming out of the volume bounded by the closed surface. If this power is negative, it means that the net power is entering the volume, which also means that the medium inside the volume has absorbing losses—electric, magnetic, or conductive. The total power absorption is obtained by integrating the losses over the whole volume: *Absorption is associated with power*.

On the other hand, in the domestic sense of the word, energy means power absorption for some time in *watt-seconds*. In electromagnetics, however, energy is interpreted as a field concentration, stored in space, expressed in *joules* (J). The total electromagnetic energy stored in a given volume is obtained by integrating the energy over the volume. The time derivative of energy yields watts, hence power.

We should now have a better understanding of the second sentence of this section: In the time domain, Poynting's theorem expresses an equality between the spatial variation of EM power and the time variation of EM energy, the sum of the electric and magnetic energies. It is often said that Poynting's theorem expresses the conservation of energy. What it precisely expresses is that, for a given volume, if there is a net flow of EM power penetrating into the volume, then the EM energy increases in the volume, a possible difference between the two quantities being the power dissipation within the volume because of the medium conductivity. In the time domain, Poynting's theorem can be expressed in either differential or integral form.

Expressed in the frequency domain, the real part of the complex Poynting vector at a point is equal to the average value of the real power flux, physically measurable, at that point. When integrated over the surface limiting a given volume, it is equal to the real power dissipated in the considered volume due to whole of the electric, magnetic, and conductive losses. Contrary to the time-domain theorem, the frequency-domain theorem shows that the imaginary part of the Poynting vector is not related to the total frequency-domain EM energy: It is related to the difference of the magnetic and electric energies. Hence, it vanishes when the two energies are equal. This situation is called *resonance*, where the power flux is entirely real. In the frequency domain also, Poynting's theorem can be expressed in either differential or integral form.

Poynting's theorem can be used in establishing a general expression for the impedance of an EM structure, for instance an antenna [3]. The structure is placed inside a virtual closed surface and the expression relates the energy stored and the power dissipated in the bounded volume.

Poynting's theorem expresses the equality between the space variation of the EM power and the time variation of the EM energy. This form of the theorem is sufficient in most cases, at least in media where the current is a conduction current. In some cases, however, a generalized form may be necessary, for instance when the current is a convection current, due to moving charges, in vacuum or other media. This may be the case in plasmas, magnetohydrodynamics, and microwave tubes. Tonks has established such a generalized form of Poynting's theorem [12], obtaining equality for the conservation of energy, where the power is the sum of EM term and a kinetic term, while the energy is the sum of EM energy and kinetic energy. On both sides of the equations, the EM and kinetic terms cannot be separated. This expresses the possibility of transforming one term into the other, for instance EM energy into kinetic power, as in a particle accelerator, as well as kinetic energy into EM power, as in solar eruptions. In most cases, however, and in biological applications in particular, the usual form of Poynting's theorem is quite satisfactory.

Some special care has to be exerted when the EM properties of a medium vary with frequency [6]. The medium is then said to be *dispersive*. It has been shown that such a material is necessarily *absorptive*. The fundamental problem is that, in this case, EM energy has no precise thermodynamic definition. When the medium has limited dispersion, it is said to be *transparent*. This is the case when permittivity and permeability vary only slowly around the operating frequency. The mean value of the total EM energy can then be calculated and used in Poynting's theorem. Such a calculation might be necessary in biological tissues and systems, but only when operating in a range of frequencies where permittivity or permeability varies with frequency. It should be remembered that, in this case, the wave energy does not propagate at the phase velocity; it propagates at the group velocity, or even at another velocity if the dispersion is high, as has been mentioned in Section 1.3.4.

Finally, the pattern of energy absorption in biological tissues or systems, as in the human body, contributes to the RF/microwave effect. This raises the question of whether a whole-body average absorption rate can be used as the only determining factor in evaluating biological effects of RF and microwaves. Other features of the radiation also need to be considered. This will be considered in Chapter 2. A radiation diagram typical of a communications antenna is shown in Figure 1.5: A paraboloid antenna is placed at the coordinate origin. As illustrated, the gain $G_i(\theta, \phi)$ of the antenna varies with the direction (θ, ϕ). The *gain* of an antenna is defined as the ratio of the power transmitted by the antenna in a given direction to that which would be transmitted by an isotropic antenna (transmitting the same power in all directions) placed in the same location. It is usually expressed in *decibels* (dB).

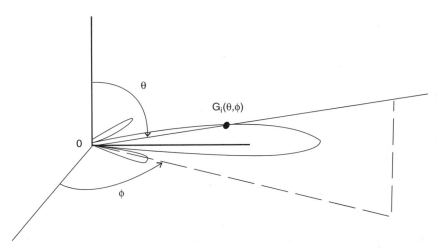

FIGURE 1.5   Radiation diagram of an antenna.

1.4.2   Influence of the Waveform

Parameters of microwave exposure are an important consideration in the production of biological effects. One key word is *dosimetry*, which takes into account the level of exposure as well as its duration. The simplest expression is the product of the level and the duration. Different durations of acute exposure lead to different biological effects and, consequently, different long-term effects occur after repeated exposure. The waveform of the radiation is also important. Differential effects have indeed been observed after exposure to pulsed-wave with respect to continuous-wave (CW) microwaves. In practice, biological effects have been observed under a variety of exposure types: CW, sinusoidal amplitude-modulated wave (AMW), pulsed wave (PW), and pulsed modulated wave (PMW) [13].

Hence, there is a difficulty in evaluating the exposure. Thermal effects are of course related to power, so that comparison of biological effects under different types of exposure should be done at constant power. What is constant power, however? Normally it should be either the CW power or the average power when the excitation is pulse modulated. In this case, however, and especially when the duty cycle is short, as in radar-type waves, the peak power may be much larger than the average power, and possible nonlinearity may induce other effects. The difficulty should even be greater for microthermal or nonthermal effects, immediately influenced by fields and not by power. This is a controversial question which shall be considered in Section 3.8. We should not consider power, however, as the only parameter able to induce effects. For instance, differential effects have also been observed after exposure to plane-*versus* circular-polarized waves.

1.4.3   Blackbody Radiation

The concept of a *blackbody* radiator is of fundamental importance for the understanding of the emission of real materials, including biological tissues, because its emission spectrum represents a reference relative to which the radiant emittance of a material can be expressed. In general, a fraction of the radiation incident upon a body is absorbed and the remainder is reflected. A blackbody is defined as an idealized, perfectly opaque material that absorbs all the incident radiation at all frequencies, with no reflection: It is a *perfect absorber*. The quantum-mechanical model of a blackbody can be described as consisting of such a large number of quantified energy levels with a correspondingly large number of allowable transitions that any photon, whatever its energy or frequency, is absorbed when incident upon the blackbody. In addition to being a perfect absorber, a blackbody is also a *perfect emitter*, since energy absorbed by a material would increase its temperature if no energy were emitted [14–16]. In other words, a blackbody is lossless. Its behavior is not to be confused, however, with that of a perfect reflector (a mirror), which is lossless too. A photon incident at a given frequency on a perfect reflector is indeed reflected back at the same frequency: The reflected ("emitted") signal is perfectly correlated with the incident signal. The perfect reflector may be called a *white body*.

Physically, blackbody radiation is produced by a body that is considered as a closed volume whose walls are in thermal equilibrium at a given temperature. An outside source maintains the whole wall at a constant temperature. It the body is a blackbody, there are no losses and the energy balance is zero. The incident radiation is transformed into thermal agitation, which in turn is transformed into emitted radiation. The emitted signal is perfectly noncorrelated with the incident signal [17]. As a consequence, a blackbody radiates at least as much energy as any other body at the same temperature. At microwave frequencies, good approximations to blackbodies are the highly absorbing materials used in the construction of anechoic chambers. Real materials, such as the earth for instance, usually referred to as *grey bodies*, emit less than blackbodies and do not necessarily absorb all the incident energy. As an example, part of the solar energy incident upon the earth is directly reflected back, so that only part of the incident energy is transformed by a blackbody process and emitted back. As a consequence, the "blackbody temperature" of the earth, that is, the temperature at which a blackbody would emit the same energy as that emitted by the earth, is lower than the physical temperature of the earth; at microwaves, this is about 254 K.

It must be noted that blackbody radiation is usually at a very low level with respect to other radiation sources, in particular power transmitted by communications systems and, especially, by radars. A few examples are worth mentioning. The sun's total power output is approximately 3.9×10^{26} W, little of which appears at wavelengths below 0.3 nm or above 3 μm. The solar energy per minute falling at right angles on an area of 1 cm^2 at a solar distance of 1 astronomical distance (equal to the mean earth's solar distance) is called the *solar constant* and at a point just outside the earth's atmosphere has a value of 0.14 W cm^{-2} = 1.4 kW m^{-2}. The planet earth reradiates a part of the received solar energy, equal to the difference between the energy received and the earth's *albedo* (relative energy absorbed is 0.34 for the earth), equal to 0.023 W cm^{-2} = 230 W m^{-2}. Cosmic noise extends from about 20 MHz to about 4 GHz. Man-made noise is an unwelcome by-product of electrical machinery and equipment operation and exists from frequencies of about 1 MHz to about 1 GHz. The peak field intensity in industrial areas exceeds the value of cosmic noise by several orders of magnitude, which draws attention to the need for judicious ground station site selection. Blackbody radiation is evaluated later in this section.

Blackbodies will be reevaluated in Chapter 2, devoted to an introduction to biological effects, more precisely hazards due to RF and microwave fields. In that chapter, energy considerations will be compared with entropy considerations while evaluating the possibility of isothermal biological effects.

Planck's Radiation Law   Planck established the mathematical expression of the energy emitted by a blackbody in 1901, in his blackbody radiation law, generally termed Planck's law. It yields the spectral energy density per unit volume, that is, the energy density per unit volume and unit bandwidth. It can easily be calculated by multiplying the energy per photon by the number of photons in a given mode and by the number of modes in a given volume, which is obtained when using semiclassical statistical mechanics [17]. Planck's law may be expressed in terms of the spectral energy per unit volume. It is often expressed, however, in terms of the *blackbody spectral intensity* per unit solid angle $I(f; \theta, \phi)$, also called *blackbody spectral brightness* $B_f(f; \theta, \phi)$, where f is frequency and (θ, ϕ) characterizes the direction in which the radiation is emitted, in watts per steradian per square meter and per hertz:

$$I(f; \theta, \phi) = B_f(f; \theta, \phi) = \left(\frac{2}{c^2}\right)\frac{hf^3}{\exp(hf/kT) - 1} \qquad \text{W m}^{-2} \text{ Hz}^{-1} \text{ sr}^{-1} \quad (1.33)$$

where f is the frequency in hertz; T the absolute temperature in kelvin; h Planck's constant $= 6.63 \times 10^{-34}$ J; k Boltzmann's constant $= 1.380 \times 10^{-23}$ J K^{-1}; and c the velocity of light in a vacuum $= 3 \times 10^8$ m s^{-1}. The only two variables in (1.33) are f and T. It should be noted that hf has the dimensions of energy, as well as kT; hence the ratio hf/kT is dimensionless. Figure 1.6 shows a family of curves of B_f plotted as a function of frequency with temperature as a parameter. It is observed that (1) as the temperature is increased, the overall level of the spectral brightness curve increases, and (2) the frequency at which the spectral brightness is maximum increases with temperature.

Rayleigh–Jeans Radiation Law   At RFs and microwaves, the product hf may be very small with respect to kT so that the denominator of the second factor on the right side of (1.33) can be written as

$$e^{hf/kT} \cong 1 + \frac{hf}{kT} - 1 \cong \frac{hf}{kT} \qquad \text{when } hf \ll kT \qquad (1.34)$$

which inserted in (1.33) reduces Planck's law to

$$B_f = 2\left(\frac{f}{c}\right)^2 kT = \frac{2kT}{\lambda^2} \qquad \text{W m}^{-2} \text{ Hz}^{-1} \text{ sr}^{-1} \qquad \text{when } hf \ll kT \qquad (1.35)$$

This is the Rayleigh–Jeans radiation law. It is a very useful low-frequency approximation of Planck's law in the radio spectrum. At millimeter waves or at very low absolute temperature, however, one needs to be careful because the condition $hf \ll kT$ may not always be satisfied. It is worth noting that the ratio hf/kT is approximately equal to

$$\frac{hf}{kT} \cong \frac{(1/20)f \text{ (GHz)}}{T \text{(K)}} \qquad (1.36)$$

Using the Rayleigh–Jeans approximation yields a definition of the *brightness temperature* of a transmitter. It is the temperature of a blackbody that emits the same brightness (1.35):

$$T_B(f; \theta, \phi) = \frac{\lambda^2}{2k} B_f(f; \theta, \phi) \qquad \text{K} \qquad \text{when } hf \ll kT \qquad (1.37)$$

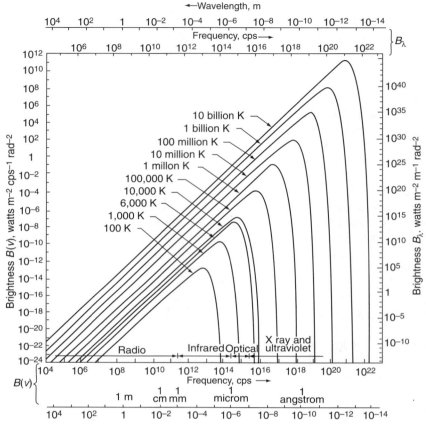

FIGURE 1.6   Planck radiation law curves (from [14]).

According to the Rayleigh–Jeans law, the brightness varies inversely as the square of the wavelength. On a log-log graph this relation appears as a straight line of negative slope. It coincides with Planck's law for the same temperature at long wavelengths. At short wavelengths, however, the brightness predicted by the Rayleigh–Jeans law would increase without limit, whereas the actual brightness reaches a peak and then decreases with decreasing wavelength, as predicted by Planck's law.

In fact, the Rayleigh–Jeans law was derived directly by Jeans from classical physics, before Planck introduced his quantum theory [14]. It should be observed that the Rayleigh–Jeans approximation does not contain Planck's constant h. It can be shown that this is because Planck's law is based on the postulate that the radiator possesses only a discrete set of possible energy values or levels, whereas the Rayleigh–Jeans approximation is based on classical physics, assuming that the radiator may possess all energy levels and that transitions are by a continuous process [17].

The classical *Johnson–Nyquist* expression for the noise emitted by a resistor can easily be derived from the Rayleigh–Jeans approximation. A circuit is formed by connecting a resistor at both ends of a lossless transmission line. Both resistors are equal to the characteristic resistance of the line; hence the line is matched and there are no reflections. The matching resistors and the line are all at the same temperature. Hence, the circuit is a one-dimensional blackbody in which the energy emitted by the resistor at the left propagates on the line and is absorbed by the resistor at the right end, while the energy emitted by the resistor at the right and propagating on the line is absorbed by the resistor at the left. As for establishing the three-dimensional Planck's law, one multiplies the energy per photon by the number of photons in a given mode and by the number of modes in the given length. The expression is reduced assuming that the Rayleigh–Jeans approximation is valid ($hf \ll kT$). Dividing by 2 to obtain the spectral power density for one resistor only yields the classical expression for the noise energy emitted by a resistor:

$$S(f) = kT \qquad \text{W Hz}^{-1} \tag{1.38}$$

It is interesting to observe that this expression is independent of the value of the resistor: Two resistors, with values $1\,\Omega$ and $1\,\text{M}\Omega$, respectively, produce the same energy. It is important to note that expression (1.38) is valid only under the Rayleigh–Jeans approximation. The power emitted in a given bandwidth B is of course

$$P = kTB \qquad \text{W} \tag{1.39}$$

From this expression one can define the noise temperature of a signal: It is the temperature at which a pure resistor should be maintained to produce the same power spectrum:

$$T(f) = \frac{S(f)}{k} \qquad \text{K} \tag{1.40}$$

In particular, this is the adequate definition for the *antenna temperature*, which is the absolute temperature at which a pure resistor must be maintained to produce the same power density as that measured at the output of the antenna. In other words, if a pure resistor maintained at the *antenna temperature* replaces the antenna, the receiver does not observe a difference. One then uses expression (1.39) for calculating the noise power emitted by a pure resistor, considered as the source of a generator with the resistor as the series resistance. A resistor of the same value to match the source terminates this source. It is well known that the generator then delivers the maximum possible power. Calculating this power yields

$$< e^2(t) > = 4RkTB \qquad \text{when } hf \ll kT \qquad (1.41)$$

where the brackets mean the average value. This is the well-known expression for the Johnson–Nyquist noise emitted by a resistor. This expression depends upon the value of the resistor and its range of application is limited because of the limitations of the Rayleigh–Jeans approximation.

It is interesting to calculate the blackbody radiation and compare it with other radiation sources evaluated earlier in this section. Equation (1.40) can be used in most cases. It shows that the noise power of a blackbody source at a physical absolute temperature of 300 K and measured in a bandwidth of 10 MHz is 41.4×10^{-15} W. If the source is at 3000 K and the receiver bandwidth is 100 MHz, the power is 41.4×10^{-13} W. It appears that the blackbody radiation is at a power level much smaller than most of the physical and industrial sources.

Stefan–Boltzmann Law Planck's law predicts the brightness curves presented in Figure 1.6 as a function of frequency and temperature. Integrating the Planck radiation law over all frequencies, which is summing the area of the Planck radiation law curve for that temperature, yields the *total brightness* B_t for a blackbody radiator:

$$B_t = \left(\frac{2h}{c^2}\right)\int_0^\infty \frac{f^3}{e^{hf/kT}-1}df \qquad \text{W sr}^{-1}\text{ m}^{-2} \qquad (1.42)$$

The integration yields the Stefan–Boltzmann relation [15]

$$B_t = \sigma T^4 \qquad \text{W sr}^{-1}\text{ m}^{-2} \qquad (1.43)$$

where B_t is the total brightness, σ a constant equal to 5.67×10^{-8} W m^{-2} K^{-4}, and T the absolute blackbody temperature in kelvin. It must be observed that the *total* brightness of a blackbody increases as the fourth power of its temperature. As an example, the total brightness of a blackbody at 1000 K should be 16 times that of at 500 K. This temperature dependence is valid only for the total brightness and not for the brightness (1.33) or (1.35).

Wien Displacement Law Planck's law represented in Figure 1.6 shows that the peak brightness shifts to higher frequencies with an increase in temperature. Maximizing Eqn. (1.33) by differentiating with respect to frequency f, that is, in terms of unit frequency, and setting the result equal to zero yield a quantitative expression for this displacement at a specific frequency f_P. Noting that the peak occurs at high values of the ratio hf_P/kT and simplifying accordingly yield the expression of the wavelength λ_P at which B is a maximum [15]:

$$\lambda_p T = \frac{hc}{3k} = 0.0048 \qquad \text{m K} \qquad (1.44)$$

This relation, in which the product of wavelength and temperature is a constant, is called the *Wien displacement law*: The wavelength of the maximum of peak brightness varies inversely with the temperature. Avoiding the simplification related to the high value of the ratio hf_P/kT yields the exact value of 0.0051 m K for the constant.

Instead of differentiating with respect to frequency, one can also differentiate with respect to wavelength, that is, in terms of unit wavelength. The brightness is then expressed in terms of unit wavelength and the wavelength for the peak brightness is not the same as when brightness is expressed in terms of unit bandwidth. The quantitative relation for the $\lambda_p T$ product when brightness is expressed in terms of unit wavelength is obtained by maximizing (1.42) and simplifying, which yield the value 0.0029 m K [15].

The frequency f_P at which the maximum radiation occurs increases as the temperature increases. Further calculations [14] yield the following expression for this frequency:

$$f_P = 5.87 \times 10^{10} T \qquad \text{Hz } (T \text{ in kelvin})\qquad(1.45)$$

and for the maximum spectral brightness B_f with respect for f:

$$B_f f_P = aT^3 \qquad \text{W sr}^{-1} \text{ m}^{-2} \text{ Hz}^{-1}\qquad(1.46)$$

where $a = 1.37 \times 10^{-19} \text{ W sr}^{-1} \text{ m}^{-2} \text{ Hz}^{-1} \text{ K}^{-3}$. Expressions (1.45) and (1.46) are valid for the spectral brightness per unit frequency. They are not valid for the spectral brightness per unit wavelength, for which similar expressions can be obtained.

Wien Radiation Law   The Rayleigh–Jeans radiation law (1.35) is a low-frequency approximation of Planck's radiation law (1.33), valid when one has $hf \ll kT$. At high frequencies, when one has $hf \gg kT$, the quantity unity in the denominator of the second factor on the right side of (1.33) can be neglected, in comparison with $e^{hf/kT}$, so that Planck's law reduces to

$$B_f = \left(\frac{2hf^3}{c^2}\right)e^{hf/kT} \qquad \text{W sr}^{-1} \text{ m}^{-2} \text{ Hz}^{-1}\qquad(1.47)$$

called the *Wien radiation law*. The Wien law coincides with Planck's law for the same temperature at frequencies considerably higher than the frequency of maximum radiation. A comparison of Planck's law with its high- (Wien) and low-frequency (Rayleigh–Jeans) approximations is provided in Figure 1.7 [14]. The coincidence of both approximations with the corresponding part of Planck's law is obvious.

1.5   PENETRATION IN BIOLOGICAL TISSUES AND SKIN EFFECT

When a conductive material is exposed to an EM field, it is submitted to a current density caused by moving charges. In solids, the current is limited by the collision of electrons moving in a network of positive ions. Good conduc-

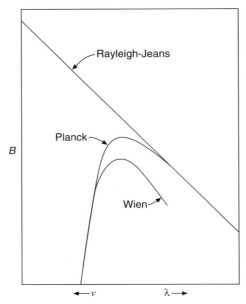

FIGURE 1.7 Comparison of Planck's law with its high- (Wien) and low-frequency (Rayleigh–Jeans) approximations at 300K (from [15], courtesy of J. D. Krause, Jr.).

tors such as gold, silver, and copper are those in which the density of free charges is negligible, the conduction current is proportional to the electric field through the conductivity, and the displacement current is negligible with respect to the conduction current. The propagation of an EM wave inside such a material is governed by the diffusion equation, to which Maxwell's equations reduce in this case. Biological materials are not good conductors. They do conduct a current, however, because the losses can be significant: They cannot be considered as lossless.

Solving the diffusion equation, which is valid mainly for good conductors, where the conduction current is large with respect to the displacement current, shows that the amplitude of the fields decays exponentially inside of the material, with the decay parameter

$$\delta = \frac{1}{(\omega\mu\sigma/2)^{1/2}} \qquad \text{m} \qquad (1.48a)$$

The parameter δ is called the *skin depth*. It is equal to the distance within the material at which the fields reduce to 1/2.7 (approximately 37%) of the value they have at the interface. One main remark is that the skin depth decreases when the frequency increases, being inversely proportional to the square root of frequency. It also decreases when the conductivity increases: The skin depth is smaller in a good conductor that in another material. Furthermore, it can be shown that the fields have a phase lag equal to z/δ at depth z.

For most biological materials the displacement current is of the order of the conduction current over a wide frequency range. When this is the case, a more general expression should then be used instead of (1.48a) [16]:

$$\delta = \left(\frac{1}{\omega}\right)\left\{\left(\frac{\mu\varepsilon}{2}\right)\left[(1+p^2)^{1/2} - 1\right]\right\}^{1/2} \qquad (1.48b)$$

where $p = \sigma/\omega\varepsilon$ is the ratio of the amplitudes of the conduction current to the displacement current. It is easily verified that Eqn. (1.48b) reduces to Eqn. (1.48a) when p is large.

The following important observations can be deduced from Eqn. (1.48a):

1. The fields exist in every point of the material.
2. The field amplitude decays exponentially when the depth increases.
3. The skin depth decreases when the frequency, the permeability, and the conductivity of the material increase. For instance, the skin depth of copper is about 10 mm at 50 Hz, 3 mm at 1 kHz, and 3 μm at 1 GHz. It is equal to 1.5 cm at 900 MHz and of the order of 1 mm at 100 GHz in living tissues.

These results are strictly valid for solids limited by plane boundaries. They are applicable to materials limited by curved boundaries when the curvature radius is more than five times larger than the skin depth. In the other cases, a correction has to be applied.

The phenomenon just described is the *skin effect*: Fields, currents, and charges concentrate near the surface of a conducting material. This is a shielding effect: At a depth of 3δ, the field amplitude is only 5% of its amplitude at the interface, and the corresponding power is only 0.25%; at a depth of 5δ, the field amplitude reduces to 1% and the corresponding power to 10^{-4}, which is an isolation of 40 dB. This shows that, at extremely low frequency, for instance at 50 Hz, it is illusory to try to shield a transformer with a copper plate: A plate 5 cm thick would be necessary to reduce the field to 1%! This is the reason why materials which are simultaneously magnetic and conducting, such as mumetal, are used for low-frequency shielding. In practice, the skin effect becomes significant for humans and larger vertebrates at frequencies above 10 MHz.

Shielding is much easier to achieve at higher frequencies. The skin effect implies that, when using microwaves for a medical application, the higher the frequency, the smaller the penetration, which may lower the efficiency of the application. Hence, the choice of frequency is important. It also implies that if a human being, for instance, is submitted to a microwave field, the internal organs are more protected at higher than at lower frequencies. As an example, the skin depth is three times smaller at 900 MHz, a mobile telephony frequency, than at 100 MHz, an FM radio frequency, which means that the fields are three times more concentrated near the surface of the body at 900 MHz than at 100 MHz. It also means that internal organs of the body are submitted to higher fields at lower than at higher frequency.

The skin effect is well known in engineering. It is also characterized by the *intrinsic* (or *internal* or *metal*) *impedance*, obtained by dividing the electric field at the surface by the current per unit width flowing into the material:

$$Z_m = \frac{E_0}{I_x} = (1+j)\sigma\delta = \sqrt{\frac{j\omega}{\sigma}} \qquad \Omega \qquad (1.49)$$

As an example, the intrinsic impedance of copper at 10 GHz is $0.026(1+j)$ Ω. It has equal real and imaginary terms. The real part R_m is sometimes called the *surface resistance*, measured in *ohms per square*.

The impedance is useful for calculating the (complex) power dissipation in the material. A property of the exponential curve is that its integration yields a result equal to the initial value times the skin depth. In other words, the intrinsic resistance of the material in which the fields decay exponentially would be the same if the current was uniformly distributed over depth δ. Hence, the total power dissipated in the material under an exponentially decaying field is equal to the power dissipated in depth δ under a field constant and equal to the value at the surface.

Table 1.2 summarizes some skin depth values for human tissues at some frequencies. The EM properties of the tissues as well as their variation as a function of frequency have been taken into account.

Figure 1.8 shows the variation of the power absorbed inside a human body as a function of the penetration depth at several microwave frequencies: We are less and less transparent to nonionizing EM radiation when the frequency increases. In the optical range, skin depth is extremely small: We are not transparent anymore. Variation of the dielectric constant as a function of frequency was taken into account in this figure.

There is a tendency to believe that RFs and microwaves exert more significant biological effects at low and extremely low frequencies. This is not necessarily true: The dielectric constant of living materials is about 10,000 times larger at ELF than at microwaves. The dielectric constant is important because

TABLE 1.2   Typical Skin Depths in Human Tissue

Parameter	Radio FM	TV Transmitter	Telephony Mobile	Telephony Mobile
Frequency (MHz)	100	450	900	1800
Skin depth (cm)	3	1.5	1	0.7
Depth at which power reduces to 1% (cm)	9	4.5	3	2

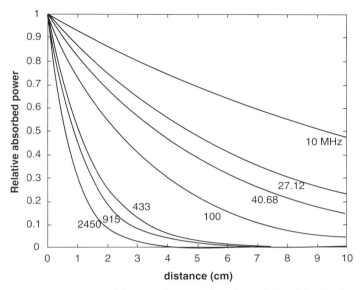

FIGURE 1.8   Power absorbed in muscles as a function of the skin depth at various frequencies.

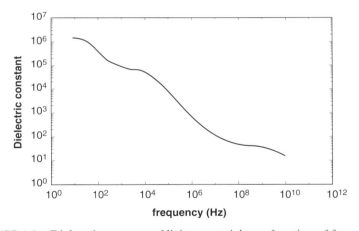

FIGURE 1.9   Dielectric constant of living material as a function of frequency.

it is the link between the source field and the electric flux density (also called the displacement field). A dielectric constant 10,000 larger implies the possibility of an electric flux density of a given value with a source field 10,000 times smaller. Figure 1.9 shows the dielectric constant of living material (muscle) as a function of frequency [18]. There is a level of about 1,000,000 at ELF up to 100 Hz, then a second level of about 100,000 from 100 Hz to 10 kHz, and, after some slow decrease, a third level of about 70–80 from 100 MHz to some giga-hertz. This last value is that of the dielectric constant of water at microwaves. One of the main constituents of human tissues is water. Hence, we have about the same microwave properties as water.

However, as has been said in Section 1.1, microwaves are in the frequency range in which the wavelength is of the order of the size of objects of common use—meter, decimeter, centimeter, and millimeter—depending of course on the material in which it is measured. One may hence wonder whether such wavelengths can excite resonance in biological tissues and systems. We shall come back with this question in Chapters 2 and 3.

1.6   RELAXATION, RESONANCE, AND DISPLAY

A good knowledge of the complex permittivity of biological media is neces-sary for evaluating biological effects as well as in medical applications. Hyper-thermia is only one example of such an application (treated in great detail in Chapter 4). A number of measured data are available for characterizing bio-logical media. It should be mentioned, however, that there are not many meas-ured data for biological and organic liquids at frequencies above 20 GHz.

1.6.1   Relaxation in Dielectrics

The dielectric constant is the real part of the complex relative permittivity. It is of primary importance when characterizing dielectrics. It should not be for-gotten, however, that the permittivity is complex in the frequency domain and that the dielectric constant gives only partial information.

Up to about 1 GHz, materials respond to *relaxation* phenomena, already illustrated in Section 1.3.1. When a rarefied *nonpolar gas* in which the mole-cules have no electric dipoles at rest is submitted to an external electric field, an electric dipole is induced. When a rarefied *polar gas* which has an electric dipole at rest is submitted to an external electric field, the dipole orientation is modified: It essentially rotates. As has been said earlier, when the density increases, classical physics almost completely fails when trying to establish quantitative models. It can, however, yield some insight on the phenomena involved with the dielectric character of materials.

The alignment of the molecule dipolar moment because of an applied field, called *dipolar polarization*, is a rather slow phenomenon. It is correctly described by a first-order equation, called after Debye [4, 5]: The dipolar polar-ization reaches its saturation value only after some time, called the *relaxation time* τ (Fig. 1.1).

This looks simple. The process is rather complicated, however. The task of dielectric theory is difficult, not so much because permanent dipoles cannot always be identified but mainly because they mutually influence one another: A dipole not only is subject to the influence of a field but also has a field of its own. The mutuality of the influence of dipoles, permanent or otherwise, on one another makes the response of the assembly a cooperative phenomenon, depending on the size and shape of the assembly [5].

The relaxation process consists in the approach to equilibrium of a system which is initially not in thermodynamic equilibrium. Such a process is irreversible and not covered by equilibrium thermodynamics. Most relaxation processes fulfill the conditions for which irreversible thermodynamics is applicable. Relaxation occurs when the free energy stored in the system is degraded into heat, in other words, if entropy is created irreversibly. The irreversibility is related to the fact that the free energy of the field is used to increase the total amount of heat stored in the dielectric plus the heat reservoir surrounding it. The thermodynamic treatment illuminates the significance of some of the concepts of relaxation and shows, in particular, that the Debye equation is the most plausible description of a relaxation process in a first approximation. A more thorough understanding requires further reading [5].

What has just been described corresponds to dielectrics with a single relaxation time. In fact, three classes of dielectric characteristics may be distinguished:

1. *Dielectrics with Single Relaxation Time.* These are rare. For instance, they are found in certain dilute solutions of large polar molecules in nonpolar solvents. Some other liquids and solids obey the Debye equation for complicated reasons.
2. *Dielectrics with Approximate Debye Behavior.* The peaks of ε'' are wider and have a somewhat different shape. This is a large class which contains many simple compounds and solutions of dipolar compounds in nonpolar solvents.
3. *Dielectrics for Which Debye Peaks Are Unrecognizable.* This class contains many practical insulating materials as well as some pure compounds.

For materials with many relaxation times, the polarizable elements are divided into groups of given relaxation times; then a principle of superposition is used to obtain an analytical expression for the complex permittivity.

Few substances have one single relaxation time. For some materials, such as water, permittivity is closely fit by a first-order Debye equation, while for others, such as biological media, where several relaxation times are involved, higher order terms are necessary [19].

1.6.2   Resonance Absorption

Relaxation refers to that part of polarization that is due to the ordering of permanent dipoles. However, matter can take energy from a field even in the absence of permanent dipoles if the field perturbs oscillations of one kind or another. The polarization caused by this mechanism is called *electronic polarization* or *optical polarization*.

Resonance can be observed in gases at frequencies below 100 GHz, although, in condensed matter, such an effect can be observed optically in the IR region of the spectrum. Besides, in gases, optical polarization may be described as a property of individual molecules, while conditions in solids and liquids are less easy to visualize.

An individual atom or molecule, whether it has a permanent dipole moment or not, consists of negative and positive charges. In terms of classical physics, it may be considered as a harmonic oscillator: Overcoming a restoring force can alter the distance between the centers of gravity of the positive and negative charges. The oscillator takes energy from an electric field, at a resonant frequency, determined by the restoring force; the energy uptake in the absence of damping tends to infinity.

The interaction of molecules and fields is best described by quantum mechanics [5]. Individual molecules have discrete energy states and they absorb energy from an alternating field of frequency f, so that

$$hf = \Delta W \qquad J \tag{1.50}$$

where ΔW is the energy difference between two quantum levels and h is Planck's constant. The molecule can take up or radiate energy only at the appropriate frequencies and in the appropriate amounts determined by Eqn. (1.50). Each individual molecule is characterized by an optical spectrum with a number of discrete frequencies. The energy differences are extremely small and the smallness of ΔW may be appreciated when it is compared with kT: For $f = 10$ GHz, ΔW is only $(6 \times 10^{-3})kT$ at room temperature while the energy per degree of freedom is $^{1}/_{2} kT$ at equilibrium. Detailed investigation of the spectra for gases at low pressures is provided by microwave spectroscopy.

When molecules exist within a gas or other assembly, they collide or otherwise interact with neighbors. These interactions result in thermal equilibrium, which implies an equilibrium distribution of energies among the constituent members of the assembly. It also implies that a given molecule retains its energy only for a certain average time τ_l. In a gas, this lifetime is shorter when the pressure is higher.

When measuring the value of an energy level in an assembly of molecules by means of some external probe, for instance microwaves, the result is unprecise according to the Heisenberg uncertainty principle. If this energy level has a lifetime τ_l, then it can only be determined with an uncertainty δW given by

$$\delta W \cong \frac{h}{\tau_l} \tag{1.51}$$

When τ_l is short, the uncertainty of the energy determination may be large compared with the magnitude of the energy difference to be determined. In this case, the magnitude of ΔW cannot be determined, and spectroscopy is impossible.

Neighboring oscillators interchange energy by collisions. Lorentz deduced a theory of line broadening which is a good approximation for the case that neither the resonance frequency nor the field frequency is very small and that line broadening is not too great, which covers most typical spectra [20]. His results can be closely approximated by introducing a damping term in the equation of the oscillator which yields the very simple solution of a second-order differential equation with a damping term, the simplest formal description of resonance absorption. The behavior of the complex dielectric constant as a function of frequency is quite different for resonance and relaxation, as may be seen from Figures 1.1 and 1.2.

The absorption spectra of crystalline solids are at short wavelengths due to electronic transitions within atoms or molecules and at longer wavelengths due to vibrations of the crystal as a whole [5]. The lattice vibrations are either optical or acoustic. Together they constitute the thermal vibrations whereby the atoms and ions, for example, within the crystal achieve thermal equilibrium. The acoustic vibrations do not involve a change of the polarization and do not interact with EM fields. The optical vibrations correspond to oscillatory displacements of charges.

The most important optical vibration in a simple ionic crystal such as KCl may be visualized as a movement of K and Cl ions in opposition to each other, as in the case of a dipole whose length is oscillating. The frequency of this vibration in the alkali halides is of the order of 10 THz, in the IR spectrum. Investigating the broadening of the IR spectra of crystalline solids is a difficult subject, although first-order approximations can be obtained rather easily.

The transition from resonant to nonresonant absorption, that is, relaxation, has been studied in gases by increasing damping, hence line broadening. It has also been approached for solids. When the resonant frequency decreases, however, optical measurements become increasingly difficult and very complex techniques have to be used. For a wavelength of about 1 mm, which corresponds to 300 GHz in a vacuum, the optical techniques of the far-IR overlap with microwave techniques.

1.6.3   Cole–Cole Display

The data provided by dielectric measurements can be presented in different ways. One classical representation is in plotting the real and imaginary parts of the permittivity as functions of frequency, most frequently as functions of $\log_{10} \omega$. The disadvantage, however, is that these two plots are then presented independently of each other while their frequency behaviors are linked through the general theoretical considerations. These were developed in Section 1.3.1 where it was shown that each part can be calculated from the variation of the other part over the whole frequency range, as indicated by the Kramer and Kronig equation (1.12) [6].

One very early representation is the plot of $\varepsilon''(\omega)$ for a certain frequency against $\varepsilon'(\omega)$ at the same frequency, which is the representation of the dielectric constant in its complex plane. This display is often called the *Cole–Cole plot* (or *complex-locus diagram* or *Argand diagram*) [5]. Using expressions (1.9), it can easily be shown that Debye's equation between the real and imaginary parts of the dielectric constant is the equation of a circle in the complex plane $\varepsilon(\omega)$:

$$[\varepsilon'(\omega) - \tfrac{1}{2}(\varepsilon_s + \varepsilon_\infty)]^2 + [\varepsilon''(\omega)]^2 = \tfrac{1}{4}(\varepsilon_s - \varepsilon_\infty)^2 \tag{1.52}$$

where ε_s and ε_∞ are the values of the real part of the relative permittivity at frequencies zero and infinity, respectively. The Cole-Cole plot therefore provides an elegant method of finding out whether a system has a single relaxation time. Figure 1.10 shows the plot of a dielectric with a single relaxation time. A given point on the semicircle corresponds to a given frequency. Frequency is zero on the right end of the diameter and infinity on the left, increasing anticlockwise. The summit corresponds to $\omega\tau = 1$, where τ is the relaxation time in Eqn. (1.7). The plot has the disadvantage that ω does not appear explic-

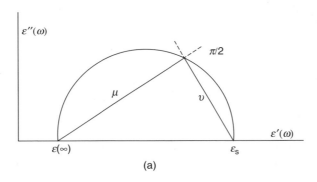

(a)

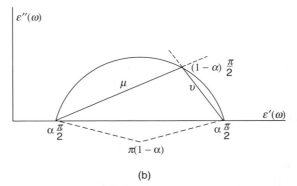

(b)

FIGURE 1.10   Cole–Cole plot for a Debye dielectric.

itly: Any material with a single relaxation time, characterized by ε_s and ε_∞, gives the same plot. Hence, in reporting results, it is essential to supply the magnitude of τ in addition to the plot. The power at which $\omega\tau$ appears in Eqn. (1.7) is not always unity. When it is smaller than unity, the plot represents only an upper portion of a semicircle.

When the steady conductivity contributes significantly to $\varepsilon''(\omega)$, the representation of dielectric data is affected in the Cole–Cole plot [5]. The presence of several relaxation times also complicates the representation. In view of this, Fröhlich has introduced a distribution function in which the relaxation time depends on the (absolute) temperature, and he assumes that the polarizable units are evenly distributed in terms of activation energy [21]. There is no reason why activation energies should be evenly distributed over a small range of values while higher or lower values should be absent. Fröhlich's distribution can be generalized by assuming that most activation energies have a median value while deviant values are increasingly unlikely, yielding some sort of a Gaussian distribution of activation energies [5].

However, Cole and Cole designed a function that is not very different in practice from a Gaussian distribution. It is a useful representation for many experimental results, and this seems reasonable in view of its similarity to a Gaussian distribution. Several other distribution functions were designed later.

1.7   DIELECTRIC MEASUREMENTS

Accurate knowledge of the complex relative permittivity of biological tissues is necessary to evaluate biological effects as well as the efficiency of medical applications. There are a variety of measurement methods, based however on a rather limited number of principles.

1.7.1   RF Measurements

The complex permittivity—dielectric constant and losses—of a material is almost always measured by inserting a specimen into a capacitor, waveguide, or cavity which forms part of an electrical circuit. The circuit is subject to an alternating voltage or wave or to step functions. Above 1 MHz, however, which covers the RF/microwave range, the source delivers most generally a CW. In all cases, errors may arise if effects due to some part of the external circuit are ascribed to the specimen. It is extremely important to point out the limits of experimental accuracy. Furthermore, some sources of error might escape the experimenter when using ready-made equipment.

For frequencies higher than a few megahertz, representation of a circuit by discrete circuit elements and by wires free of inductance and capacitance becomes gradually unrealistic. For higher frequencies, circuits are considered as distributed impedances, and the specimen is incorporated accordingly.

For frequencies up to 1 GHz resonant circuits may be used, the specimen being combined with a known inductance in the form of a Q-meter using essentially a series combination of C, L, and R [5]. The energy absorption exhibits a resonant peak and the width of this peak $2\Delta\omega$ is measured, where $\Delta\omega$ is the difference between the frequency where the energy loss is a maximum and the frequency where it is reduced to half the maximum. For a simple series circuit the Q value is given by

$$Q = \frac{\omega_0}{2\Delta\omega} = \frac{\omega_0 L}{R} = \frac{1}{\omega_0 CR} \tag{1.53}$$

Insofar as the capacitor is the only lossy component, Q is inversely proportional to tan δ of the dielectric at ω_0. Q-meters which use leads from instrument to specimen may be inexpensive and versatile, but they are generally not suitable for precision measurements of loss angle. Leads can be avoided, for instance, by using a capacitance whose value can be modified by micrometer screws. This, however, makes the method difficult to use over more than a narrow range of temperatures.

1.7.2 Microwave Measurements

At frequencies above 300 MHz, the shape and size of the measuring assembly may become important in relation to the wavelength. Because of this, different specimens have to be used for not so very different frequencies. Measurements in this microwave region are discussed in a number of textbooks, essentially with nonautomated equipment however [22–24].

In the early days of microwave measurements, accurate measurements were made with the use of a slotted line. In practice, this reduces to measuring the amplitude and phase of an impedance at a given frequency as a function of the position on the line. The slotted line is still the most accurate measurement device. It is not in use anymore, however, because it is limited to one single frequency at a time and making measurements over a range of frequencies necessitates a number of tedious adjustments. Because of transmission line properties, however, the important propagation term is the product of the propagation constant and the position on the line. Hence, operating at constant frequency as a function of the position on the line yields the same type of results as operating at a given point on the line as a function of frequency.

This is what most measuring devices have been able to accomplish since about 1990. They are called *vector analyzers*, and they yield the amplitude and phase of an impedance in a specific point of a line as a function of frequency, for instance from 40 MHz to 40 GHz and higher. The calibration of these instruments is extremely important: It is a key element in doing good measurements. It takes into account all the elements included in the measurement process of the various components outside of the device under test: mismatches, losses, defects, imperfections, and so on. Calibration compensates all these, so that the actual measurements are not degraded. Some devices with known properties are necessary for calibration purpose, such as a short circuit, an open circuit, and transmission lines with different lengths.

A poor calibration may degrade the measurements without making the experimenter aware of this. Because there are a variety of devices to be measured, there are also a number of calibration procedures. They become more complicated when the number of imperfections to be corrected increases. For instance, the number of terms needed for calibrating the measurement setup depends upon the fact that the physical quantity to be measured is a reflection or a transmission, on a device with small or high losses, reciprocal or not, and so on.

Scattering matrix parameters are most often used for calculating or estimating the errors. They describe a two-port by its reflection and transmission characteristics. When measuring the electric properties of a dielectric, the container in which the specimen is placed is considered as the electrical two-port.

A number of precautions have to be taken, especially at the higher frequencies, above 20 GHz [17]. In particular, connectors can be a source of inaccuracy if not adequately chosen, especially the miniature connectors introduced around 1960. Better connectors are necessary when operating at higher frequencies.

Measuring the complex permittivity of biological tissues fortunately does not offer the same variety as can be found with electronic circuits, passive and active. However, one needs to be careful sometimes: Anisotropy may be present because of the inhomogeneity of the structure. In such a case, calibration has to take the possibility of anisotropy into account.

1.7.3   Liquids

Measuring the dielectric properties of liquids raises specific difficulties. In particular, possibly the container must be reusable while being tight. The usual practice for liquids is to measure the reflection coefficient of an open-ended coaxial probe [25]. Two difficulties, however, relate to this practice, especially when applied in the high range of microwave frequencies: The structure radiates and the size of the coaxial probe is very small; hence the mechanical accuracy may not be very good.

A new procedure for measuring the complex permittivity of liquids was presented in 1997 [26]. It is based on the measurement of the scattering parameters of waveguide two-ports and on an original calibration method developed by the authors, called line–line (LL) [27]. The complex permittivity is obtained explicitly. Hence the method avoids the difficulties related to the inversion of implicit expressions through tedious numerical iterations.

A waveguide transmission method was used on a vector analyzer, yielding the four complex scattering parameters: reflection at both ends and transmission in both ways. A waveguide spacer is placed between the two waveguide ends of the vector analyzer. The connection between the waveguides and both ends of the analyzer is made by coax-to-waveguide transitions specific to each frequency band. To use the LL method for calibration purposes, two wave-guide spacers that are identical except for their thickness are necessary. The LL method requires a "short" and a "long" line, respectively. A synthetic film is placed at both ends between waveguides and spacers, containing the liquid in a known volume. Reference planes are placed at both ends in the plane of each film. The calibration method ensures that the transition flange–film–flange has no effect on the result, provided this transition is reproducible when using both spacers. Commercially available simple configurations are adequate. In practice, the film is put on both sides of the waveguide spacer and soldered together. Holes are made in the film, in the place of the holes of the waveguide flanges, to ensure tightness. Then, using a syringe, the liquid is inserted into the spacer. Air bubbles must be avoided. The two wave-guides are then screwed together, through the holes of the spacer. This oper-ation has to be done for each of the two spacers and for each waveguide band.

There are several advantages to these methods. First, the test cell is made of commercially available devices. Second, only two operations are needed to extract the required experimental data, and explicit expressions provide the values of the complex permittivity. This is especially helpful for characterizing highly dispersive and lossy substances. Finally, the influence of the air–liquid interface is removed from the procedure, so that mismatch, radiation, and higher order mode generation have no significant influence on the result.

The measured liquids were dioxane, methanol, blood, and axoplasm. Dioxane is a perfect dielectric, with a dielectric constant of about 2.25 and no losses up to 110 GHz. It is extremely useful as a calibration liquid dielectric to check the validity of the measurement setup and, in particular, the quality of matching the various waveguides. Results showed for the first time the complex permittivity of biological and organic liquids at frequencies from 8 to 110 GHz in five waveguide bands. The comparison has been made between the measurements and the first-order Debye equation. For some materials, such as water, permittivity is closely fit by a first-order Debye equation, while for others, such as biological media, where several relaxation times are involved, higher order terms are necessary [19]. Two relaxation times were exhibited by the measurements on blood.

The configuration just described is completely closed, which provides very accurate results. One may need or wish to use open structures. For instance, the complex permittivity of live and dead neurological cell cultures has been measured on a setup made of open microstrip transmission lines, from 20 to 40 GHz [28]. One advantage of this configuration is that it is open. Hence, the biological medium can be submitted to a microwave exposure and measure-ment of the complex permittivity can be an indicator of the activity of the medium.

1.7.4 Applicators

The applicators usually operate in the near-field region of the antenna, where the biological tissue, or part of the human body, is placed. One needs to be very careful about how to match an applicator to the part of interest of the human body. Furthermore, the dielectric parameters of the tissue exposed to the antenna are probably not too well known, because they have most certainly been measured in another configuration. It would be very difficult to measure with some accuracy the parameters of a flat living surface, in particular because it would be extremely difficult, if not impossible, to design a calibration method to be used in this configuration. As a consequence, matching an applicator to a biological surface is a very specific subject. The reader is referred to specific further reading, for instance [29].

1.8 EXPOSURE

The importance of evaluating correctly the electric parameters of biological tissues has been stressed in this chapter. This is because such knowledge is necessary in evaluating biological effects, in particular in medical applications. This will be considered in detail in the last chapters. One must, however, realize already at this stage that biological effects are to be evaluated with respect to fields that are inside the biological material, and this implies a good knowledge of both the field at the interface with the material and the material properties. In Section 1.7 were exposed the main guidelines for measuring the electric properties of biological materials, together with the precautions to be taken to ensure a good knowledge at the interface.

A totally different situation occurs in real life, when human beings are exposed to a surrounding RF or microwave field due to a variety of sources: AM radio, FM radio, television, mobile telephony, radars, and so on. Similarly, animals prepared for epidemiological studies are submitted to a specific field. In both cases it is necessary to have an accurate knowledge of the field amplitude, for instance, for comparing the human exposure to standards or evaluating possible hazards, the risk of interference in a hospital, the field amplitude at which the animals will be submitted in the epidemiological study, and so on.

Standards for human beings are based on two kinds of limitation:

1. *Basic limitations* that should always be respected.
2. *Reference levels* that could be exceeded if the basic limitations are not exceeded.

The basic limitations are expressed in terms of absorption by the human body, more precisely by the part of the body that is exposed since, because of the skin effect, only the most external layer is absorbing the power. The biological aspects of this will be considered in Chapter 2. It is quite appropriate, however, to note that this power absorption is expressed in *watts per kilo* (W/kg) of absorbing matter. This quantity is called the *specific absorption rate* (SAR). The SAR is of course not measured on a living person.

This is why there are reference levels, which are expressed in measurable physical quantities, namely the fields, electric or magnetic. At RF/microwaves, the electric field is mostly used as a reference. Standards then specify the value of the electric field, for instance, which should not be exceeded in some place or for some time. This creates a situation that should be well understood: The reference level is a level that the reference, for instance the electric field, should not exceed. This value is measured in the *absence* of the person. The corresponding basic limitation then indicates the power absorbed by a person that should be *present* in this electric field, measured in the absence of the person.

Hence, correctly measuring the electric field or the associated EM power is of prime importance. The basic theory has been exposed in the first sections of this chapter. It should be stressed, however, that in most circumstances the electric field can vary quite significantly from one place to another, not far away from the first: The *variability in space* may be quite important. This may happen, for instance, because of standing waves caused by steady obstacles. If the obstacles are moving, for instance cars in a street, *variability in time* is added. The difficulty of correctly evaluating a living system should not be underestimated.

REFERENCES

[1] R. S. Elliot, *Electromagnetics*, New York: IEEE Press, 1993.

[2] P. Lorrain, D. Corson, *Electromagnetic Fields and Waves*, San Francisco: Freeman, 1970.

[3] A. Vander Vorst, *Electromagnétisme. Champs et Circuits*, Brussels: De Boeck, 1994.

[4] A. Matveyev, *Principles of Electrodynamics*, New York: Rheinhold, 1966.

[5] V. V. Daniel, *Dielectric Relaxation*, New York: Academic, 1967.

[6] L. D. Landau, E. M. Lifschitz, *Electrodynamics of Continuous Media*, Oxford: Pergamon, 1960.

[7] S. Ramo, J. R. Whinnery, T. Van Duzer, *Fields and Waves in Communication Electronics*, New York: Wiley, 1965.

[8] J. Van Bladel, *Relativity and Engineering*, Berlin: Springer-Verlag, 1984.

[9] A. Vander Vorst, *Transmission, Propagation et Rayonnement*, Brussels: De Boeck, 1995.

[10] C. Balanis, *Antenna Theory*, New York: Harper & Row, 1982.

[11] H. G. Booker, *Energy in Electromagnetism*, Stevenage: P. Peregrinus, 1982.

[12] L. Tonks, "A new form of the electromagnetic energy when free charged particles are present," *Phys. Review*, Vol. 54, p. 863, Nov. 1938.

[13] A. Vander Vorst, F. Duhamel, "1990–1995 Advances in investigating the interaction of microwave fields with the nervous system," *IEEE Trans. Microwave Theory Tech.*, Vol. 44, pp. 1898–1909, Oct. 1996.

[14] F. T. Ulaby, R. K. Moore, A. K. Fung, *Microwave Remote Sensing*, Vol. I, Norwood, MA: Artech, 1981.

[15] J. D. Kraus, *Radio Astronomy*, New York: Mc-Graw-Hill, 1966.

[16] E. Jordan, *Electromagnetic Waves and Radiating Systems*, Englewoods Cliff, NJ: Prentice-Hall, 1950.

[17] A. Vander Vorst, D. Vanhoenacker-Janvier, *Base de l'Ingénierie Micro-onde*, Brussels: De Boeck, 1996.

[18] A. Gérin, B. Stockbroeckx, A. Vander Vorst, *Champs Micro-ondes et Santé*, Louvain-la-Neuve: EMIC-UCL, 1999.

[19] H. P. Schwan, K. R. Foster, "RF-field with biological systems: Electrical properties and biophysical mechanisms," *Proc. IEEE*, Vol. 68, No. 1, pp. 104–113, Jan. 1980.

[20] H. A. Lorentz, *The Theory of Electrons*, New York: Dover, 1915.

[21] H. Fröhlich, *Theory of Dielectrics*, London: Oxford University Press, 1958.

[22] C. G. Montgomery, *Technique of Microwave Measurements*, New York: Dover, 1966.

[23] E. L. Ginzton, *Microwave Measurements*, New York: McGraw-Hill, 1957.

[24] M. Sucher, J. Fox, *Handbook of Microwave Measurements*, New York: Polytechnic, 1963.

[25] D. Misra, M. Chabbra, B. R. Epstein, M. Mirotznik, K. R. Foster, "Noninvasive electrical characterization of materials at microwave frequencies using an open-ended coaxial line: Test of an improved calibration technique," *IEEE Trans. Microwave Theory Tech.*, Vol. 38, No. 1, pp. 8–14, Jan. 1990.

[26] F. Duhamel, I. Huynen, A. Vander Vorst, "Measurements of complex permittivity of biological and organic liquids up to 110 GHz," *IEEE MTT-S Microwave Int. Symp. Dig.*, Denver, June 1997, pp. 107–110.

[27] I. Huynen, C. Steukers, F. Duhamel, "A wideband line-line dielectrometric method for liquids, soils, and planar substrates," *IEEE Trans. Instrum. Meas.*, Vol. 50, No. 5, pp. 1343–1348, Oct. 2001.

[28] M. R. Tofighi, A. S. Daryoush, "Characterization of the complex permittivity of brain tissues up to 50 GHz utilizing a two-port microstrip test fixture," *IEEE Trans. Microwave Theory Tech.*, Vol. 50, No. 10, pp. 2217–2225, Oct. 2002.

[29] J.-C. Camart, D. Despretz, M. Chivé, J. Pribetich, "Modeling of various kinds of applicators used for microwave hyperthermia based on the FDTD method," in A. Rosen and A. Vander Vorst (Eds.), Special Issue on Medical Applications and Biological Effects of RF/Microwaves, *IEEE Trans. Microwave Theory Tech.*, Vol. 44, No. 10, pp. 1811–1818, Oct. 1996.

PROBLEMS

1.1. In a nonmagnetic and homogeneous lossy medium, the electric field intensity of a plane wave is $E = E_0 e^{-\gamma z}$, where $\gamma = \alpha + j\beta$. The quantities α, β, and γ are the phase, attenuation, and propagation constants, respectively. Knowing that the phase velocity $\gamma = j\omega\sqrt{\mu_0\varepsilon'}$ where μ_0 (= $4\pi \times 10^{-7}$ H m^{-1}) is the permeability of free space, ω is the radian frequency, and ε' is the real part of the complex permittivity ε of the medium:

$$\varepsilon_0 \varepsilon = \varepsilon_0 (\varepsilon' - j\varepsilon'') = \varepsilon_0 \left(\varepsilon' - \frac{j\sigma}{\omega \varepsilon_0} \right) = \varepsilon_0 \varepsilon' (1 - j \tan \delta)$$

In this relation, $\varepsilon_0 = 8.854 \times 10^{-12}\,\mathrm{F\,m^{-1}}$ and σ and $\tan\delta$ $(= \sigma/\omega\varepsilon_0\varepsilon'')$ are the conductivity and the loss tangent of the medium, respectively.

(a) Prove the following expressions:

$$\alpha = \frac{\omega}{c} \sqrt{\frac{\varepsilon'}{2}} \sqrt{\sqrt{1 + \tan \delta^2} - 1} \qquad \mathrm{Np\,m^{-1}}$$

$$\beta = \frac{\omega}{c} \sqrt{\frac{\varepsilon'}{2}} \sqrt{\sqrt{1 + \tan \delta^2} + 1} \qquad \mathrm{rad\,m^{-1}}$$

Hint: Calculate γ^2, which yields $\beta^2 - \alpha^2$ and $\alpha\beta$.

(b) Show that the wavelength $(= 2\pi/\beta)$ is

$$\lambda = \lambda_0 \left(\sqrt{\frac{\varepsilon'}{2}} \sqrt{\sqrt{1 + \tan \delta^2} + 1} \right)^{-1} \qquad \mathrm{m}$$

1.2. The complex permittivity of skin, fat, and muscle at 915 MHz, 2.45 GHz, and 10 GHz, respectively, is given in the following table:

	915 MHz		2.45 GHz		10 GHz	
	ε'	ε''	ε'	ε''	ε'	ε''
Skin	41.5	17	38	11	31	14.5
Fat	11.3	2.2	10.9	2	8.8	3.1
Muscle	55	19	53	12.5	43	19

(a) Calculate the loss tangent and the parameters α, β, and λ for these media at the specified frequencies.

(b) The penetration depth δ is the inverse of the attenuation constant α. Find δ for these media at the specified frequencies.

1.3. A plane wave is incident from semi-infinite medium 1 to semi-infinite medium 2 and propagates normal to the boundary between the two media. The reflection coefficient of a plane wave with normal incidence on a flat boundary is given as

$$\Gamma = \rho e^{j\varphi} = \frac{\sqrt{\varepsilon_1} - \sqrt{\varepsilon_2}}{\sqrt{\varepsilon_1} + \sqrt{\varepsilon_2}}$$

where ρ and φ are the magnitude and phase of the reflection coefficient, respectively. Find the reflection coefficients of the air–skin, skin–fat, and fat–muscle interfaces at the specified frequencies of problem 1.2.

1.4. The electric field intensity of the plane wave inside medium 1 of problem 1.3 is given by $E = E_{01}^+ e^{-\gamma z} + \Gamma E_{01}^+ e^{\gamma z}$. The first term characterizes the wave propagating in the $+z$ direction, and the second term characterizes the wave propagating in the $-z$ direction. The interface determining the reflection coefficient Γ is located at $z = 0$. Show that the absorbed power in the unit volume is

$$P = \tfrac{1}{2}(\sigma|E_{01}^+|)[e^{-2\alpha z} + \rho^2 e^{2\alpha z} + 2\rho \, \cos(2\beta z + \varphi)]$$

1.5. Assume that a wave with $E_0\,(z = 0^-) = 200\,\text{V m}^{-1}$ is propagating in the z direction inside a fat medium and is incident upon a semi-infinite muscle medium that fills $z > 0$.

  (a) Write explicit relations for the power absorbed per unit volume P in both media at 915 MHz, 2.45 GHz, and 10 GHz, respectively.

  (b) Plot the values of power absorbed as a function of z ($-0.1\,\text{m} < z$).

1.6. Repeat problem 1.5 for power absorbed in medium 2, assuming that medium 1 is air and medium 2 is skin. Assume that the skin is thick enough to ensure there is no reflecting wave present in it. Plot the results for $0 < z < 0.003$ m.

1.7. In multilayered cases, the reflection coefficient at the interface of mediums i and $i + 1$ is

$$\Gamma_i = (Z_{\text{in}(i+1)} - Z_{0i})/(Z_{\text{in}(i+1)} + Z_{0i})$$

where $Z_{0i} = 120\pi/(\varepsilon_0\varepsilon_i)^{1/2}$

and

$$Z_{\text{in}(i+1)} = Z_{0(i+1)}\frac{1 + \Gamma_{(i+1)}e^{-2\gamma_{1+i}d_{i+1}}}{1 - \Gamma_{(i+1)}e^{-2\gamma_{1+i}d_{i+1}}}$$

is the input impedance of medium $i + 1$ seen in this boundary, with d_i being the thickness of medium i. Assume that a wave at 915 MHz and $E_0^+ = 1$ V/m is propagating in air ($i = 1$) and is incident upon three layers of skin ($d_2 = 1\,\text{mm}$), fat ($d_3 = 5\,\text{mm}$), and muscle ($d_4 = \infty$).

  (a) Find the reflection coefficient at each interface.

  (b) Find the electric field intensity and absorbed power at the two sides of each interface and 10 mm into the muscle. Note that the electric field inside medium i can be written as $E_i = E_{0i}^+ e^{-\gamma_i(z-z_i)} + \Gamma_i E_{0i}^+ e^{+\gamma_i(z-z_i)}$, where E_{0i}^+ is the electric field in the $+z$ direction wave at $z = z_i$.

  (c) Plot the profile of the absorbed power as a function of the penetration into these media.

1.8. Repeat problem 1.7 for $f = 2.45$ GHz and comment on the results.

1.9. Repeat problem 1.7 for $f = 10$ GHz and comment on the results.

1.10. The dimensions of some standard 50-Ω semiflexible coaxial cables are given in the following table (dimensions in millimeters):

Outer diameter	2.20	3.58	1.19	0.58
$2b$	1.68	2.98	0.94	0.42
$2a$	0.51	0.92	0.29	0.13

The dielectric is Teflon ($\varepsilon_c = 2.1$ and $\tan \delta = 0.0001$). At 915 MHz and 2.4 GHz, find the dielectric, conductive, and total loss if the conductors are copper.

1.11. The 2.2-mm coaxial cable of problem 1.10 is terminated to muscle (as shown in Fig. P1.11).

An approximate model of the system is shown in Figure P1.11*b*, where C_f is the fringing capacitance on the left side of the $z = 0$ boundary and C_0 is the open-end capacitance on the right side of the $z = 0$ boundary, where the line is air terminated. The values $C_0 = 0.05$ pF and $C_f = 0.01$ pF are given.

(a) Show that when the cable is terminated to the tissue, the ratio of the real to the imaginary part of the admittance associated with the external capacitance is the loss tangent of the medium.

(b) Calculate the input admittance and the reflection coefficient at $z = 0$ for f values of 915 MHz and 2.45 GHz.

(c) Calculate the input reflection coefficient seen at a source located at the input of a 100-cm coaxial line terminated at the tissue

(d) If the source provides an available power of 10 W (50 Ω source impedance), how much is the power absorbed in the tissue?

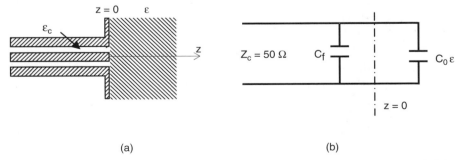

(a)   (b)

FIGURE P1.11   Coaxial probe terminated to lossy medium with complex permittivity ε: (*a*) schematic of the probe; (*b*) equivalent circuit model.

1.12. Repeat problem 1.11 for fat.

1.13. A monopole antenna is fabricated by removing length l of the outer conductor and the dielectric of a 2.2-mm Teflon-filled coaxial cable. For a sufficiently small value of l, the impedance of the antenna in free space is approximated as

$$Z_{in}^0 = 10(kl)^2 - \frac{j60[\ln(2l/a)-1]}{kl}$$

where $k\,(= 2\pi/\lambda = \omega\sqrt{\mu_0\varepsilon_0})$ is the wavenumber in free space and a is the inner conductor radius (Fig. P1.13).

(a) Use what is called Deschamps' theorem

$$Y_{in}(\omega,\varepsilon\varepsilon_0) = \sqrt{\varepsilon}\,Y_{in}^0(\omega\sqrt{\varepsilon},\varepsilon_0)$$

to find a general relation for Y_{in} and Z_{in}, the admittance and impedance of the monopole in the lossy medium, if the antenna is immersed in a tissue with complex permittivity ε.

(b) For 915 MHz and 2.45 GHz, respectively, find the input impedance and reflection coefficient of a monopole with $l = 10$ mm made from a 2.2 mm coaxial cable immersed in muscle.

(c) Repeat part (b) if the antenna is in a fat medium.

1.14. Suppose that the available power in the input to the monopole of Problem 1.13 is 2 W. Find the power absorbed in the tissue.

1.15. An insulated dipole with half-length h and radius a is rounded by two cylindrical concentric dielectric layers with radii b and c, and permittivity ε_2 and ε_3, respectively. The antenna is inside a tissue medium with complex permittivity ε_4. The dipole is fed by a source V_0^e placed at the middle of it (Fig. P1.15a).

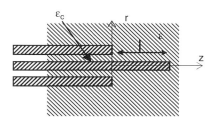

FIGURE P1.13   Coaxial probe monopole applicator immersed in lossy medium with complex permittivity ε.

Given the conditions $|k_4/k_2|^2 \gg 1$, $|k_4/k_3|^2 \gg 1$, $(k_2b)^2 \ll 1$, and $(k_3c)^2 \ll 1$ are satisfied, it is known that the current distribution on the center conductor can be written as[1]

$$I(z) = I(0) \frac{\sin k_L (h - |z|)}{\sin k_L h}$$

where $I(0) = V_0^e/Z_0$ is the input current to the antenna and Z_0 is the antenna input impedance with

$$Y_0 = 1/Z_0 = -(j/2Z_c)\tan k_L h$$

The parameters k_L and Z_c are complex quantities that follow the relations

$$k_L = k_{2e}[\ln(c/a) + F]^{1/2}[\ln(c/a) + n_{24}^2 F]^{-1/2}$$

and

$$Z_c = (\omega\mu_0 k_L / 2\pi k_{2e}^2)[\ln(c/a) + n_{2e4}^2 F]$$

with

$$k_{2e} = k_2 \left[\frac{\ln(c/a)}{\ln(b/a) + n_{23}^2 \ln(c/b)} \right]^{1/2}$$

$$\varepsilon_{2e} = \varepsilon_2 n_{2e4}^2$$

the effective wavenumber and dielectric constant, respectively, of an equivalent dielectric for regions 2 and 3, where $n_{2e4}^2 = k_{2e}^2/k_4^2$, $n_{23}^2 = k_2^2/k_3^2$, $n_{24}^2 = k_2^2/k_4^2$, and $F = H_0^{(1)}(k_4c)/k_4cH_1^{(1)}(k_4c)$. Note that k_2, k_3, and k_4 are the wavenumbers in regions 2, 3, and 4 respectively, and $H_n^{(1)}$ is the nth order Hankel function of the first kind.

A dipole is built with $h = 20\,\text{mm}$, $a = 0.255\,\text{mm}$ and is covered with a dielectric layer (region 2, $b = 0.84\,\text{mm}$, $\varepsilon_2 = 2.1$) followed by an outer layer of plastic tube (region 3, $c = 1.2\,\text{mm}$, $\varepsilon_2 = 4$). The dipole is surrounded by heart tissue ($\varepsilon_4 = 60 - j24.5$ at 915 MHz).

(a) Find ε_{2e}, k_{2e}, Z_c, and k_L.

(b) Find the input impedance of the antenna at 915 MHz.

[1] R. W. King, B. S. Trembly, J. W. Strohbehn, "The electromagnetic field of an insulated antenna in a conducting or dielectric medium," *IEEE Trans. Microwave Theory Tech.*, Vol. MTT-31, No. 7, pp. 574–583, July 1983.

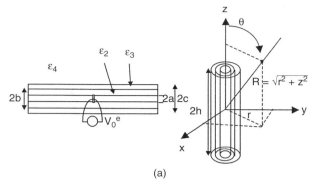

(a)

FIGURE P1.15a   Insulated dipole in lossy medium.

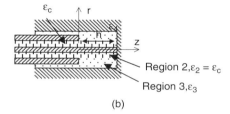

(b)

FIGURE P1.15b   Catheter in lossy medium that is similar to insulated monopole corresponding to dipole of Figure 1.15a.

(c) Plot the current distribution $I(z)$ as a function of axial distance z.

(d) A monopole antenna is built that corresponds to half of the above dipole (Fig. P1.15b). This monopole is fed by a 2.2-mm Teflon-filled coaxial probe ($a = 0.255$ mm, $b = 0.84$ mm, $\varepsilon_c = 2.1$). Find the input impedance of the monopole. Assume that the absence of the flange (ground plane) has a negligible effect on the monopole input admittance.

1.16. Repeat problem 1.15 for $h = 31$ mm, $a = 0.47$ mm, $b = 0.584$ mm, $c = 0.8$ mm, $\varepsilon_2 = 1$, $\varepsilon_3 = 1.78$, $\varepsilon_4 = 42.5 - j17.3$, and $f = 915$ MHz.[2]

1.17. According to the Wien displacement law, the wavelength–temperature product is a constant [Eqn. (1.44)]. In Section 1.4.3 a value of 0.0048 mK

[2] R. W. King, B. S. Trembly, J. W. Strohbehn, "The electromagnetic field of an insulated antenna in a conducting or dielectric medium," *IEEE Trans. Microwave Theory Tech.*, Vol. MTT-31, No. 7, pp. 574–583, July 1983.

is obtained. The derivation involves an approximation. It has been mentioned that a more accurate value of 0.0051 mK can be obtained without this approximation. Prove this assertion.

1.18. Calculate the brightness of a blackbody radiator at a temperature of 6,000 K and a wavelength of 0.5 μm.

1.19. When the brightness is expressed in terms of unit wavelength, the wavelength for the peak brightness is not the same as when brightness is expressed in terms of unit bandwidth. The quantitative relation for the $\lambda_p T$ product when brightness is expressed in terms of unit wavelength is obtained by maximizing (1.42) and simplifying. Prove that it yields the value 0.0029 mK [15].

1.20. Show that the arc plot of a dielectric with a single relaxation time (Cole-Cole diagram, Eqn. 1.52), obtained by representing $\varepsilon''(\omega)$ as a function of $\varepsilon'(\omega)$, is a semicircle. Verify that a given point on the semicircle corresponds to a given frequency while the summit corresponds to $\omega\tau = 1$ and find the points for $\omega = 0$ and $\omega = \infty$. Observe the disadvantage: ω does not appear in it. Any material with a single relaxation time, characterized by ε_s and ε_∞ gives the same arc plot. Observe that it is therefore essential to supply the magnitude of τ in addition to the arc plot.

CHAPTER TWO

RF/Microwave Interaction Mechanisms in Biological Materials

2.1 BIOELECTRICITY

2.1.1 Fundamentals

The effects of the interaction of RF and microwave radiation with biological tissues can be considered as the result of three phenomena:

1. The penetration of EM waves into the living system and their propagation into it.
2. The primary interaction of the waves with biological tissues.
3. The possible secondary effects induced by the primary interaction.

The word *interaction* is important. It stresses the fact that end results not only depend on the action of the field but also are influenced by the reaction of the living system. Living systems have a large capacity for compensating for the effects induced by external influences, in particular EM sources. This is very often overlooked while it is one main reason for which conclusions derived from models have to be taken with precaution. *Physiological compensation* means that the strain imposed by external factors is fully compensated and the organism is able to perform normally. *Pathological compensation* means that the imposed strain leads to the appearance of disturbances within the functions of the organism and even structural alterations may result. The bor-

RF/Microwave Interaction with Biological Tissues, By André Vander Vorst, Arye Rosen, and Youji Kotsuka
Copyright © 2006 by John Wiley & Sons, Inc.

derline between these two types of compensation is obviously not always easy to determine.

The radiation mechanism considered consists of a source that emits EM energy. Part of the incident energy is absorbed and transformed within the biological system. Hence, there is the sequence source–radiation–target. The physical laws of EM field theory, reflection, diffraction, dispersion, interference, optics, and quantum effects, must be applied to investigate and explain the observed phenomena. This is true in general for the whole spectrum of EM radiation. In this book, however, the study is limited to RF and microwaves.

The increasing industrialization of the world and the tendency to increase the power of equipment raised the question of health risks first for personnel, then for the general public. It gave an impetus to carry out large research projects and collect a vast amount of experimental data and clinical observations. Before starting any interpretation of the results obtained, however, it is necessary to survey the basic phenomena involved in the interaction of RF and microwave radiation with living systems. The first step is to review basic bioelectricity. Further reading on this subject can be found in a number of textbooks [e.g., 1]. An excellent summary is by Reilly et al., on which this section is largely based [2].

Natural bioelectric processes are responsible for nerve and muscle function. Externally applied electric currents can excite nerve and muscle cells. The nervous system is concerned with the rapid transfer of information through the body in the form of electrical signals. It is conveniently divided into the central nervous system (CNS) and the peripheral nervous system. The CNS consists of the brain and spinal cord. The peripheral nervous system consists of *afferent* neurons, which convey information inward to the CNS, and *efferent* neurons, which convey information from the CNS to the body. The efferent system is subdivided into a somatic nervous system and an autonomic nervous system. The autonomic nervous system consists of neurons that convey impulses to smooth muscle tissue, cardiac muscle tissue, and glands, which are usually considered involuntary, that is, not under conscious control. The autonomic nervous system has sympathetic and parasympathetic divisions. Because they control opposite effects in various organs, they are usually considered antagonistic to each other. The sympathetic nervous system tends to mobilize the body for emergencies (secretion of adrenaline), whereas the parasympathetic nervous system is concerned with vegetating functions of the body, such as digestion. One responsibility of the autonomic nervous system is to maintain body homeostasis.

Muscles can be stimulated directly or indirectly through the nerves that enervate the muscle. Thresholds of stimulation of nerve are generally well below thresholds for direct stimulation of a muscle. Hence, an understanding of neuroelectric principles is a valuable foundation for investigating both sensory and muscular responses to electrical stimulation.

2.1.2 Cells and Nerves

The functional boundary of a biological cell is a bimolecular lipid and protein structure termed the *membrane*, a semipermeable dielectric that allows some ionic interchange. The membrane is thin, of the order of 10 nm. Electromechanical forces across the membrane regulate chemical exchange across the cell. The medium inside the cell (*plasma*) and outside the cell (*interstitial fluid*) is composed largely of water containing various ions. Ions particularly involved with the electrical response of nerve and muscle are sodium and potassium, Na^+ and K^+, respectively. The concentrations of these ions inside and outside the cell dictate the electric potential, termed the *Nernst potential*. The difference in the concentrations of ions inside and outside the cell causes an electromechanical force across the membrane. Example concentrations for a nerve fiber are $[Na^+]_i = 50$, $[Na^+]_o = 460$, $[K^+]_i = 400$, and $[K^+]_o = 10\,\mu M\,cm^{-3}$ [2].

The term *nerve* usually refers to a bundle of nerve fibers. *Neurons* are nerve cells. Sensory neurons carry information from sensory receptors in the peripheral nervous system to the brain; motor neurons carry information from the brain to the muscles. The conducting portion of the nerve fiber is a long, hollow structure termed an *axon*. A neuron can be classified in the presence or absence of *myelin*, which is a fatty substance. When a neuron is covered with myelin, it is termed a myelinated nerve or fiber. A myelinated nerve fiber has periodically exposed unmyelinated gaps, termed *nodes of Ranvier*. The length of the gaps is in the range 1–2.5 μm and the distance between the nodes in the range 0.2–2 mm. Myelinated nerve cells have diameters typically in the range 2–20 μm and conduction velocities of 5–120 m/s. Unmyelinated nerves have diameters of 0.3–1.3 μm and conduction velocities of 0.6–2.3 m s^{-1}. Fiber lengths may be up to 1 m.

The body is equipped with a vast array of sensors (receptors) for monitoring its internal and external environment. Specialized receptors are associated with the visual and auditory systems. Chemical receptors make neurons communicate with one another. The *somatosensory* system is found in the skin and internal organs. It is generally involved in electrical stimulation. The somatosensory system includes a variety of sensors, for instance, for mechanical stimulation, chemical stimulation, and heat and cold stimuli. It also includes nociceptors that are usually associated with pain and respond only when the stimulus reaches the point where tissue damage is imminent. They respond to a broad spectrum of noxious levels of mechanical, heat, and chemical stimuli. The muscles are equipped with specialized receptors to monitor and control muscle posture and movement.

When a sensory receptor is stimulated, it produces a voltage change termed a *generator potential*. The generator potential is graded: Squeezing a pressure sensor, for instance, produces a voltage, while squeezing it harder produces a higher voltage. The generator potential initiates a sequence of events that leads to a propagating *action potential*, which in common parlance is usually termed a "nerve impulse." The action potential is a propagating change in the conductivity and potential across a nerve cell's membrane. More specifically, it is a rapid change in the membrane potential that involves a depolarization followed by a repolarization. *Evoked potentials* are modifications of the electric brain activity due to the application of sensory stimulation [3]. Depending on the type of stimulation, they can be visual, auditive, or somesthetic.

In a quiescent state, nerve and muscle cells maintain a membrane potential typically around –60 to –90 mV, with the inside of the cell negative relative to the outside. This is termed the *resting* potential. The potential is referred to a reference plane that is the outside solution in which the nerve cell is immersed. The cell remains in a state of electrochemical disequilibrium, because the Na^+ potential is of the order of +60 mV while that of K^+ is somewhat more negative than the resting potential. The energy that maintains this force derives from the metabolic process of the living cell, which is the method by which cells use oxygen and produce carbon dioxide and heat. Considering the value of the transmembrane potential ($\cong 100$ mV) and the membrane thinness ($\cong 10$ nm), the electric field across the membrane is enormous ($\cong 10$ MV m^{-1}). Hence, a cell membrane resembles a charged capacitor operating near the breakdown voltage. The transmembrane charge is the result of a metabolic process. A typical value for the capacitance is about $1\,\mu F\,cm^{-2}$.

The membrane is a semipermeable dielectric that allows some ionic interchange. The ionic permeability of the cell membrane, however, varies substantially from one ionic species to another. Also, the ionic channels in the excitable membrane will vary their permeability in response to the transmembrane potential; this property distinguishes the excitable membrane from the ordinary cellular membrane and supports propagation of nerve impulses.

The electrodynamics of the excitable membrane of unmyelinated nerves was first described in detail by Hodgkin and Huxley, a work for which they obtained the Nobel Prize [4]. They established an electrical model for the membrane; this model was extended later, in particular to the myelinated nerve membrane [5]. The model is represented in Figure 2.1. It includes potential sources, which are the Nernst potentials for the particular ions. The dielectric membrane separating the conductive media on either side forms a capacitance C_m. The nonlinear conductances g_{Na} and g_K apply to Na^+ and K^+ channels, respectively. They highly depend on the voltage applied across the membrane as described by a set of nonlinear differential equations. The conductance g_L is a leakage channel that is not specific to any particular ion.

In the resting state one has $g_{Na} \ll g_K$, and the membrane potential moves toward the Nernst potential for Na^+. This is the depolarized state and the membrane is said to be excited. The transition between the resting and excited conditions of the membrane occurs rather abruptly when the membrane potential has been depolarized by about 15 mV. The duration of the excited state lasts for about 1 ms. After excitation, the ionic channel conductances vary again, causing the membrane to revert back to its resting potential. Then, it cannot be reexcited until a recovery period, termed the *refractory period*, has passed.

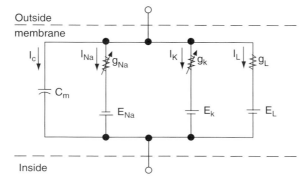

FIGURE 2.1   Hodgkin–Huxley membrane model.

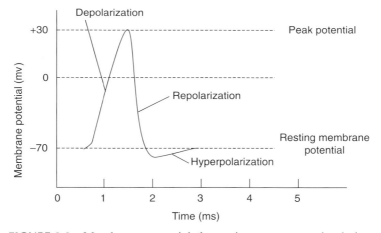

FIGURE 2.2   Membrane potential change in response to stimulation.

The *absolute* refractory period refers to the period of time during which a second action potential cannot be initiated, even with very strong stimulus; it is of the order of 0.4 ms for large fibers up to possibly 4 ms for small fibers [6].

Figure 2.2 shows a typical evolution of the membrane potential responding to stimulation. The *relative* refractory period refers to the period of time during which a second action potential can be initiated, but only by a stronger than normal stimulus. It corresponds roughly to the period of increased K^+ permeability resulting in a brief hyperpolarization, which means that the cell is more negative than the resting level.

In other words, when considering the membrane in the resting state, the concentration of Na^+ is high outside and low inside the cell while, in contrast, the concentration of K^+ is higher inside the cell. Therefore, the K^+ ions tend to diffuse (leak) out of the cell while the Na^+ ions tend to diffuse into the cell, down both a concentration and an electric field gradient. If an excitatory stimulus higher than the threshold stimulus is applied to a polarized membrane, the membrane's permeability to Na^+ ions greatly increases at the point of stimulation. This is because voltage-gated Na^+ channels open and permit the influx of Na^+ ions by diffusion. Because positively charged Na^+ ions are entering the cell, the membrane potential starts to change. The potential inside the membrane shifts from about $-70\,mV$ first toward $0\,mV$, then to a positive value; the membrane loses its polarity and becomes depolarized.

A voltage-gated ion channel, for instance for Na^+, has two separate gates: an activation and an inactivation gate. For a resting membrane, the activation gate is closed and the inactivation gate is open. When a threshold stimulus is applied to a polarized membrane, the same voltage change that causes the activation gate to open also causes the inactivation gate to close. The closure, however, occurs about $10^{-4}\,s$ after the opening of the activation gate. When the membrane becomes depolarized, the voltage change causes a slow opening of voltage-sensitive K^+ channels. They open simultaneously with the closure of the voltage-sensitive Na^+ channels. The outward diffusion of K^+, down both the concentration and electric field gradients, causes the outer surface of the membrane to become electrically positive. The loss of positive ions leaves the inner surface of the membrane negative again, and the membrane is repolarized.

The action potential is the progression of the membrane voltage during the period of excitation and recovery. A nerve impulse that is generated at any point of the membrane usually excites adjacent portions of the membrane, causing the impulse to propagate. The body's natural condition, however, is to initiate an action potential at the end of the axon, which then propagates in only one direction. The propagation relies on local electrical currents along the membrane and is termed *continuous* conduction. In myelinated fibers the activation of the membrane occurs only at nodes of Ranvier. A nerve impulse jumps from node to node, where the impulse is regenerated. This type of impulse conduction is termed *saltatory* conduction.

The *propagation velocity* of a nerve impulse is independent of the stimulus strength. Once a neuron reaches its threshold of stimulation, the propagation velocity is normally determined by the diameter of the fiber, the temperature, and the presence or absence of myelin. Conduction may also be altered by conditions such as the presence of toxic material in the cells, fatigue, and temperature. In principle, however, the action potential is transmitted or not. If a stimulus is strong enough to generate a nerve action potential, the impulse is conducted along the entire neuron at a constant and maximum strength for the existing conditions. Even if the stimulus intensity is increased, the amplitude of the action potential will remain unchanged. A propagation velocity of about $40\,m\,s^{-1}$ is typical of a 20-μm fiber [2]. Nerve excitation is investigated by nonlinear differential equations, which relate the transmission line equations to the driving function, that is, the electric field in the biological medium, depending of course on how the current is distributed in the medium.

2.1.3   Bioelectric Phenomena

All living cells exhibit bioelectric phenomena. However, a small variety only produce potential changes that reveal their physiological function. There are familiar bioelectric recordings. The three most prominent bioelectric effects, those of heart, skeletal muscle, and brain, are recorded by the following: the electrocardiogram (ECG), reflecting the excitation and recovery of the whole heart; the electromyogram (EMG), reflecting the activity of skeletal muscle; and the electroencephalogram (EEG), reflecting the activity of the outer layers of the brain, the cortex. In these cases, the action potentials are used for diagnostic purposes, and extracellular electrodes are used that are both large and distant from the population of cells that become active and recover. The depolarization and repolarization processes send small currents through the conducting environmental tissues and fluids, resulting in a time-varying potential field. Appropriately placed electrodes can record the electrical activity of the bioelectric generators. However, the waveforms of such recordings are vastly different from those of the transmembrane action potentials. It has been shown by Geddes and Baker that such extracellular recordings resemble the second derivative of the excursion in the transmembrane potential [7]. Despite the difference in waveform, extracellular recordings identify the excitation and recovery processes very well [2]. Furthermore the eye, ear, sweat glands, and many types of smooth muscles also produce action potentials that are used for their diagnostic value [7].

Bioelectricity is extremely important in a living body. It has long been shown that direct application of an externally generated voltage may have an effect on bone and cartilage repair. Considerable animal and in vitro experimentation suggests the clinical usefulness of electric currents for soft tissue repair and possibly to enhance repair of nerve fibers that have sustained crush or transsection injury [2]. There is no doubt that bioelectricity has to be taken into account seriously when investigating possible medical applications of RFs and microwaves as well as when wondering about possible hazards on human beings and animals due to RF or microwave exposure. These issues will be examined in more detail in subsequent sections of this chapter.

2.2   TISSUE CHARACTERIZATION

The classical book by Michaelson and Lin, reviewing in 1987 the entire area of biological effects due to RF radiation, is still a reference for those who want to acquire a good knowledge of the field [8]. Thuery reviewed, in 1992, the industrial, scientific, and medical applications of microwaves, with a number of basic information sources on the interaction between microwave fields and the nervous system [9]. In 1996, Polk and Postow reviewed the whole field of biological effects of EM fields [10]. More specifically, a summary of the action of microwave EM fields on the nervous system has been published recently [11].

2.2.1 Ionization and Nonionization

Radio frequencies and microwaves are nonionizing radiation. To appreciate the difference between nonionizing and ionizing radiation, one must note that the energy of EM waves is quantified with the quantum of energy (in joules) being equal to Planck's constant times frequency [Eqn. (1.50)]. This energy can also be expressed in electron volts, that is, in multiples of the kinetic energy acquired by an electron accelerated by a potential difference of 1 V, so that $1\,eV \cong 1.6 \times 10^{-19}\,J$. The energy is proportional to frequency: While the energy of one quantum of energy is 4.12×10^{-7} at 100 MHz, it is 4.12×10^{-5} at 10 GHz, 41.2 at $10^{16}\,Hz$ (ionizing UV rays), and 4.12×10^5 at $10^{20}\,Hz$ (penetrating X-rays) [10].

Ionization can be brought about not only by absorption of EM energy but also either by collision with foreign (injected) elementary particles of the requisite energy or by sufficiently violent collision among its own atoms. Ionization by any outside agent of the complex compounds that make up a living system leads to profound and often irreversible changes in the operation of the system. Ionization is due to the possible coupling of appropriate frequencies to vibrational and rotational oscillations. If the incident energy quantum has sufficient magnitude, it can tear an electron away from one of the constituent atoms. The "ionization potential" is the energy required to remove one electron from the highest energy orbit. Typical ionization potentials are of the order of 10 eV. Since chemical binding forces are essentially electrostatic, ionization implies profound chemical changes.

At RF/microwaves, and even at millimeter waves, the quantum energies are well below the ionization potential of any known substance. Excitation of coherent vibrational and rotational modes requires considerably less energy than ionization and it can occur at RF. Many other possible biological effects require energies well below the level of ionizing potentials, such as heating, dielectrophoresis, depolarization of cell membranes, and piezoelectric transduction. In all these cases, one has of course to estimate the rates at which energy has to be delivered to produce specific effects.

2.2.2 Dielectric Characterization

The dielectric properties of materials have been characterized in Sections 1.6 and 1.7. The characterization, however, was limited to bulk homogeneous materials. The present section will investigate dielectric properties of tissues in more detail, in particular for heterogeneous materials. The interested reader may find more information in a number of books, in particular in [8–10].

Three relaxation processes are mainly responsible for the dielectric properties of tissues: dipolar orientation, interfacial polarization, and ionic diffusion. The theories summarized below apply to linear responses to weak fields. As the field intensity is increased, at some level the response will no longer be linear. The threshold at which nonlinearity becomes noticeable depends on the system and the investigated dielectric effect. Nonlinearity thresholds have been evaluated elsewhere [12].

Dipolar Orientation   Dipolar orientation has been described in Sections 1.6 and 1.7. The basic theory is macroscopic and does not strictly apply to molecular systems. It applies poorly to liquids, especially to water. In tissues, several dipolar effects may be anticipated. Globular proteins typically exhibit total dielectric increments of the order of 1–10 per gram of protein per 100 g of solution, with relaxation frequencies in the range 1–10 MHz. Polar side chains on protein relax at some higher frequencies. However, these are likely to be relatively small effects in tissues at RFs and below for which charging phenomena or counterion effects dominate the dielectric properties.

Water is the major constituent in most tissues, and in several respects the dielectric properties of tissues reflect those of water. At RFs, the conductivity of tissue is essentially that of its intracellular and extracellular fluids. At microwave frequencies the dielectric dispersion arises from the dipolar relaxation of the bulk tissue water. Dipolar relaxation of water is a dominant effect in tissues at microwave frequencies. Pure water undergoes nearly single time constant relaxation centered at 20 GHz at room temperature and 25 GHz at 37°C. Water associated with protein surfaces has a lower relaxation frequency than that of the bulk liquid, and this water fraction contributes noticeably to the dielectric dispersion at frequencies near 1 GHz.

Interfacial Relaxation   If a material is electrically heterogeneous, charges do appear at the interfaces within the material because of boundary conditions at the interfaces. Interfacial effects typically dominate the dielectric properties of colloids and emulsions. Evaluating the effect of interfaces within a material is a classical EM problem, in particular for evaluating the transmission and reflection of EM waves, solved in a number of textbooks [e.g., 13–15]. Simple models can easily be analyzed.

Cartesian configurations are most easy to evaluate, for instance two semi-infinite media in contact with each other, one slab of a given thickness inserted into an infinite medium separated in two parts by the slab, two slabs in contact with each other, and so on. The bulk permittivity and conductivity of the composite material can easily be calculated. When the bulk material properties of the constituent phases vary with frequency, the frequency dependence of the heterogeneous material can no longer be characterized by a single relaxation time.

The conductivity of a dilute suspension of spherical particles has been calculated by Maxwell as a function of the volume fraction of the suspended particles. His analysis has been extended by Wagner to the case of alternating currents, which led to a set of dispersion equation in the form of Eqn. (1.9). These results clearly show that the presence of heterogeneity introduces dispersion in the material. This is often termed the Maxwell–Wagner theory. Fricke extended it to the case of prolate or oblate spheroids for which a shape factor has to be introduced, yielding what is sometimes termed the Maxwell–Fricke mixture theory. It has also been extended to cover the cases of a spherical particle contained within a larger sphere and of a suspension of spheres surrounded by a thin shell of a material with different dielectric properties. These structures are membrane-covered spheres, such as cells suspended in physiological fluids. It is shown that the dielectric properties of the suspension are rather insensitive to the membrane conductance at RFs. An approximate analysis might not be valid for small cells or concentrated cell suspensions or at high frequencies. A more detailed analysis shows that there is a second dispersion, generally at frequencies in the range of 100–1000 MHz. An equivalent circuit for the Maxwell–Wagner dispersion has been established and a physical interpretation of these results has been suggested. At low frequencies, the current flows around the cells; at high frequencies the current flows without restriction through the cells. This gives rise to a dispersion whose relaxation frequency corresponds to the midpoint of the transition between these two limiting situations and is determined by the time constant for charging the cell membrane capacitance through the extracellular and intracellular fluids. This time constant is of the form RC, where R is a resistance, inverse to the conductivity, and C the capacitance of the permittivity. Maxwell–Wagner effects typically are dominant at RFs [12].

Ionic Diffusion: Counterion Polarization Effects   Counterion phenomena are due to ionic diffusion in the electrical double layers adjacent to charged surfaces. Counterion polarization effects have been reported in a variety of systems containing charged surfaces: emulsions, suspensions of charged polystyrene spheres, microorganisms, and long-chain macromolecular polyions such as DNA. The time constants for such effects are of the form L^2/D, where L is the length over which diffusion occurs and D is a diffusion coefficient. [In contrast, the time constant for the Maxwell–Wagner effect is of the form of RC or $\varepsilon\varepsilon_0/\sigma$, where R (or $1/\sigma$) is a resistance and C (or $\varepsilon\varepsilon_0$) a capacitance]. Counterion effects contribute to the alpha dispersion that is found in tissues at low frequencies. They are dominant in biological systems at kilohertz frequencies and below: Suspensions of submicrometer-sized polystyrene spheres have permittivity values approaching 10^4 with relaxation frequencies in the kilohertz range [12].

The recent model by Grosse assumes that the counterion layer is very thin and the diameter of the particle much larger than the Debye screening length of the electrolyte [16, 17]. The model also assumes that the counterion layer contains only ions that are opposite in sign to that of the fixed charges on the particles, and the counterions can exchange freely only with ions of the same sign in the bulk electrolyte. When an electric field is applied, the ions redistribute under the influence of both the field and diffusion. A set of coupled differential equations are obtained for the ion concentrations and current densities, which yields a broad, asymmetrical, low-frequency dispersion in the permittivity of the particle. The dispersion is broad and can be fitted to a Cole–Cole relaxation function with $\alpha = 0.4$ [see Eqn. (1.37)].

The physical interpretation of counterion polarization is rather simple [10]. The motion of an ion in the bulk electrolyte near the particle depends greatly on whether its sign is the same or opposite than of the ions in the counterion layer. Ions of the same sign can enter the counterion layer and their charge is quickly conducted to the opposite side of the particle: The particle acts as a good conductor. Those of opposite sign are excluded from the layer and must travel around the particle in the bulk electrolyte surrounding the particle: The particle acts as an insulator. As a result, charges accumulate in the electrolyte near the particle, giving rise to an induced dipole moment of the system and hence a large permittivity of the suspension.

2.2.3 Dielectric Dispersion in Tissues

The Kramer and Kronig equations [Eqn. (1.12)] relate the real (imaginary) part of the permittivity to the imaginary (real) part of the permittivity through an integral over the whole frequency range. One may also say that they relate the conductivity of materials, including biological tissues, to their permittivity. It has been observed in Section 1.3.1 that for the imaginary part to exist, the permittivity has to vary as a function of frequency. When a material has parameters such as permittivity, conductivity, and/or permeability varying as a function of frequency, it is said to be *dispersive*; this property is called *dispersion*. Evaluating dielectric dispersion consists in considering the variation of the properties as a function of frequency. On the other hand, when a material has a nonzero imaginary part of the permittivity (or of the permeability), it exhibits losses and is said to be *dissipative* (or *lossy*); this property is called *dissipation*. It can easily be proven that dissipation induces dispersion and reciprocally, but in a universe satisfying causality [18].

In Section 1.3.1, Eqns. (1.15) and (1.16) formulated the correspondence between the conductivity σ and the imaginary part of permittivity ε'' of a material; in particular,

$$\sigma_{eff} = \sigma' + \omega\varepsilon'' \qquad S\ m^{-1} \qquad [Eq.\ (1.15)]$$
$$\sigma''_{eff} = \sigma/\omega + \varepsilon'' \qquad F\ m^{-1} \qquad [Eq.\ (1.16)]$$

This shows that in tissues at low frequencies one has $\varepsilon''_{eff} \gg \varepsilon'$, and for most practical purposes only their conductivity must be taken into account. This is not true at higher frequencies, as will be shown below. Equations (1.15) and (1.16) show that an increase of conductivity with frequency is associated with a decrease of permittivity.

It has already been said that water is the major constituent in most tissues. The water contained in tissues is sometimes called *biological water*. It is obviously difficult to evaluate the differences between bulk tissue water and bulk water. Dielectric relaxation, conductivity, and diffusion have however been investigated for the sake of comparison. For instance, Cole–Cole representations show a relaxation time of the water in muscle tissue 1.5 times longer than that of pure liquid water, which is not much of a difference [12]. The question of relaxation time of tissue water is, more precisely, that of the distribution or relaxation times, as described in Section 1.6.3.

Conductivity   At low frequencies, below 100 kHz, a cell is poorly conducting compared to the surrounding electrolyte, and only the extracellular fluid is available to current flow. A typical conductivity of soft, high-water-content tissues at low frequencies is 0.1 or $0.2 \, S \, m^{-1}$. It varies strongly on the volume fraction of extracellular fluid, which can be expected to vary with physiological changes in the cells.

At RFs, from 1 to 100 MHz, the cell membranes are largely shorted out and do not offer significant barrier to current flow. The tissues can be considered to be electrically equivalent to suspensions of nonconductive protein (and other solids) in electrolyte. The conductivity of most tissues approaches a plateau between 10 and 100 MHz.

At microwaves, above 100 MHz, three effects add significantly to the conductivity:

1. *Maxwell–Wagner Effect Due to Interfacial Polarization of Tissue Solids through Tissue Electrolyte.* It introduces some dispersion at several hundred megahertz because of the difference in the electrical properties of tissue protein and electrolyte. Assuming a tissue–electrolyte conductivity of about $1 \, S \, m^{-1}$ and a relative permittivity of around 10 with no conductivity for the protein predicts increases in conductivity of only a few hundredths of siemens per meter with a relaxation frequency of about 300 MHz.

2. *Dielectric Loss of Small Polar Molecules and Polar Side Chains on Proteins.* The time constant for dipolar relaxation of globular proteins varies as the cube of the protein radius [12]. Thus, small molecules relax at higher frequencies than larger ones. Since the maximum conductivity increase is proportional to the mean relaxation frequency, a small component with a high relaxation frequency may contribute noticeably to the total conductivity at these high frequencies.

3. *Dielectric Relaxation of Water.* As has been said in Section 2.2.2, water is the major constituent in most tissues, and in several respects the dielectric properties of tissues reflect those of water. Dipolar relaxation of water is a dominant effect in tissues at microwave frequencies. Pure water undergoes nearly single time constant relaxation centered at 20 GHz at room temperature. Using Eqn. (1.9), it can easily be calculated that its conductivity will rise nearly quadratically with frequencies below about 10 GHz. For typical high-water-content tissues, the conductivity at 3–5 GHz is due in roughly equal proportions to ionic conductivity and to dipolar loss of water.

Permittivity   The relative permittivity of tissues at frequencies as low as 50–60 Hz is extraordinary large, being of the order of 10^5–10^7, typically 10^6. Figure 1.9 illustrates the variation of tissue permittivity as a function of frequency. There are three major dispersion regions where the value of permittivity is varying strongly with frequency, called alpha, beta, and gamma, at relaxation frequencies in the ranges of kilohertz, hundreds of kilohertz, and gigahertz, respectively. The permittivity undergoes an almost monotonous decrease over the entire frequency range.

The astonishingly high value of the permittivity at very low frequency (ELF), due to alpha dispersion, can be largely ascribed to counterion diffusion effects. Indeed, theory predicts a dielectric increment of the order of 10^6 [16, 17]. The alpha dispersion also results from other contributions: active membrane conductance phenomena, charging of intracellular membrane-bound organelles that connect with the outer cell membrane, and perhaps frequency dependence in the membrane impedance itself [19]. Although alpha dispersion is very striking in the permittivity, it does not appear in the conductivity. Assuming a dielectric increment of 10^6 and a relaxation frequency of 100 Hz, the Kramer and Kronig relations yield a total increase in conductivity associated with alpha dispersion of about $0.005 \, \mathrm{S \, m^{-1}}$ while the ionic conductivity is about 200 times larger. Thus, at low frequencies, tissues are essentially resistive despite their tremendous permittivity values.

Beta dispersion occurs at RFs. It arises principally from the capacitive charging of cellular membranes in tissues. A small contribution might also come from dipolar orientation of tissue proteins at high RFs. As an example, blood exhibits a total dielectric increment of 2000 and a beta relaxation frequency of 3 MHz [12]. The associated increase in ion conductivity is about $0.4 \, \mathrm{S \, m^{-1}}$. For tissues, the static permittivity and relaxation times of this dispersion are typically larger than in blood.

Gamma dispersion occurs with a center frequency near 25 GHz at body temperature. It is due to the dipolar relaxation of the water, which accounts for 80% of the volume of most soft tissues, yielding a total dielectric increment of 50. These values of dielectric increment and relaxation frequency yield a total increase in conductivity of about $70 \, \mathrm{S \, m^{-1}}$ [12].

Some authors have called delta dispersion a small dispersion occurring in tissues and other biological materials between 0.1 and 3 GHz. The lack of a single dominant mechanism makes the interpretation of this dispersion region difficult.

2.2.4   Measurements

Tissues   A good knowledge of the complex permittivity of biological media is necessary for adequately determining the action of EM fields, in biological effects as well as in medical applications. Hyperthermia is only one example. Dielectric properties of a variety of selected tissues can be found in three main handbooks [8–10]. In particular, blood has been investigated for many years [20].

The low-frequency conductivity of muscle tissues has been reviewed in detail [21]. Muscle exhibits an extreme anisotropy in its configuration and hence in its electrical properties: There is a 7- to 10-fold variation in conductivity and permittivity (of dog skeletal muscle) at low frequencies. Similar variations have been reported in skeletal and heart muscle from many other species. For muscle tissue oriented perpendicular to the external field, the plateau between alpha and the beta dispersion is at about 10^5 and the relaxation frequency is around 250 kHz. In the longitudinal orientation, the conductivity at low frequencies is higher and varies much less as a function of frequency. At microwaves, there is not really a distinct separation between beta and higher frequency dispersions.

There has been interest also on other soft tissues with high water content, such as liver or breast tissue, either normal or tumoral. The main nonwater component of soft tissue is protein. At RFs, beta dispersion is related to a broad distribution of relaxation times, which arises principally from the presence of membrane-bound structures of widely varying dimensions. A simple model for liver, assuming that the contributions to permittivity from each major tissue structure are additive, yields remarkably good agreement with the measured properties of the tissue above 1 MHz [22]. One main result is that the dielectric dispersion in liver tissue in the range of 1–100 MHz represents the high-frequency end of the beta dispersion of the cells together with the organelles. Dielectric properties have been parameterized at RF and microwaves and empirical equations have been obtained from data over the frequency range 0.01–18 GHz [12]:

$$\frac{\varepsilon'}{\varepsilon} = 1.71 f^{-1.13} + \frac{\varepsilon_s^m - 4}{1 + (f/25)^2} + 4$$

$$\sigma = 1.35 f^{0.13} \sigma_{0.1} + \frac{0.00222(\varepsilon_s^m - 4)f^2}{1 + (f/25)^2} \tag{2.1}$$

where f is in gigahertz and σ in siemens per meter. These properties may be used to estimate the dielectric properties at somewhat higher frequencies. However, they are not valid below this range because they do not include the effects of beta dispersion.

Tumor tissues have often been found noticeably different from normal tissues, but not in all the experimental data [12]. They have significantly higher water content than normal tissues. Necrosis in the tumor leads to breakdown of cell membranes, so that a larger fraction of the tissue can carry current at low frequencies. Experimental data showed that necrosis yields 5- to 10-fold higher conductivity while the permittivity is generally smaller than that of normal tissue. The infiltration in normal tissue of neoplastic tissues of high water content leads to pronounced changes in the dielectric properties, for instance in breast tissues. However, the practical significance of these differences is unclear. For instance, they should be taken into account when selecting an optimal frequency for RF/microwave hyperthermia treatment (see Chapter 4).

The dielectric properties of bone have also been investigated under near-normal physiological conditions. The conductivity of bone at low frequencies is associated with fluid-filled channels that permeate the tissue and is proportional to the conductivity of the medium surrounding the tissue. The DC conductivity varies by a factor of 3 with orientation. The corresponding variation of permittivity is largely unknown. Fluid-saturated bones exhibit a permittivity of about 1000 at audio frequencies, decreasing to 10–20 at 100 MHz. The conductivity increases by about 0.05 S m^{-1} over the same frequency range. This dispersion can be fitted by a Cole–Cole function with distribution parameter $\alpha = 0.5$ and a center relaxation frequency of a few kilohertz [23, 24].

Adipose tissues such as fat and bone marrow are distinguished by their low water content and by cells largely filled with lipids. Fat and bone marrow show a large alpha dispersion between 10^4 and 10^5 Hz. Beta dispersion is either absent or small in comparison with that of soft tissues. It has been observed that the conductivity of fat samples is higher than that of a soft tissue such as liver, probably because of its larger extracellular fluid fraction [25].

The electric parameters of a number of biological tissues may be found in handbooks [8–10]. Table 2.1 summarizes some parameters for a few tissues.

Liquids   There are not many measured data for biological and organic liquids at frequencies above 20 GHz. Such measurements require some experimental skill. A method for measuring the complex permittivity of liquid dielectrics at microwaves is described in Section 1.7.3 in detail. Some results obtained when using this method are presented in this section.

The usual practice for liquids is to measure the reflection coefficient of an open-ended coaxial probe [26]. As has been said in Section 1.7.3, two difficulties are related to this practice, especially when using it at high microwave frequencies: The structure radiates and the size of the coaxial probe is very small.

TABLE 2.1 Some Biological Parameters

Tissue	Volume Fraction of Water	Extrapolated Microwave Relative Permittivity	Conductivity at 0.1 GHz (S m^{-1})	Extrapolated Microwave Conductivity (S m^{-1})
Brain (gray matter)	0.84	44	0.70	1.13
Brain (white matter)	0.74	34	0.48	0.75
Skeletal muscle	0.795	47	0.70	2.40
Fat	0.09	10	0.005	0.10
Liver	0.795	43	0.67	2.30

Source: From [12].

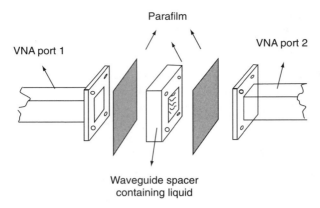

FIGURE 2.3   Measurement configuration.

A new procedure for measuring the complex permittivity of liquids was recently described [27]. It is based on the measurement of the scattering parameters of waveguide two-ports, up to 110 GHz, and on an original calibration method [28, 29]. These measurements led to measured values of the complex permittivity of biological and organic liquids in the range from below 10 to above 100 GHz. The liquids were methanol, dioxane, blood, and axoplasm, measured at frequencies from 8 to 110 GHz in five waveguide bands.

Figure 2.3 illustrates the measurement configuration. A waveguide spacer is placed between the two waveguide ends of the vector analyzer. The connection between the waveguides and both ends of the analyzer is made by coax-to-waveguide transitions, appropriate for each frequency band. For calibration purpose and for using the LL method, two waveguide spacers that are identical except for their thickness are necessary. The LL method [29] requires a "short" and a "long" line. A synthetic film (Parafilm) is placed at both ends between waveguides and spacers, hence containing the liquid in a known volume. Reference planes are placed at both ends in the plane of each film. The LL method ensures that the transition flange–Parafilm–flange has no effect on the result, provided this transition is reproducible when using both spacers. In practice, the Parafilm is put on both sides of the waveguide spacer and soldered together. Holes are made in the film, in the place of the holes of the waveguide flanges, to ensure tightness. Then, using a syringe, the liquid is inserted into the spacer. Air bubbles must be avoided. The two waveguides are then screwed together, through the holes of the spacer. This operation has to be done for each of the two spacers and for each waveguide band.

As an example of obtained results, Figure 2.4 presents the value of dioxane ($C_4H_8O_2$) measured between 50 and 100 GHz. Only the real part of the relative permittivity is presented, because the imaginary value is absolutely indiscernible. Dioxane is a lossless liquid dielectric up to at least 100 GHz and, as

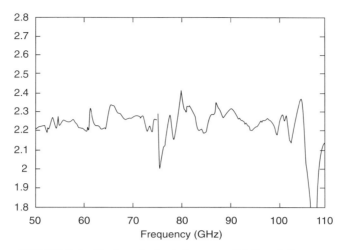

FIGURE 2.4   Real part of relative permittivity of dioxane.

such, is extremely useful as a calibration dielectric to check if a procedure is correct, in particular when changing the frequency from one waveguide band to the next one. These measurements complete results published earlier up to 20 GHz [8]. They show that the dielectric constant of dioxane is equal to 2.25 over the whole microwave range and part of the millimeter waves, up to 110 GHz.

The two solid curves in Figure 2.5 show the real and imaginary parts of the relative permittivity of methanol (CH_3OH), respectively, from 8.2 to 110 GHz. (Results in the frequency band 12–18 GHz are missing.) The two dashed curves result from Debye's equation. A discrepancy between Debye's law and the measurements is visible in the high-frequency range, above 20 GHz. These results complete results published earlier up to 20 GHz [26].

Figure 2.6 presents (solid curves) the real and imaginary parts of the complex permittivity of beef blood from 50 to 110 GHz. The blood was refrigerated and regularly stirred. When starting the measurements, the blood temperature was 13°C. After preparation and during measurements, it reached an estimated value of 20°C. These new measurements follow well Debye's law, based on earlier published results at microwave frequencies, up to 20 GHz (dashed curves in Fig. 2.4), despite a possible temperature effect [30].

The measurement configuration is completely closed, which provides very accurate results. Open structures may however be necessary. For instance, the complex permittivity of live and dead neurological cell cultures has been measured on a setup made of open microstrip transmission lines, from 20 to 40 GHz [31]. One advantage of this configuration is that it is open. Hence, the biological medium can be submitted to microwave exposure and the meas-

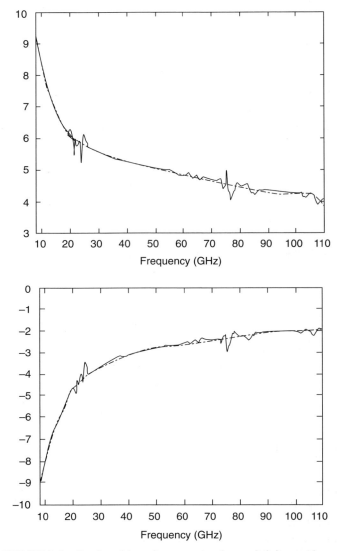

FIGURE 2.5   Real and imaginary parts of permittivity methanol.

urement of the complex permittivity can be an indicator of the activity of the medium.

Influence of Temperature   Information about the influence of temperature on the electric parameters of the material may be needed for some applications. At low frequencies, below beta dispersion, the conductivity reflects the

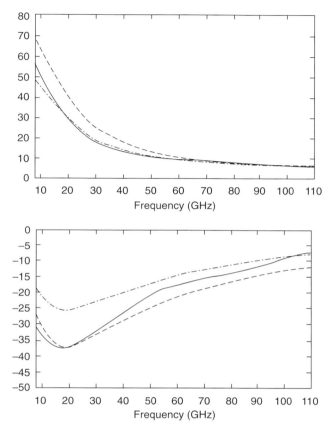

FIGURE 2.6   Real and imaginary parts of permittivity of beef blood.

volume fraction of extracellular space. For small temperature increases, it has been reported that the conductivity properties of tumor and normal tissue, at 44 kHz and 1 MHz, varied reversibly, with a temperature coefficient of about 2%/°C. These changes reflect thermally induced changes in the conductivity of tissue electrolyte. For large temperature increases (above about 44.5°C), however, the dielectric properties exhibit abrupt and irreversible changes, reflecting thermal damage to the tissue. In excised tissue maintained at 44°C, the low-frequency conductivity decreased initially by approximately 10% during the first hour, then gradually increased by approximately 50% during the next 8 h; 10-fold changes were reported in the permittivity over similar periods [12]. Hence, monitoring temperature is accompanied with a lot of uncertainty.

At RF/microwaves, the conductivity of tissues reflects that of the tissue electrolyte. There are two (sometimes opposing) effects. Increasing temperature increases the ionic conductivity and also the relaxation frequency. At frequencies below the relaxation frequency this will result in a decrease in the loss and, hence, in a decrease in the dipolar contribution to the conductivity. At low frequencies, this contribution varies inversely as the relaxation frequency and will show a temperature coefficient of about –2%/°C. In tissues, the contributions of dipolar loss and ionic conduction to the total conductivity are comparable at about 2 GHz. Thus, the temperature coefficient of the conductivity should change sign close to this frequency. Dielectric properties of muscle and brain have been reported [12]. They confirm that the conductivity below 1 GHz has the same temperature coefficient as that of simple electrolytes, about 2%/°C, while there is a sign change in the temperature coefficient near 2 GHz, to about –2%/°C above this frequency. At microwave frequencies the permittivity of tissues reflects that of water, the static permittivity of which varies little with temperature.

At RFs, the situation is more complicated. In this frequency range, Maxwell–Wagner effects associated with the charging of cell membranes determine the dielectric properties. The low-frequency permittivity increase depends on the cell size and membrane capacitance; it is insensitive to temperature. The relaxation frequency, however, scales as the conductivity of tissue electrolyte; it should exhibit temperature coefficients of about 2%/°C. Hence, temperature coefficients of the order of 0–2%/°C may be expected, depending on frequency.

For many practical applications these temperature coefficients are negligible. They must, however, be considered when applying data from tissues at 20°C to predict effects at 37°C, for instance microwave absorption.

2.3 THERMODYNAMICS

Section 1.4 was devoted to RF and microwave energy, discussing the relation between power and energy and evaluating the RF and microwave properties of matter. Electromagnetic properties, however, are only one way to characterize materials. Other relations are necessary to calculate, for instance, the thermal and mechanical properties of matter: Energies other than EM have to be defined. These energies will be reviewed in Section 2.4. Their definition, however, requires having recourse to thermodynamics and the definition of entropy.

The classical theory of thermodynamics deals with the time-averaged properties of systems containing large numbers of particles. It avoids description of the individual motions of the particles. By manipulations of thermodynamic variables, many thermodynamic relations can be developed for calculating many of the thermal, electrical, and mechanical properties of matter from a relatively limited set of measurements.

Thermodynamics as a new science has its roots in the early nineteenth century, when Carnot invented the steam engine [32]. It had to do with the conversion of heat into work. The efficiency, the ratio of useful mechanical work to the used heat, became the principal characteristic of this conversion. Then, in 1854, Clausius introduced the concept of entropy [33], defined by

$$S = \frac{dQ}{T} \qquad (2.2)$$

where S is the entropy, Q is the heat, and T is the temperature in kelvin. He formulated the second law of thermodynamics as

$$dS \geq 0 \qquad (2.3)$$

where equality is applicable for reversible processes and inequality for irreversible processes. This will be developed in Section 2.4. In 1877, Boltzmann published the statistical character of the second law of thermodynamics by determining the relation between the entropy and the probability of the system state, Γ [34]:

$$S = k \ln \Gamma \qquad (2.4)$$

where k is Boltzmann's constant (1.380×10^{-23} J K^{-1}). In 1947, Prigogine showed that the entropy of a system could be split in two parts [35]:

1. The flow of entropy due to interactions with the environment.
2. The entropy generation inside the system.

This second part is characteristic of the thermodynamics of irreversible processes.

Thermodynamics has no connection with the structure of the system and ignores the inside processes. It handles four parameters: volume, pressure, temperature, and entropy. It considers the interaction of the system with the environment, taking into account entropy and energy at the system input and output, dealing only with three different types of systems:

1. An *isolated* system, which has no exchange at all with the environment.
2. A *closed* system, which can exchange energy with the environment.
3. An *open* system, which can exchange both energy and mass with the environment.

At the turn of the nineteenth and twentieth centuries, the theory of blackbody radiation was elaborated. Fundamentals have been exposed in Section 1.4.3. A few reminders are summarized in this section. The concept of blackbody radiation is of fundamental importance for understanding the emission of materials, including biological tissues. Blackbody radiation is produced by a body that is considered as a closed volume whose walls are in thermal equilibrium at a given temperature. A source at the outside maintains the whole wall at a constant temperature. A blackbody is defined as an idealized material that absorbs all the incident radiation at all frequencies, reflecting none: It is a perfect absorber. In addition, a blackbody is also a perfect emitter, since energy absorbed by a material would increase its temperature if no energy were emitted. There are no losses and the energy balance is zero: The incident radiation is transformed into thermal agitation, which in turn is transformed into emitted radiation. A perfect reflector is also lossless. However, it reflects the incident energy, the reflected photons are at the same frequency as the incident photons, and there is no inside thermal agitation; it could be called a *white body*.

Planck established the mathematical expression of the energy emitted by a blackbody in 1901, in a law that is called Planck's law [Eqn. (1.33)], yielding the spectral energy density per unit volume, that is, the energy density per unit volume and bandwidth. The Rayleigh–Jeans radiation law is a very useful low-frequency approximation of Planck's law, obtained when the product hf is very small with respect to kT [Eqn. (1.35)]. It is generally valid at RFs and microwaves, except in special circumstances. According to the Rayleigh–Jeans law, the brightness varies inversely as the square of the wavelength. On the other hand, integrating Planck's law over all frequencies yields the total brightness B_t for a blackbody radiator, which is the Stefan–Boltzmann law [Eqn. (1.43)]. This law shows that the total brightness of a blackbody increases as the fourth power of its temperature. Also, maximizing Planck's law [Eqn. (1.33)] with respect to frequency yields an expression for the peak brightness shifts to higher frequency with increase in temperature, in which the wavelength–temperature product is a constant. This is called the Wien displacement law [Eqn. (1.44)], showing that the wavelength of the maximum of peak brightness varies inversely with the temperature. Finally, simplifying Planck's law at high frequencies, when one has $hf \gg kT$, yields the Wien radiation law [Eqn. (1.47)]. Figure 1.7 shows a comparison of Planck's law with its high- (Wien) and low-frequency (Rayleigh–Jeans) approximations.

Most biological materials are inhomogeneous: They are composed of several parts. Dealing with inhomogeneous systems requires considering the various parts of the system. The *separable* parts of a thermodynamic system are termed *phases*. For example, an iced drink contains two phases, the liquid phase and the solid phase (the ice). Undissolved sugar in the drink would constitute a second solid phase, a stirring spoon a third solid phase. The dissolved sugar and the molecules of H_2O, however, are intimately mixed and constitute only a single phase: This phase has two *chemical constituents* or *species*. While the phases of a system must be distinguishable, this does not preclude the transfer of atoms of molecules of the constituents from phase to phase: Ice melts and sugar dissolves or recrystallizes. Often each phase contains some quantity of all chemical constituents present, but some of these may be small; for instance, a metallic stirring spoon does not dissolve into a regular drink.

2.4 ENERGY

A simple medium, such as a lossless dielectric medium, exhibits no dispersion; that is, its electrical characteristics do not vary with frequency. In this case, the EM energy has an exact thermodynamic significance: It is the difference between the internal energy per unit volume with and without the field, respectively, with unchanged density and entropy. In the presence of dispersion, however, no such simple interpretation is possible. Moreover, in the general case of arbitrary dispersion, the EM energy cannot be rationally defined as a thermodynamic quantity [36]. This is because the presence of dispersion in general signifies a dissipation of energy, which is the reason why a dispersive medium is also an absorbing medium.

To determine this dissipation under a single-frequency EM excitation, one averages with respect to time the rate of change of energy in unit volume of the body. This yields the steady rate of change of the energy, which is the mean quantity Q of heat evolved per unit time and volume. It has been shown in Section 1.4 that the imaginary parts of ε and μ determine the absorption (dissipation) of energy. These imaginary parts are associated with the electric and magnetic losses, respectively. On account of the law of increase of entropy, the sign of these losses is determinate: The dissipation of energy is accompanied by the evolution of heat, that is, $Q > 0$. It therefore follows from Poynting's theorem that the imaginary parts of ε and μ are always positive,

$$\varepsilon'' > 0 \qquad \mu'' > 0 \tag{2.5}$$

for all substances in thermal equilibrium and at all frequencies. If the body is not in thermal equilibrium, then Q may in principle be negative. The second law of thermodynamics requires only a net increase in entropy as a result of the effects of the variable EM field and the absence of thermodynamic equilibrium, the latter effect being independent of the presence of the field. On the other hand, the signs of the real parts of ε and μ for $\omega \neq 0$ are subject to no physical restriction.

Any nonsteady process in an actual body is to some extent thermodynamically irreversible. The electric and magnetic losses in a variable EM field therefore always occur to some extent, however slight. That is, the functions $\varepsilon''(\omega)$ and $\mu''(\omega)$ are not exactly zero for any frequency other than zero, which does not exclude the possibility of only very small losses in certain frequency ranges. These ranges are then called *transparency ranges* and the material is said to be *transparent* in these frequency ranges. It is possible to neglect the absorption in these ranges and to introduce the concept of *internal energy* of the body in the EM field in the same sense as in a constant field. To determine this, it is not sufficient to consider an EM source of only one single frequency, since the strict periodicity results in no steady accumulation of EM energy. One then considers a field whose components have frequencies in a narrow range around some mean value of frequency. This yields the mean value of the EM part on the internal energy per unit volume of a transparent medium [36].

These considerations will now be generalized, following closely the excellent presentation by Beam [37]. Any thermodynamic system may be described by a total energy U, sometimes termed the *internal energy*. Since it is not really possible to define absolute zero-energy conditions, energy is generally measured with respect to some arbitrary reference conditions, for instance zero temperature. It is, however, possible to define, precisely, *changes* in energy. For example, if a change in volume dV of a system occurs while a given pressure is applied, the change in internal energy will be $dU = -p\, dV$. Other sources of energy are electric fields, magnetic fields, stresses, and sources of heat energy. Each means of altering the system energy involves some external force function whose magnitude is independent of the size of the system: electric field $\overline{E}$, magnetic field $\overline{T}$, stress $\overline{T}$, and temperature T. These variables are termed *intensive variables*. The changes in the system are proportional to the number of particles in the system; they are termed *extensive variables*: volume V, electric polarization $\overline{P}$, magnetic polarization $\overline{M}$, and strain $\overline{S}$. A differential change of system energy may be expressed as

$$dU = dQ + \overline{E} \cdot \overline{dP} + \mu_0 \overline{H} \cdot \overline{dM} + \overline{T} \cdot \overline{dS} - p\, dV \qquad (2.6)$$

for a system of unit volume, because $\overline{P}$, $\overline{M}$, $\overline{T}$, and $\overline{S}$ are so defined. It is valid for any subsystem in which these quantities are uniform. The quantity dQ represents *heat input* to the system.

One should expect that dQ is also the product of an intensive variable with the differential of an extensive variable, similar to the other terms. This is indeed the case when the heat input takes places slowly so that the system remains at *thermal equilibrium*, which means that there is *no temperature gradient* in the system. There are also no electrical conduction currents moving steadily through the material, for such currents imply conversion of externally supplied energy into heat (except in superconductors).

When there is heat input to a system maintained at thermal equilibrium, the differential dQ equals $T\, dS$, where T is the temperature and S the entropy, a complementary extensive variable. Suppose that a system is composed of two parts, A and B, each of which is at thermal equilibrium but having different temperatures T_A and T_B, respectively. If a quantity of heat energy dQ passes out of part A into part B, the change of entropy of part A must be $-dQ/T_A (= dS_A)$. The change of entropy of part B is $dQ/T_B (= dS_B)$. The change in entropy of the entire system is $dQ(1/T_A - 1/T_B)$. The heat will flow from the hotter part to the colder part; hence, if $T_A > T_B$, the system will experience a net *increase* in entropy *without* the addition of heat energy from an external source. The entropy of the overall system will continue to increase until the temperatures of the two parts become equal. This type of change is termed an *irreversible* change, for there is nothing one can do to the entire system from outside to reestablish the temperature difference that existed initially.

Only when heat is added to the system slowly, so that the equilibrium is maintained constantly, is the change in entropy given by dQ/T. Accordingly, one can write, in general,

$$dQ \le T\,dS \tag{2.7}$$

where equality is for reversible changes in the system and inequality for irreversible changes. Accordingly, (2.6) should be stated:

$$dU \le T\,dS + \overline{E}\cdot\overline{dP} + \mu_0\overline{H}\cdot\overline{dM} + \overline{T}\cdot\overline{dS} - p\,dV \tag{2.8}$$

This indicates that, if the entropy, polarizations, strains, and volume are kept constant, one has $dU \le 0$: The energy will seek a *minimum* value in keeping with the constraints on the system. These constraints, however, are difficult to maintain. For instance, it is very difficult to maintain constant the volume of a solid.

One can form, however, new *energy functions*, sometimes termed *thermodynamic potentials*, which are more readily handled. For example, what is termed the *Gibbs free energy* is the function

$$G = U - TS - \overline{E}\cdot\overline{P} - \mu_0\overline{H}\cdot\overline{M} - \overline{T}\cdot\overline{S} + pV \tag{2.9}$$

Its differential is

$$dU - T\,dS - S\,dT - \overline{E}\cdot\overline{dP} - \overline{P}\cdot\overline{dE} - \mu_0\overline{H}\cdot\overline{dM}$$
$$- \mu_0\overline{M}\cdot\overline{dH} - \overline{T}\cdot\overline{dS} - \overline{S}\cdot\overline{dT} + p\,dV + V\,dp$$

Substituting into (2.8) yields

$$dG \le -S\,dT - \overline{P}\cdot\overline{dE} - \mu_0\overline{M}\cdot\overline{dH} - \overline{S}\cdot\overline{dT} + V\,dp \tag{2.10}$$

The advantage of this formulation is that G is unambiguous because the parameters $U, T, S, p, \ldots$, involved in the definition of G, are all uniquely definable functions of the system. Furthermore, if all *intensive* variables are held constant—and these are the ones one generally has control of—the condition on dG is simply

$$dG \le 0 \tag{2.11}$$

which means that, as equilibrium is approached in a system under these conditions, G seeks for a minimum. It is therefore a valuable function to use in equilibrium calculations.

Other thermodynamic functions can be defined. The more useful ones are the *enthalpy*,

$$H = U + pV \tag{2.12}$$

and the *Helmholtz free energy*,

$$F = U - TS \tag{2.13}$$

These functions are more generally used in thermodynamic description of gaseous systems.

Particular functions are most valuable in dealing with particular thermal processes. For example, presuming that no fields, stresses, or strains are involved, when heat ΔQ is added to a system at constant pressure to raise temperature by an amount ΔT, one does not directly obtain a measure of the change of energy U. Equation (22.12) shows that one has

$$dH = dQ + V \, dp \tag{2.14}$$

Hence, at constant pressure ($dp = 0$),

$$\left(\frac{\partial H}{\partial T} \right)_p \left(= \frac{\Delta H}{\Delta T} \right) = c_p \tag{2.15}$$

which is termed the *specific heat at constant temperature*. Likewise, the *specific heat at constant volume c_v* is defined as

$$c_v = \left(\frac{\partial U}{\partial T} \right)_p \tag{2.16}$$

While c_v is useful in dealing with gaseous systems, the specific-heat values for solids are generally taken at constant pressure.

An inhomogeneous system, as are most of the biological systems, is composed of several parts. The *separable* parts are termed *phases*. Then, the extensive parameters of the entire system are the *sums* of the extensive parameters of the separate phases. For example, the volume of the liquid plus the volume of the solid phases equals the total volume in the iced drink. Likewise, the energy U is the sum of the energies of the two parts. The same applies for S, F, G, ... In one phase, however, one cannot measure extensive parameters for each of the chemical constituents present. (Dissolution of sugar in coffee does not increase the volume of the sweetened coffee by an amount equal to the volume of the dissolved sugar.)

At equilibrium, in the absence of field- and stress-dependent terms, temperature is the same in all phases, as generally is pressure. In addition to changes of the entropy and volume of a phase, there may also be changes in composition. The addition of an atom or molecule, or even an electron, will generally raise its internal energy by a small amount.

In a one-constituent, one-phase system at equilibrium, temperature and pressure are the only variables (dispensing with fields and stresses). In a more complex system, however, molecules or atoms move from one phase to another, or chemical reactions may take place to minimize the free energy G. At equilibrium (constant temperature and pressure), passage of one molecule from one phase to another, or of an atom from one chemical compound to another, should *not* further lower the value of G. Thus, if one molecule of a specific constituent goes from one phase to another, the increase of G in this second phase must exactly equal the decrease of G in the first phase. The chemical potential *at equilibrium* is therefore *the same in all phases* for each constituent.

An important principle that can be inferred from energy functions is the so-termed *Gibbs phase rule*, which states the number of degrees of freedom (variables) available in a system. Description of the thermodynamic state of a system of P phases ($\alpha = 1, 2, \ldots, P$) and C chemical constituents ($i = 1, 2, \ldots, C$) includes the pressure and temperature of the system and the *composition* of each phase. The latter requires in general $C - 1$ variables; one needs specify only the fraction of $C - 1$ constituents, the final one being given automatically. In P phases, allowing the possibility that each phase contains at least a small quantity of each constituent, one requires $P(C - 1)$ composition measures with temperature and pressure, the total number of variables is $PC - P + 2$. All are not independent, however: There are constraints relating the *chemical potentials*. These are chemical potentials for each constituent in each phase; they have the dimensions of energy per molecule and are the energy that must be given to one molecule of a specific constituent to move it *into* a particular phase. There are $C(P - 1)$ of these constraints. As an example, three phases have two such constraint relations for *each* constituent. Each constraint reduces the number of independent system variables required by *one*. The total number of independent variables required to describe the system is therefore

$$P(C - 1) + 2 - C(P - 1) = C - P + 2 = \text{degrees of freedom} \qquad (2.17)$$

This is the Gibbs phase rule. It helps to explain *phase diagrams*, but this is essentially related to metal mixtures.

REFERENCES

[1] J. P. Reilly, *Electrical Stimulation and Electropathology*, New York: Cambridge University Press, 1992.

[2] J. P. Reilly, L. A. Geddes, C. Polk, "Bioelectricity," in R. C. Dorff (Ed.), *The Electrical Engineering Handbook*, Boca Raton, FL: CRC Press, 1993.

[3] J.-M. Guérit, *Les Potentiels Évoqués*, 2nd ed., Paris: Masson, 1993.

[4] A. L. Hodgkin, A. F. Huxley, "A quantitative description of membrane current and its application to conduction and excitation in nerve," *J. Physiol.*, Vol. 117, pp. 500–544, 1952.

[5] B. Frankenhaeuser, A. F. Huxley, "The action potential in the myelinated nerve fiber of *Xenopus laevis* as computed on the basis of voltage clamp data," *J. Physiol.*, Vol. 171, pp. 302–315, 1964.

[6] E. Cocherova, "The effects of microwave radiation of nerve fibers," Ph. D. Thesis, Bratislava: Elec. Eng. Inf. Tech., 2001.

[7] L. A. Geddes, L. E. Baker, *Principles of Applied Biomedical Instrumentation*, 3rd ed., New York: Wiley, 1989.

[8] S. M. Michaelson, J. C. Lin, *Biological Effects and Health Implications of Radiofrequency Radiation*, New York: Plenum, 1987.

[9] J. Thuery, *Microwaves: Industrial, Scientific and Medical Applications*, Boston, MA: Artech House, 1992.

[10] C. Polk, E. Postow, *Handbook of Biological Effects of Electromagnetic Fields*, Boca Raton, FL: CRC Press, 1996.

[11] A. Vander Vorst, F. Duhamel, "1990–1995 advances in investigating the interaction of microwave fields with the nervous system," in A. Rosen and A. Vander Vorst (Eds.), Special Issue on Medical Applications and Biological Effects of RF/Microwaves, *IEEE Trans. Microwave Theory Tech.*, Vol. 44, No. 10, pp. 1898–1909, Oct. 1996.

[12] K. R. Foster, H. P. Schwan, "Dielectric properties of tissues," in C. Polk and E. Postow (Eds.), *Handbook of Biological Effects of Electromagnetic Fields*, Boca Raton, FL: CRC Press, 1996.

[13] E. C. Jordan, *Electromagnetic Waves and Radiating Systems*, Englewood Cliffs, NJ: Prentice-Hall, 1950.

[14] S. Ramo, J. R. Whinnery, T. Van Duzer, *Fields and Waves in Communication Electronics*, New York: Wiley, 1965.

[15] A. Vander Vorst, *Transmission, Propagation et Rayonnement*, Brussels: De Boeck, 1995.

[16] G. Grosse, K. R. Foster, "Permittivity of a suspension of charged spherical particles in electrolyte solution," *J. Phys. Chem.*, Vol. 91, p. 3073, 1987.

[17] G. Grosse, "Permittivity of a suspension of charged spherical particles in electrolyte solution. II. Influence of the surface conductivity and asymmetry of the electrolyte on the low and high frequency relaxations," *J. Phys. Chem.*, Vol. 92, pp. 3905–3910, 1988.

[18] A. Vander Vorst, *Electromagnétisme. Champs et Circuits*, Brussels: De Boeck, 1994.

[19] H. P. Schwan, "Electrical properties of cells: Principles, some recent results and some unresolved problems," in W. S. Aldeman and D. Goldman (Eds.), *The Biophysical Approach to Excitable Systems*, New York: Plenum, 1981.

[20] E. D. Trautman, R. S. Newbower, "A practical analysis of the electrical conductivity of blood," *IEEE Trans. Biomed. Eng.*, Vol. 30, p. 141, 1983.

[21] L. A. Geddes, L. E. Baker, "The specific resistance of biological material—a compendium of data for the biomedical engineer and physiologist," *Med. Biol. Eng.*, Vol. 5, p. 271, 1967.

[22] R. D. Stoy, K. R. Foster, H. P. Schwan, "Dielectric properties of tumor and normal tissues at radio through microwave frequencies," *Phys. Med. Biol.*, Vol. 27, p. 107, 1981.

[23] J. D. Kosterich, K. R. Foster, S. R. Pollack, "Dielectric permittivity and electrical conductivity of fluid saturated bone," *IEEE Trans. Biomed. Eng.*, Vol. 30, p. 81, 1983.

[24] J. D. Kosterich, K. R. Foster, S. R. Pollack, "Dielectric properties of fluid saturated bone: Effect of variation in conductivity of immersion fluid," *IEEE Trans. Biomed. Eng.*, Vol. 31, p. 369, 1984.

[25] S. R. Smith, K. R. Foster, "Dielectric properties of low-water-content tissues," *Phys. Med. Biol.*, Vol. 30, p. 965, 1985.

[26] D. Misra, M. Chabbra, B. R. Epstein, M. Mirotznik, K. R. Foster, "Noninvasive electrical characterization of materials at microwave frequencies using an open-ended coaxial line: Test of an improved calibration technique," *IEEE Trans. Microwave Theory Tech.*, Vol. 38, No. 1, pp. 8–14, Jan. 1990.

[27] F. Duhamel, I. Huynen, A. Vander Vorst, "Measurements of complex permittivity of biological and organic liquids up to 110 GHz," *IEEE MTT-S Microwave Int. Symp. Dig.*, Denver, June 1997, pp. 107–110.

[28] I. Huynen, C. Steukers, F. Duhamel, "A new line-line calibration method for measuring the permittivity of liquids, soils, and planar substrates over a wide frequency range," *IEEE Trans. Instrum. Meas.*, Vol. 50, No. 5, pp. 1343–1348, Oct. 2001.

[29] M. Fossion, I. Huynen, D. Vanhoenacker, A. Vander Vorst, "A new and simple calibration method for measuring planar lines parameters up to 40 GHz," *Proc. 22nd Europ. Microwave Cont.*, Helsinki, Sept. 1992, pp. 180–185.

[30] Y. Wei, S. Sridhar, "Biological applications of a technique for broadband complex permittivity measurements," *IEEE MTT-S Int. Microwave Symp. Dig.*, Albuquerque, June 1992, pp. 1271–1274.

[31] M. R. Tofighi, A. S. Daryoush, "Study of the activity of neurological cell solutions using complex permittivity measurement," *IEEE MTT-S Microwave Int. Symp. Dig.*, Seattle, June 2002, pp. 1763–1766.

[32] S. Carnot, *Réflexions sur la Puissance Motrice du Feu et sur les Machines Propres à Développer Cette Puissance,* Paris: Bachelart, 1824.

[33] R. Clausius, "Über eine verwandte Form des zweiten Haupsatzes," *Poggendorf Annalen*, CXIII, 1854.

[34] L. Boltzmann, "Über die Beziehung zwischen dem zweiten Hauptzatze des mechanischen Wärmetheorie und der Warscheinlichkeitsrechnung respektive den Sätzen Über das Wärmegeleichgewicht," *Wiener Berichte*, Vol. 76, pp. 373–435, 1877.

[35] I. Prigogine, *Introduction to Thermodynamics of Irreversible Processes*, Springfield: C. C. Thomas, 1954.

[36] L. Landau, E. Lifshitz, *Electrodynamics of Continuous Media*. Oxford: Pergamon 1960.

[37] W. R. Beam, *Electronics of Solids*, New York: McGraw-Hill, 1965.

PROBLEMS

2.1. Plot Cole–Cole diagrams for the following (1) a series circuit with capacitance ε_1 and conductance σ_1; (2) a capacitance ε_∞ in parallel with a series circuit with capacitance ε_1 and conductance σ_1; (3) a capacitance ε_∞ in parallel both with a series circuit with capacitance ε_1 and conductance σ_1 and with a conductance σ_0; (4) a capacitance ε_∞ in parallel with two series circuits with capacitance ε_1 and conductance σ_1 and capacitance ε_2 and conductance σ_2. A general assumption is that the values of conductivity do not contribute significantly to $\varepsilon''(\omega)$. Observe that a steady conductivity distorts the permittivity plot.

2.2. The representation of complex dielectric data is convenient where the conductivity does not contribute significantly to $\varepsilon''(\omega)$. Where the conductivity is appreciable, Grant uses the complex conductivity, plotting its imaginary part versus its real part [5]. Draw Grant's diagrams for the four cases described in problem 2.1 and compare with the corresponding Cole–Cole diagrams. Observe that the presence of ε_∞ distorts the conductivity plot.

2.3. In a system of several phases the extensive parameters of the entire system are the sums of the extensive parameters of the separate phases. In one phase, however, one cannot measure extensive parameters for each of the chemical constituents present. (Dissolution of sugar in coffee does not increase the volume of the sweetened coffee by an amount equal to the volume of the dissolved sugar.) In the absence of field- and stress-dependent terms, show that one has for the system of several phases

$$dU = \sum_\alpha T\,dS_\alpha - \sum_\alpha p\,dV_\alpha + \sum_\alpha \sum_i \mu_{i\alpha}\,\Delta m_{i\alpha}$$

The variable $m_{i\alpha}$ is the total number of "molecules" of chemical constituent i in phase α and $\Delta m_{i\alpha}$ the change in this number. The parameter $\mu_{i\alpha}$ is known as the *chemical potential* for constituent i in phase α. It has the dimensions of energy per molecule and is the energy that must be given to one molecule of i to move it *into* a particular phase α. This equation permits description of an *open system*, for one can add molecules from an *external* supply as well as exchange them between phases. Substitute in the above equation the definition of G [Eqn. (2.9)], again in the absence of fields and stresses, and obtain the expression for dG, similar to that shown for dU.

CHAPTER THREE

Biological Effects

3.1 ABSORPTION

3.1.1 Fundamentals

It should be reminded that only the fields inside a material can influence it. Similarly, only the fields inside the tissues and biological bodies can possibly interact with these: The biological effects of microwaves do not depend solely on the external power density; they depend on the dielectric field inside the tissue or the body. Hence, the internal fields have to be determined for any meaningful and general quantification of biological data obtained experimentally. Dosimetric studies attempt to quantify the interactions of RF fields with biological tissues and bodies. An excellent introduction to radio and microwave dosimetry will be found in Stuchly and Stuchly [1].

The theoretical approach is a formidable task. A biological body is an inhomogeneous lossy dielectric material. Both the inhomogeneity of the dielectric properties and the complexity of the shape make the calculation of a solution extremely difficult, if not impossible. Only very simplified models can be analyzed. Experimental approaches are also subject to enormous difficulties. The intensity of the internal fields depends on a number of parameters: frequency, intensity, and polarization of the external field; size, shape, and dielectric properties of the body; spatial configuration between the exposure source and the exposed body; and the presence of other objects in the vicinity. As mentioned by Stuchly [1, p. 296]: "With a complex dependence on so many parameters, it is apparent that the internal fields in a mouse and a man exposed to the same external field can be dramatically different, and so will be their biological response, regardless of physiological differences. Conversely, different expo-

RF/Microwave Interaction with Biological Tissues, By André Vander Vorst, Arye Rosen, and Youji Kotsuka
Copyright © 2006 by John Wiley & Sons, Inc.

sure conditions, that is, different frequencies, may induce similar fields inside such diverse shapes as a mouse and a man.

It should be stressed that very complex field distributions can occur, both inside and outside biological systems exposed to EM fields. Refraction within these systems can focus the transmitted energy, resulting in markedly nonuniform fields and energy deposition. Different energy absorption rates can result in thermal gradients causing biological effects that may be generated locally and are difficult to anticipate and perhaps unique [2]. The geometry and electrical properties of biological systems will also be determining factors in the magnitude and distribution of induced currents at frequencies below the microwave range.

When EM fields pass from one medium to another, they can be reflected, refracted, transmitted, or absorbed, depending on the complex conductivity of the exposed body and the frequency of the source. Absorbed RF energy can be converted to other forms of energy and cause interference with the functioning of the living system. Most of this energy is converted into heat: This is *absorption*. However, not all EM field effects can be explained in terms of energy absorption and conversion to heat. At frequencies well below 100 kHz, it has been shown that induced electric fields can stimulate nervous tissue. At the microscopic level, other interactions have been observed, as will be shown in the next sections of this chapter.

It has been shown in Section 1.5 that waves do not necessarily penetrate the entire body. Penetration is limited by what is termed the *skin effect*, characterized by the *skin depth*. The depth to which microwaves can penetrate tissues is primarily a function of the electric and magnetic properties of the tissues and of the microwave frequency. In general, at a given frequency, the lower the water content of the tissue, the deeper a wave can penetrate. Also, at the frequencies of interest, the lower the frequency, the deeper is the depth of penetration into tissues with given water content. The medical devices using microwaves depend on the ability of microwaves to deeply penetrate into living tissues.

3.1.2 Dosimetry and SAR

As mentioned in Chapter 1, exposure levels in the microwave range are usually described in terms of *power density*, reported in watts per square meter or derived units (e.g., milliwatts per square meter or microwatts per square meter). However, close to RF sources with longer wavelength, the values of both the electric volts per meter and magnetic amperes per meter field strengths are necessary to describe the field.

For this reason two different, although interrelated, quantities are commonly used in dosimetry. At lower frequencies (below approximately 100 kHz), many biological effects are quantified in terms of the current density in tissue, and this parameter is most often used as a dosimetric quantity. At higher frequencies, many (but not all) interactions are due to the rate of energy deposition per unit mass. This is why the *specific absorption rate* (SAR) is used as the dosimetric measure at those frequencies. It is expressed in watts per kilogram or derived units (e.g., milliwatts per kilogram). In some respect, the generalized use of the SAR at microwaves is unfortunate because the SAR is based on absorption only, which raises questions about using this parameter for evaluating effects that may be of another nature than absorption. The possibility of nonthermal effects is a controversial issue and having only the SAR for evaluating all the biological effects does not help in reducing the controversy. In fact, the SAR may be a valid quantitative measure of interaction mechanisms other than absorptive ones when the mechanism is dependent on the intensity of the E field, except, however, when the direction of the *E* field is of importance with respect to the biological structure. Similarly, the SAR concept may not be sufficient for the interactions directly through the *H* field.

With these limitations in mind, the SAR concept has proven to be a simple and useful tool in quantifying the interactions of RF/microwave radiation with living systems, enabling comparison of experimentally observed biological effects for various species under various exposure conditions, providing the only means, however imperfect, of extrapolating the animal data into potential hazards to humans exposed to RF radiation, and facilitating, planning, and effectively executing the therapeutic hyperthermic treatment [1].

The SAR is defined as the time derivative of the incremental energy dW absorbed by or dissipated in an incremental mass dm contained in a volume element dV of a given density ρ [2, 3]:

$$\text{SAR} = \left(\frac{d}{dt}\right)\left(\frac{dW}{dm}\right) = \left(\frac{d}{dt}\right)\left[\frac{dW}{\rho(dV)}\right] \tag{3.1}$$

or, using the Poynting vector theorem for sinusoidal varying electromagnetic fields:

$$\text{SAR} = \left(\frac{\sigma}{2\rho}\right)|\overline{E_i}|^2 = \left(\frac{\omega\varepsilon_0\varepsilon''}{2\rho}\right)|\overline{E_i}|^2 \tag{3.2}$$

where $\overline{E_i}$ is the peak value of the internal electric field (in volts per meter). The average SAR is defined as the ratio of the total power absorbed in the exposed body to the mass in which it is absorbed, which is not necessarily that of the total body: The SAR is the *ratio of absorbed power by absorbing mass*. The local SAR refers to the value within a defined unit volume or unit mass, which can be arbitrarily small.

When extrapolating the results of, for instance, animal experiments to human exposure, the conditions of EM similitude can be applied [4, 5]. These conditions are often used in a reduced form termed *frequency scaling*. They enable us to use results obtained with a given object and adapt them to predict the results to be obtained with another object, similar in form to the first and differing only by a scale factor. Using reduced scale models is of common use in a variety of fields, such as hydraulic and mechanical engineering. It is of common practice also in electrical engineering, for instance when constructing a reduced-scaled antenna for predicting the behavior of a huge antenna to be designed and built. With this in mind, it is necessary to establish the conditions to be satisfied for ensuring that the EM situation of the reduced-scale model is identical to that of the normal-scale model. This is accomplished by transforming the basic equations of the specific technical area, Maxwell's equations in electrical engineering, to obtain dimensionless equations. This isolates dimensionless factors that have to remain constant when going from one model to another and indicate how a parameter has to be modified when another parameter is submitted to a change.

The frequency-scaling application is well known. It is often said that the conditions of EM similitude are satisfied when a smaller size model is submitted to a correspondingly smaller wavelength, that is, a correspondingly higher frequency. From this, it is frequently understood that a two-times smaller model will yield the same results as the original one if the frequency is doubled. It should be strongly emphasized that *this is true only if there are no losses* [4, 5]. The reason is in the Maxwell's equations: In the presence of losses there is one more term to take into account. This modifies the factors that have to remain constant when shifting from one model to another to keep the equations dimensionless. As a consequence, keeping the ratio size–wavelength constant is not a sufficient condition anymore. There are circumstances where frequency scaling is very inaccurate, if not inadequate. This may indeed be the case in biological systems because biological losses can be very important.

The frequency-scaling principle can be used, for instance, to determine the equivalence between the exposure of a man of a height l_m and an animal of a height l_a, with both man and animal having the same orientation with respect to the exposure field, using the relationship

$$f_m l_m = f_a l_a \tag{3.3}$$

where f_m and f_a are the frequencies at which man and animal are exposed, respectively [6]. It must be emphasized that the SAR distributions will only be similar in the two bodies, not equal, and that the equivalence is essentially valid for low losses, which may not be the case in biological structures.

3.1.3 Thermal Considerations

Thermal effects have been investigated for some time. It should never be forgotten that to obtain reliable temperature measurements, only probes suitable for operation in RF fields should be used. Thermal measurements are important, in particular on human beings. A comprehensive database is available on effects of a thermal nature, but it mainly concerns animal studies and in vitro studies. Several methods of biological effect determination are based on thermal measurements:

1. *Calorimetric methods* particularly suited for in vitro measurements, in which heating and cooling data can be analyzed to estimate the energy absorbed by an exposed sample.
2. *Thermometric methods*, used to measure the temperature due to microwaves with particular types of nonperturbing thermometers, with only a few commercially available.
3. *Thermographic techniques*, used to measure temperature with particular thermographic cameras.

This last technique is noninvasive and, as a consequence, especially attractive for thermal measurements on human beings. As an example, human thermal effects have been measured on faces exposed to the emission of a mobile telephone handset using a thermographic camera to measure the thermal distribution over the surface of the face and obtain quantitative thermometric data. The camera was a forward-looking IR camera, measuring and imaging the emitted IR radiation from the head, through a pattern of 10,000 thermocouples located in the lens [7].

The rate of temperature change in the subcutaneous tissue in vitro exposed to RF or microwave energy is related to the SAR as [1]

$$\frac{\Delta T}{\Delta t} = \frac{(\text{SAR} + P_m - P_c - P_b)}{C} \tag{3.4}$$

where ΔT is the temperature increase, Δt the exposure duration, P_m the metabolic heating rate, P_c the rate of heat loss per unit volume due to thermal conduction, P_b the rate of heat loss per unit volume due to blood flow, and C the specific heat. If before the exposure a steady-state condition exists such as

$$P_m = P_c + P_b \tag{3.5}$$

then during the initial period of exposure one has

$$\frac{\Delta T}{\Delta t} = \frac{\text{SAR}}{C} \tag{3.6}$$

and the SAR can be determined from measurements of an increase in the tissue temperature over a short period of time following the exposure. For tissue phantoms and tissues in vitro, Eqn. (3.6) may be used as long as the thermal conductivity can be neglected, that is, for short durations of exposure. Several methods of SAR determination are based on thermal measurements and utilization of this equation. The specific heat and density of various tissues are summarized in Table 3.1.

More generally, from a macroscopic point of view, thermal effects resulting from the absorption of EM waves inside biological tissues are described in terms of the bioheat equation. Time-domain modeling should be necessary when the exposure is pulsed microwave energy. Spatial approximations usually

TABLE 3.1   Specific Heat Capacity and Density

Tissue	Specific Heat Capacity $(J\,kg^{-1}\,{}^{\circ}C^{-1})$	Density $(kg\,m^{-3})$
Skeletal muscle	3470	70
Fat	2260	940
Bone, cortical	1260	1790
Bone, spongy	2970	1250
Blood	3890	1060

Source: From [1].

have to be made. As an example, when investigating the effect of microwaves on the spinal cord, because of the narrowness of the medium and imperfect knowledge of the power deposition in the nearby tissues of the spinal cord (bone, skin, and muscle), only the longitudinal two-dimensional section of the spinal cord has been considered, assuming a homogeneous medium [8]. With this approximation, the bioheat equation can be written as

$$\rho(x,y)c(x,y)\frac{\partial T(x,y,t)}{\partial t} = k_t(x,y)\,\nabla^2 T(x,y,t) + v_s(x,y)[T_a - T(x,y,t)]$$
$$+ Q_m(x,y) + Q(x,y,t) \tag{3.7}$$

where T is the temperature inside the medium at the considered point (°C), $\rho(x,y)$ the volume mass of the tissues ($1020\,kg\,m^{-3}$), $c(x,y)$ the specific heat of the tissues ($3500\,J\,kg^{-1}\,{}^{\circ}C^{-1}$), k_t the thermal conductivity of the tissues ($0.6\,W\,m^{-1}\,{}^{\circ}C^{-1}$), v_s the blood heat exchange coefficient ($= 14560\,W\,m^{-3}\,{}^{\circ}C^{-1}$), T_a the arterial temperature (37°C), Q_m the heat generated by the metabolic activity ($W\,m^{-3}$), Q the heat generated by the absorption of the electromagnetic energy ($W\,m^{-3}$), t the time (s), and

$$Q(x,y,t) = \tfrac{1}{2}\sigma(x,y)|\overline{E}(x,y)|^2 g(t) \tag{3.8}$$

where $g(t) = 1$ during the microwave exposure and $g(t) = 0$ otherwise. In the absence of adequate information about the thermal constants of the spinal cord, the thermal constants of white matter have been used. The terms c, v_s, k_t, and Q_m depend on the type of tissues. They were assumed to be independent of temperature. The usual practice of assuming that Q_m may be neglected when compared to Q has been followed. Finally, it has been assumed that the spinal cord is evenly perfused. As can be seen, solving the bioheat equation even in a simplified case requires skill.

Possible *nonthermal effects* are under question. This is not a new question [9, 10]. It will be examined in more detail in Section 3.8. In fact, some of the so-termed nonthermal effects might actually be microthermal effects [11]. Also, one needs to be very careful about the conditions of an experimental study to take into account all the components of power. As an example, an in vivo experiment was conducted on rabbits to measure evoked potentials in the presence and absence of microwave exposure of the spinal cord, respectively. An electrical stimulus was applied to the peripheral nervous system of a rabbit while an electrode in the cortex measured the impulse response—the evoked potential. The spinal cord was exposed at 4.2 GHz by an implanted microantenna. The purpose of the experiment was to distinguish between thermal and possible nonthermal effects. A statistical treatment of the recorded data showed that there was a microwave effect after a long period of exposure and that the induced variations were reversible. Calculating the power deposition, the bioheat equation in the time domain indicated that microwave exposure resulted in a temperature increase within the spinal cord. The conclusion was that no nonthermal effect was observed in the experiment [8].

A number of specific thermal effects have been observed [12–14]. They depend, of course, on the spatial distribution of the SAR [15, 16]. It is estimated that a SAR of about $1\,W\,kg^{-1}$ produces an increase of $1°C$ in the human body temperature, taking thermoregulation into account [9]. Behavior is the most sensitive measure of effects [17]. In some experiments, it has been observed that pulsed fields produce a detectable effect at power levels smaller than continuous waves [8, 17].

On the other hand, several modeling methods and tools are now available. Finite-difference time-domain (FDTD) techniques offer broadband analysis and computational efficiency. Recent anatomical models of the head yield a better accuracy. Such models have greatly improved data visualization [18].

The level of SAR influences the thermal effects [19–23]:

- One watt per kilogram yields an increase of $1°C$ in the human body, taking thermal regulation into account.
- Ocular damage (cataracts) has been observed at $100\,mW\,cm^{-2}$ and more above $1\,GHz$.
- Corneal damage has been observed in monkeys for a SAR of $2.6\,W\,kg^{-1}$ at $2.45\,GHz$; it is interesting to observe that, with drug pretreatment such as phenylthiazines and miotics [8], damage has been observed for a ten times lower value of SAR, in the range $0.26–0.5\,W\,kg^{-1}$.
- Retinal damage has been observed on monkeys with a SAR of $4\,W\,kg^{-1}$ in the range $1.25–2.45\,GHz$ with pulsed fields.
- The SAR above $15\,W\,kg^{-1}$ produces malformations, with more than $5°C$ temperature increase.

For mobile communication fields, the interaction between a handset antenna and the human body is of special interest [18, 24–26].

Hyperthermia is the name given to one of the medical treatments for cancer to heat tumors up to therapeutic temperatures (43–45°C) without overheating the surrounding normal tissue. Key issues are heating, measuring the temperature, and controlling the system [27]. For about 20 years, quite an effort has been devoted to hyperthermia and the measurement of temperature, including by radiometry. A number of types of applicators, in particular microstrip antennas, have been investigated. Two different ways have been used to evaluate heating devices: SAR and thermal distribution. To evaluate thermal distribution, however, solving the heat equation is necessary. This is difficult, especially because it requires imposing thermal boundary conditions. Hyperthermia is treated in great detail in Chapter 4.

The present controversy about possible hazards due to mobilophony is due to the fact that heating processes are now well known and there are a number of scientific and technical data which have led to recommendations, while effects other than heating are much less known and do not allow numbers to be established on a scientific basis. As an example, RF/microwave safety standards based on dosimetry have long been established, following some breakthrough papers [28, 29]. However, these standards do not offer general protection, since other biological effects might affect health. They will be reviewed in Section 3.10.

Although there are still a number of uncertainties about the biological effects of microwaves, there is today a substantial database on biological effects of a thermal nature. Furthermore, research is underway on new issues: investigations on body-average and local SAR per 10 g or per 1 g tissue, improvements for reliable evaluations; human studies, long-term animal studies, cancer-related studies, epidemiology, and the impact of new technologies.

As a consequence, there are now many applications of heating through absorption at RF/microwaves. Medical applications are reviewed in the next chapters of this book. It must be emphasized that the application of RF/microwaves in medicine, essentially in cardiology, urology, and surgery, has been made possible because of the huge amount of research devoted for many years to the understanding and modeling of microwave absorption in living materials.

3.2 NERVOUS SYSTEM

For years, the effect of microwave radiation on the CNS and on behavior has been the subject of a great controversy. This is due to a lack of unification in the methodological approaches and an inadequacy of the database for selecting the most appropriate methods [29]. Moreover, there have been few attempts to evaluate and reproduce the findings of other laboratories. For these reasons, a program of cooperation between the United States and the Soviet Union was undertaken as early as 1975 [30]. It was agreed that duplicate projects should be initiated with the general goal of determining the most sensitive and valid test procedures. Male rats of the Fisher 344 strain were exposed or sham exposed to a 2.45-GHz continuous wave [31]. Animals were subjected to behavioral, biochemical, or electrophysiological (EEG frequency analysis) measurements during and/or immediately after exposure. Behavioral tests used were passive avoidance ("learned" behaviors) and activity in an open field ("naturalistic" behaviors).

The results of these studies are interesting. Neither group observed a significant effect on open field activity. Both groups observed changes in variability of the data obtained using the passive avoidance procedure, but not on the same parameters. The US group, and not the USSR group, found significantly less ionic activity in the microwave-exposed animals compared to the sham-exposed animals. Both groups found incidence of statistically significant effects in the power spectral analysis of EEG frequency, but not at the same frequency. The failure of both groups to observe effects in the behavioral tests attributable to microwave exposure was unexpected because previous experience with these tests, using Wistar rats [32], had detected differences between microwave- and sham-exposed rats. The failure of each group to substantiate the results of the other reinforced the contention that such duplicate projects were important and necessary.

The present section is partly based on a paper reviewing the action of microwave fields on the nervous system [17]. Section 2.1.2, reviewing cells and nerves from a bioelectric point of view, constitutes an adequate introduction to the subject. In this section, however, the point of view will be that of an investigation of the system as a whole.

3.2.1 General Description

The nervous system is the body's main control and integration network. It serves three functions: sensing the changes both within the body and with the outside environment, interpreting and integrating these changes, and responding to the interpretation by initiating actions in the form of muscular contractions or glandular secretions.

The nervous system of vertebrates is composed of two primary parts: the *central nervous system*, composed of the brain and spinal cord, and the *peripheral nervous system*, composed of all nerves outside the CNS. The nervous system is further organized into sensorimotor and autonomic components.

The *sensory system* virtually has millions of sense organs, whose basic functions involve gathering information about the environment. The sensory nerves of the skin, for example, transmit impulses to the appropriate portions of the CNS, where the signals are interpreted as sensations of pressure, pain, temperature, or vibration. Other sensory receptors exist for sight, hearing, taste, and smell.

The *autonomic nervous* system has its part in the process. It operates involuntarily to control many bodily functions. It is divided into the sympathetic and parasympathetic systems. These components are usually considered to be antagonistic to each other, because they control opposite effects in various organs. They are distinguished by the anatomic distribution of their nerve fibers. One major difference between them is of prime interest in biomicrowaves: The postganglionic fibers secrete different neurohormones. The parasympathetic system secretes substances such as acetylcholine from fibers termed "cholinergic," while the sympathetic system secretes epinephrine and adrenaline from fibers termed "adrenergic."

Many of the functions of the autonomic system are regulated by the *hypothalamus*, located in the brain between the cerebrum and the midbrain. The hypothalamus affects the cardiovascular system, body temperature, appetite, and the endocrine system, among its many functions. Body temperature and endocrine functions are under direct influence of EM fields.

The *nerve impulses* pass to or come from the brain via 12 pairs of cranial nerves or via the spinal cord. Some cranial nerves, such as the optic nerve, are sensory nerves interfacing with sense organs. Others, like the facial nerve, are motor nerves, connecting to muscles and glands. Interneurons carry impulses between the sensory and motor nerves.

In the CNS, the *spinal cord*, which is continuous with the brain, is the center for spinal reflexes and is also a two-way communication system between the brain and the body. From an electrical engineering point of view, it is interesting to wonder about the equivalent electrical parameters of such a system.

The *neuroglial cells* are within the tissues of the brain and spinal cord. They fill in the space and support the neurons by a process known as scaffolding. Unlike neurons, neuroglial cells can reproduce.

If the EM fields are active in altering the activity and function of the CNS, then one may expect to see those changes reflected in the concentrations of *neurotransmitters* in various regions of the brain [33]. Hence, total or local exposure to EM fields at levels and frequencies where the CNS could be influenced will yield changes in neurotransmitter concentrations. It has been shown that the excitation of acupuncture points by microwaves (from 0.2 to 3 GHz) may produce an efficient analgesic effect, as shown by the corresponding increase in the pain threshold, measured by a dolormeter [34]. Furthermore, respective variations in pain threshold and neurotransmitter release in the center of pain reception in the brain are proportional [35].

Data about the effects of microwaves form a heterogeneous ensemble of facts that are not readily classified in terms of thermal versus nonthermal interactions [13]. The first category includes interactions with the peripheral nervous system and certain neurovegetative functions, alterations in EEG, changes in animal behavior and, possibly, the permeability of the blood–brain barrier (BBB). The second category could include membrane interactions that affect the ion fluxes, the modulation of neuronal impulse activity, and, possibly, induced arrhythmia in isolated hearts. Also included in this category could be the somewhat confused collection of dystonias and behavioral effects that are often referred to as the microwave syndrome. (See Sections 3.2.5 and 3.5.1.) It is also possible that there are some subtle very low-level effects of microthermal or nonthermal origin, which are masked by the more apparent thermal effects, when the power absorbed is of the same order of magnitude as the basic metabolic rate.

3.2.2   Effects on Brain and Spinal Cord

Molecular processes involved with the *brain energy metabolism* of the rat may be affected by radiation in the RF range. It has been indicated that there are 27–29 divalent iron atoms and two divalent copper atoms in molecules of the respiratory chain [36]. Each of these is essential to the function of the respiratory chain at the molecular level. Charged particles in an electric field undergo translational motion if not otherwise constrained. Therefore, some of the divalent metal ions in these key chains could be affected by the oscillating electric field of the RF exposure, at least until the critical relaxation frequency for each structure is exceeded. This concept has been tested by examining brain metabolism of rats during exposure at CW frequencies of 200, 591, and 2450 MHz and by developing dose–response relationships for the effects [37]. The presence of specific molecular interaction(s) would indeed be supported by a frequency specificity based on the macroscopic dielectric properties of biological systems. The measurement techniques included time-sharing fluorescence at the brain surface to determine relative levels of reduced nicotinamide adenine dinucleotide (NADH) and biochemical assays for adenosine triphosphate (ATP) and creatine phosphate (CP). The NADH, ATP, and CP are key compounds in brain energy metabolism. The ATP is a key compound in energy metabolism because it is the carrier of energy to the processes in living cells. The NADH is oxidized to produce ATP in the mitochondria, while brain ATP concentration is maintained at the expense of CP. When demand for ATP is higher than the mitochondrial production capacity, CP is rapidly converted to ATP to sustain ATP levels, and significant decreases in CP levels are observed prior to any decrease in ATP.

Frequency-dependent changes have been found for all three key compounds. The measured temperature in the brain of the rat was essentially constant for all exposures. At 200 and 591 MHz, NADH fluorescence increased in a dose-dependent manner between approximately 1 and 10 mW cm^{-2} and then became constant at higher exposures, up to at least 40 mW cm^{-2}. There was no effect at 2450 MHz. Levels of ATP and CP were measured in the whole brain after exposure. The CP levels decreased only at 591 MHz. The effect was investigated for all compounds at 200 and 2450 MHz exposure up to 5 min and at 591 MHz up to 20 min.

Microwave exposure at 591 MHz resulted in a decrease in ATP levels to 75% of controls, while CP levels are no lower than 60% of controls. During the 200-MHz exposures, brain ATP levels decreased to 80–90% of controls even though CP levels were not significantly decreased. These results are not consistent with normal energy metabolism where ATP levels are maintained at the expense of CP and suggest that at 200 MHz there is RF inhibition of the reaction converting CP to ATP. Furthermore, these results occur at RF exposure levels that do not increase brain temperature, thus suggesting a direct inhibition of metabolic processes by RF exposure. The authors of the study proposed the following speculative mechanism, which is consistent with the results obtained. Radio-frequency exposure at 200 and 591 MHz inhibits specific enzymes or electron transport proteins important in maintaining the cell's ATP pool. Such inhibition can be the result of RF-radiation-induced dipole oscillation involving the divalent metal ion in the active site during catalytic or transport activity. During such oscillations, the ability of the enzyme to perform its function could be decreased. For a given molecular species, the induction of dipole oscillation and change in catalytic or transport activity would be frequency dependent and should be responsive to the local electric field, that is, the field in the tissue. The frequency dependence would be determined by the detailed structure of the segment of the molecule and its freedom of movement in relation to the other parts of the active site. This proposed mechanism is consistent with Pethig's observation that RF-induced dipole oscillations in proteins in solution are found precisely in the 30–1000-MHz frequency range [38]. It is also consistent with the delta dispersion of the dielectric constant for proteins in solution, which exists between 10 and 1000 MHz [39].

To measure the relationship between pain and the *release of neurotransmitters*, a push–pull cannula has been inserted into the center of pain reception in the brain of rabbits [40]. Artificial cerebrospinal fluid (CSF) was injected at a very low and very constant speed. The CW microwave stimulation at 2.45 GHz was applied by a coaxial cable in an adequate acupuncture point. The sample of fluid was collected from the pull cannula. The pain threshold was measured three times per stimulation, before, during, and after, using an especially designed dolormeter. The radioenzymatic assay was used to measure the levels of norepinephrine release in the sample perfusates. It was found that the respective variations in pain threshold and neurotransmitter release were proportional [35].

In an in vivo experiment set up to compare evoked potentials in both the presence and absence of microwaves illuminating directly the spinal cord, an electrical stimulus was applied on the peripheral nervous system of rabbits while the impulse response (evoked potential) was measured by an electrode in the cortex. The spinal cord was exposed to 4.2-GHz pulses by a microantenna implanted in the spinal cord within the dorsal column. The purpose of the experiment was to distinguish between thermal and possible nonthermal effects. A statistical analysis of the recorded data clearly illustrated a microwave effect. Power deposition was calculated and used in the bioheat equation, which showed that the microwave effect resulted in an increased temperature within the spinal cord. No nonthermal effect was observed in this experiment [8].

3.2.3 Blood–Brain Barrier

A series of investigations on BBB permeability changes at a very low level of microwave exposure has captured increasing attention. The BBB protects the mammalian brain from potentially harmful compounds in the blood. It is a selectively permeable, hydrophobic barrier that is readily crossed by small, lipid-soluble molecules. It serves not only to restrict entry of toxic polar molecules into the brain but also as a regulatory system that stabilizes and optimizes the fluid environment of the brain's intracellular compartment. A dysfunctioning BBB allows influx of normally excluded hydrophilic molecules into the brain tissue. This might lead to cerebral edema, increased intracranial pressure, and, in the worst case, irreversible brain damage.

The BBB is an anatomic/physiological complex associated with the cerebral vascular system. It is composed of a network of astrocytic pseudopodia which envelops the tight junctions of the vascular endothelium. It is a natural defense system that maintains the physiochemical environment of the brain within certain narrow limits that are essential for life. It functions as a differential filter that permits the selective passage of biological substances from blood to brain. For instance, amino acids, anesthetics, and glucose may gain access to brain cells, while carbohydrates, proteins, and most microorganisms and antibiotics are excluded from brain tissues by the BBB. Unintentional opening of the BBB may subject the CNS to assault from extraneous microorganisms.

This selective permeability has the disadvantage that agents and drugs that are effective in treating diseases in other parts of the body may not be able to gain entry into the brain to combat infection. The ability to selectively open the BBB suggests the possibility of using microwave regional hyperthermia to facilitate chemotherapy for brain tumors and facilitate the delivery of anti-cancer drugs such as methotrexate. This substance is the drug most often used for high-dose chemotherapy, with BBB permeability; however, it is among the lowest of the agents currently used clinically and is normally excluded from the brain.

About 30 investigations on the effect of microwave radiation on BBB permeability were reported until 2002 [41]. The assay methods employed include

- visual dye markers, such as Evans blue, sodium fluorescein, and rhodamine-ferritin
- radioactive tracers
- horse radish peroxidase and electron microscopy
- endogenous albumin.

Visual dye markers are very easy to use. This is, however, a qualitative method. Radioactive tracer methods offer a quantitative analysis.

Among these investigations, the studies showing increased BBB permeability in experimental animals are about equal in number with those that do not report BBB disruption at high as well as at low SARs. Some of the apparent discrepancies undoubtedly stemmed from the complexity of the BBB and from differences in microwave exposure conditions, such as frequency power level and SAR distribution, and from differences in the use of a variety of assays and procedures to detect changes in BBB permeability. The physical and biological phenomena are highly complex, requiring sophisticated experimental procedures.

The first investigations exhibited changes in BBB permeability at high SAR. The use of markers and tracers indicated that when the absorbed microwave power is high enough (165 W kg^{-1} or more) to elevate the temperature of the rat brain to about 42°C, BBB permeability increases for substances normally excluded [42]. More recent reports, however, using the leakage of serum albumin, suggest that microwave exposure can alter BBB permeability at SARs that are well below the maximal permissive level for cellular telephone, for instance 1.6 W kg^{-1}, including extremely low levels (0.016 W kg^{-1}). A plausible question is, then, under repeated exposures of the human brain to microwaves from cellular mobile telephones, could albumin and other toxic molecules leak into and accumulate around and in the brain cells? [41].

As a specific example, effects of microwave fields on the BBB have been investigated by exposing male and female Fisher 344 rats in a TEM line chamber to 915-MHz CW microwaves as well as pulse-modulated waves with repetition rates of 8, 16, 50, and 200 per second. The SAR varied between 0.016 and 5 W kg^{-1}. The rats were not anesthetized during the 2 h exposure. All animals were sacrificed by perfusion fixation of the brains under chloral hydrate anesthesia about 1 h after the exposure. The brains were perfused with saline and fixed in 4% formaldehyde. Central coronal sections of the brains were dehydrated and embedded in paraffin. Albumin and fibrinogen were demonstrated immunohistochemically. The results show albumin leakage in 5 of 62 of the controls and in 56 of 184 of the animals exposed. Continuous waves resulted in 14 positive findings of 35, which differ significantly from the controls ($P = 0.002$). With the pulsed microwaves, 42 of 149 were positive, which is highly significant at the $P = 0.001$ level. This reveals that both CW and pulsed microwaves have the potential to open up the BBB for albumin passage, with no significant difference between continuous and pulsed microwaves in this respect [43].

Specific results were the following:

1. Exposed animals are at risk for opening of the BBB (odds ratio 3.8; $P = 0.0004$).
2. The response is independent of pulse repetition rate, and the response is the same for CW as compared with pulse modulation.
3. The response is independent of SAR in the interval 0.016 > SAR > 2.5 W kg^{-1} (odds ratio 3.3) but rises for SAR < 2.5 W kg^{-1}.

The question of whether the opening of the BBB can be induced by microwave exposure of the general public to television, radio, and mobile telephony transmitters and, hence, if it constitutes a health hazard demand further investigation. Such investigations are attracting the attention of many in the field [44].

3.2.4 Influence of Parameters of Microwave Exposure

The parameters of microwave exposure are very important to consider when investigating biological effects. As an example, different *durations* of acute exposure lead to different biological effects; consequently, different long-term effects may occur after repeated exposure. The *waveform* of the radiation is also important. This has been observed when comparing pulsed- versus continuous-wave exposure and plane- versus circularly polarized wave exposure. Furthermore, the *pattern of energy absorption* in the body also contributes to the microwave effect. These findings raise the question of whether the whole-body average SAR can be used as the only determining factor in evaluating biological effects of low-level microwaves. Other features of the radiation also need to be considered [45, 46].

The effects of CW, sinusoidal-amplitude-modulated, and pulsed square-wave-modulated 591-MHz microwave exposure on *brain energy metabolism* in male Sprague-Dawley rats (175–225 g) have been compared [47]. Brain nicotinamide adenine dinucleotide reduced form (NADH) fluorescence, ATP concentration, and CP concentration were determined as a function of modulation frequency. Brain temperatures of animals were maintained within –0.1 and –0.4°C from the preexposure temperature when subjected to as much as 20 mW cm^{-2} (average power) CW, pulsed, or sinusoidal-amplitude-modulated 591-MHz radiation for 5 min. Sinusoidal-amplitude-modulated exposures at 16–24 Hz showed a trend toward preferential modulation frequency response in inducing increased brain NADH fluorescence. The pulse-modulated and sinusoidal-amplitude-modulated at 16-Hz microwaves were not significantly different from CW exposures in inducing increased brain NADH fluorescence and decreased ATP and CP concentrations. When the pulse modulation frequency was decreased from 500 to 250 pulses per second, the average incident power density threshold for inducing an increase in brain NADH fluorescence increased by a factor of 4, that is, from about 0.45 to about 1.85 mW cm^{-2}. Since brain temperature did not increase, the microwave-induced increase in brain NADH and decrease in ATP and CP concentrations were not due to hyperthermia. These data suggested to the authors a direct interaction mechanism that is consistent with the hypothesis of microwave inhibition of mitochondrial electron transport chain function of ATP production.

Microwave *evoked body movements* are one of the biological effects associated with high-peak-power although low- to average-power microwaves. These evoked body movements were studied in mice [48]. A resonant cavity was used to provide head and neck exposure of the mouse to pulsed and gated 1.25-GHz CW. No difference in response to pulsed and gated CW stimuli of equal average power was found. The incidence of the microwave evoked body movement increased, however, proportionally with specific absorption (dose) when the whole-body average SAR was at a constant level ($7300\,\text{W}\,\text{kg}^{-1}$). A single microwave pulse could evoke body movements. The lowest whole-body specific absorption (SA) tested was $0.18\,\text{kJ}\,\text{kg}^{-1}$, and the corresponding brain SA was $0.29\,\text{kJ}\,\text{kg}^{-1}$. Bulk heating potentials of these SAs were less than $0.1°\text{C}$. For doses higher than $0.9\,\text{kJ}\,\text{kg}^{-1}$, the response incidence was also proportional to subcutaneous temperature increment and subcutaneous heating rate. The extrapolated absolute thresholds (0% incidence) were $1.21°\text{C}$ temperature increment and $0.24°\text{C}\,\text{s}^{-1}$ heating rate. Due to high subcutaneous heating rates, the mouse must have perceived the radiation as an intense thermal sensation but not a pain sensation. The temperature increment was indeed well below the threshold for thermal pain. Results of the study could be considered in promulgation of personnel protection guideline against high-peak-power but low- to average-power microwaves.

It has been observed that rats acutely exposed (45 min) to pulsed $2.45\,\text{GHz}$ (2-ms pulses, 500 pps, power density $1\,\text{mW}\,\text{cm}^{-2}$, average whole-body SAR $0.6\,\text{W}\,\text{kg}^{-1}$) showed *retarded learning* while performing in the radial-arm maze to obtain food rewards [49]. Deficits in memory functions, even transient ones, can lead to serious detrimental consequences. A new series of experiments was carried out to better understand this behavioral effect of microwaves and, especially, the underlying neural mechanisms involved [50]. It was shown that this behavioral deficit was reversed by treatment, before exposure, with the cholinergic agonist physostigmine or the opiate antagonist naltrexone, whereas pretreatment with the peripheral opiate antagonist naloxone methiodide showed no reversal of effect. These data indicate that both cholinergic and endogenous opioid neurotransmitter systems in the brain are involved in the microwave-induced spatial memory deficit. This would imply that reversal of cholinergic activity in the hippocampus alone is sufficient to reverse the behavioral deficit.

The effects of absorption of EM energy, when modulated, on brain tissue and cell membranes depend on the frequency and type of modulation. They appear to be especially important at *frequency and amplitude modulation below 300 Hz*, although few results are available which illustrate a definite effect. These modulation frequencies are comparable to the EEG wave frequency spectra. Electrical coupling of both frequencies is however unlikely [51]. Whole-body irradiation at $30\,\text{mW}\,\text{cm}^{-2}$ (SAR $25\,\text{W}\,\text{kg}^{-1}$) of rats caused an increase in EEG activity (sum of EEG wave frequency bands) immediately at the first stage after exposure and a slight increase in delta waves (0.5–$4\,\text{Hz}$) of the EEG recordings [52]. Simultaneously, the rheoencephalogram increased, but there was no significant change in heart rate. A $10\,\text{mW}\,\text{cm}^{-2}$ exposure with SAR = $8.4\,\text{W}\,\text{kg}^{-1}$, however, did not cause change in any frequency bands of the EEG. It produced a slight temperature rise in the brain. The rheoencephalogram amplitude increased after both exposure levels as a consequence of the increase of cerebral blood flow. The ECG records and heart rate did not show any change after the radiation exposure. Brain localized exposure showed that microwave field interactions affect the electrical activity of not only brain tissue and cerebral blood flow but also blood vessels. The recorded brain tissue impedance changes of vascular pulsation are explained as an effect not only on the cerebral blood flow and the wall of blood vessels but also on the electrical conductivity of the tissue and the rheological characteristics of the blood [53].

A *nonlinear effect of modulated waves* on chicken cerebral tissue has been demonstrated. The samples were impregnated with radioactive $45Ca^{2+}$ and were exposed to $0.8\,mW/cm^{-2}$ at $147\,MHz$, amplitude modulated by a sinusoidal signal of 0.5–$35\,Hz$. A statistically significant increase in net $45Ca^{2+}$ transport was observed for modulating frequencies of 6–$16\,Hz$ followed by a fall in the range 20–$35\,Hz$ [54]. The existence of *frequency windows* has been confirmed [55]: For an incident power flux of $1\,mW\,cm^{-2}$, there was a positive response when the modulation lay between 6 and $12\,Hz$ and little response at 0.5 and $20\,Hz$. A *power window* was also shown to exist at constant frequency: When the chicken cerebral tissue was submitted to a 450-MHz carrier wave modulated at $16\,Hz$ there was a significant increase in $45Ca^{2+}$ transport for power levels of 0.1 and $1\,mW\,cm^{-2}$, while no effect was observed for power levels of 0.05 and $5\,mW\,cm^{-2}$. The limits of frequency and power windows seem to be 6–$20\,Hz$ and 0.1–$1\,mW\,cm^{-2}$, respectively [29]. The carrier frequency should be less than $1\,GHz$, but itself has little effect. The optimum frequency appears to lie between 150 and $450\,MHz$. For example, for a 450-MHz carrier modulated by a 16-Hz sine wave with an average power density of $0.5\,mW\,cm^{-2}$, the amplitude of the phenomenon reaches 38% of its value at rest in $10\,min$ [13].

Low-level effects are thought to be due to the direct interaction between neuron membranes and the local electric field. In the vicinity of a membrane, the local field is not very different from its value in free space, that is, 61 and $194\,V\,m^{-1}$ for 0.1 and $1\,mW\,cm^{-2}$, respectively. This field is negligible in comparison with the value of $2 \times 10^7\,V\,m^{-1}$ characterizing the static transmembrane potential ($90\,mV$ across $4\,nm$) but is nevertheless large in comparison with slow brain waves ($1\,V\,m^{-1}$) or with terrestrial ELF fields (from 10^{-3} to $10^{-6}\,V\,m^{-1}$) picked up by fish, birds, and mammals and used by them for navigation, detection of prey, and regulation of the circadian rhythm. One model has examined the possibility of induced transmembrane potential on the order of 10–$100\,mV$ [56]. Other studies [57] have shown that, because of the spike or edge effect, the interface field may locally exceed the macroscopic field by up to two orders of magnitude, which is sufficient to reduce by a factor of 10^2 the threshold of sensitivity to the EM flux. This phenomenon is confined to a very narrow ELF modulation band and is believed unlikely to cause any damage. The American National Standards Institute (ANSI) subcommittee in charge of the revision of the American safety standards did not judge it necessary to take this effect into account [13].

Frequency and intensity windows of Ca^{2+} have been observed in the presence of *weak* fields below $100\,Hz$, similar to the ELF-modulated RF fields [58, 59]. These findings suggest further interactions, resulting in specific EEG changes that are affected by modulated RF fields. Furthermore, different effects of modulated and CW microwave exposure were found on morphology and cell surface negative charges. As static experimental models (i.e., biochemical models) do not give information about the *transient effects* and functional changes induced by EM exposure, especially in the nervous system, an electrophysiological approach is more appropriate to explain the phenomena in the CNS. Such experiments [52] were performed on F_1 hybrid male anesthetized rats. They addressed the following questions: (a) Does a weak modulated field cause measurable effects on systemic and/or localized regulation mechanisms in the CNS? (b) Is there any correlation between the field activated systemic and/or localized regulatory mechanisms and the changes in CNS activity? (c) Does evaluation of measured biopotentials provide more information about the effects on the CNS of modulated and CW field exposure?

In two series of experiments on 40 anesthetized rats (a) before and after 10 min whole-body exposure to 2.45-GHz CW and (b) during 30 min exposure to 16-Hz amplitude modulation 4 GHz, the effects on the CNS were observed simultaneously with those of the cardiovascular system by quantitative polygraphic measurements. *Transient effects* of absorbed CW and modulated microwave radiation in the brain were examined in order to clarify the involvement of physiological modalities during local and/or systemic exposure. The local cerebral modalities (EEG, rheoencephalogram (REG), DC impedance, and brain temperature) were recorded simultaneously with the systemic regulation modalities (ECG, respiration, and rectal temperature). Slight changes in cerebral metabolism and cerebral blood were observed during exposure. These transient alterations did not exceed the range of normal physiological variations, although various compensating factors with different speeds were involved.

The total power of EEG spectra increased after whole-body 30 mW cm^{-2}, 2.45-GHz CW exposure. No changes occurred below 10 mW cm^{-2}. The cerebral blood flow (CBF) increased above 10 mW cm^{-2}. The power of EEG delta waves (0.5–4 Hz) was increased by brain localized 4-GHz CW exposure at SAR = 42 W kg^{-1}, simultaneously with the REG amplitude as an index of CBF. Amplitude modulation at 16 Hz and SAR = 8.4 W kg^{-1} were associated with increased power of EEG beta waves (14.5–30 Hz). Changes were not observed in the CBF. Continuous-wave exposure at SAR = 8.4 W kg^{-1} increased the CBF but did not modify EEG spectra. Thus, the *correlation between physiological modalities* in microwave-activated systemic or localized regulatory mechanisms and changes in the CNS do not seem to be identical. It has not been clarified whether the mechanisms of the observed microwave-induced modifications are due to energy absorption in the brain.

The necessity for quantitative evaluation of these phenomena concerning CNS functions is obvious.

3.2.5   Nervous System Modeling and Simulation

Models and computer simulations of nervous fibers and of the nervous system have developed rather recently. McNeal introduced a first step in computer simulation of neural reactions to electrical stimuli in 1976. He used a popular spatial model consisting in a network of lumped circuit models [60]. An excellent discussion of various models is found in [61].

The *propagation of neural activity* in an axon is a consequence of the electrical properties of the axonal cell membrane. An electric circuit consisting of a capacitance, a voltage source, and nonlinear resistances, which represent the gating of the ionic channels, can model a small segment of the membrane. Potentials within and on the surface of a finite cylindrical volume conductor due to a single-fiber along its center have been calculated by solving Laplace's equation using a relaxation model [62]. The results have led to the estimation of the variation of the single-fiber surface potential (SFSP) that would be recorded from a surface electrode for differing nerve depths and conduction velocities. A conduction velocity of $60\,\mathrm{m\,s^{-1}}$ was chosen, which results in an instantaneous active length of 30 mm on the fiber. A length of 150 mm was chosen for the model to reduce end effects. The model radius was taken as 10 mm, which corresponds to the approximate depth of the ulnar nerve at superficial sites such as the elbow and axilla. Results are that the SFSP is more elongated than the transmembrane current waveform and the relative amplitude of its first positive phase is considerably decreased. The SFSP wave shape is independent of fiber velocity and its amplitude is proportional to the velocity squared.

The computation of the steady-state *field potentials and activating functions* led to the evaluation of the effect of electrical stimulation with several electrode combinations on nerve fibers and different orientations in the spinal cord [63]. The model comprises gray matter, white matter, CSF, epidural fat, and a low-conductivity layer around the epidural space. This layer represents the peripheral parts, like the vertebral bone, muscle, fat, and skin. Dura mater, pia mater, and arachnoid were not incorporated in the model, estimating that these membranes have only negligible influence on the field potential distribution in the spinal cord [64].

First, an infinite homogeneous model was used, solving Laplace's equation. Second, the spinal cord and its surrounding tissues were modeled as an inhomogeneous anisotropic volume conductor using a variational principle by which a functional representing the power dissipation of the potential fields must be minimized to obtain the solution. Applying this method, inhomogeneities, anisotropy, and various boundary conditions can easily be incorporated and only first-order derivatives are used, while in direct discretization methods of Laplace's equation second-order derivatives appear. The effect on spinal nerve fibers was approximated using the activating function that, for myelinated fibers, is the second-order difference of the extracellular potentials along the fiber. The effect of mediodorsal epidural stimulation was calculated. It was concluded that with cathodal simulation, mediodorsally in the epidural space, longitudinal fibers are depolarized while dorsoventral ones are hyperpolarized. The opposite occurs with anodal stimulation. It was found that parameters substantially affecting the potential distribution in the dorsal column are the conductivity of the white matter and the width and conductivity of the CSF layer. The CSF has a major influence on the potential distribution on the spinal cord. Although limited to the analysis of electrode configurations and the design of alternative electrodes for spinal cord DC stimulation, these results are important in that they give an insight on the field potential distribution in the dorsal column.

Interest in *neural excitation by EM induction* has grown in recent years. Magnetic stimulation of the CNS is a painless alternative to electrical stimulation and is finding increasing use. The spatial distribution of the currents induced by the stimulating coil has been calculated from a computer model [65]. Two configurations of a plane circular coil are considered: parallel to the tissue surface and perpendicular to the surface. The surface is assumed planar and infinite in extent. The tissue is modeled as a uniform, isotropic volume conductor. A quasi-static approximation is made in calculating the electric field. The current density is mapped as a function of position, including depth. In both configurations, it is always parallel to the surface and maximum at the surface. There is no perpendicular (vertical) current. The stimulation of a nerve fiber requires that the component of the surrounding electric field, and hence bulk current flow, is parallel to the fiber and should exceed a particular threshold value. These results suggest that nerve fibers running parallel to the skin surface are more likely to be stimulated than those running obliquely and that it is extremely difficult to stimulate nerve fibers running perpendicularly. On the other hand, it was found that, for a given coil, the current density at a particular point in the tissue is only dependent on its distance from the coil. This current density is independent of the distances of the point and of the coil from the tissue surface. In this analysis, the magnetic field produced by the induced current in the tissue has been ignored because it is much smaller than that produced by the primary current in the coil. The frequency components of the coil current waveform are at about 10 kHz. At this frequency, the skin depth is approximately 10 m, that is, much larger than the object.

A model of magnetic stimulation of an unmyelinated nerve fiber that predicts where and when excitation occurs has been proposed [66]. It consists of a one-dimensional cable equation that is forced by a term analogous to the activating function for electrical stimulation with extracellular electrodes. While neural stimulation is caused by a three-dimensional electric field distribution, a one-dimensional cable model generally describes the response. These one- and three-dimensional representations have been reconciled, and an activating function for magnetic stimulation was derived which was consistent with both [67]. From a three-dimensional volume conductor model of magnetic stimulation, the induced electric field and its resultant transmembrane potential distribution along an axon were derived analytically. This model validated several simplified assumptions on which the one-dimensional model was based: (1) The electric field within the axon is axial, (2) the field in the membrane is radial, (3) the electric field in the membrane due to induction is negligible compared to the electric field due to charge separation, and (4) the extracellular potential is negligible, so that the transmembrane potential equals the intracellular potential.

These linear models assume an infinitely long fiber, that is, a fiber whose length is large relative to the distance to the electrode. In a number of usual applications, this condition does not apply and it is necessary to consider the *termination conditions of fibers* in the field of a focal electrode. An analytical model of a fiber terminal in the field of a monopolar, time-varying, spherical or point source has been presented, and the effects of the termination imped-ance of an axon on the membrane polarization induced by extracellular stimuli have been shown [68]. The significance of any termination current is deter-mined by the ratio of the termination impedance to the axon's input imped-ance. If the ratio is large, the fiber is effectively "sealed" and the termination current is insignificant. If it is small, the fiber is "unsealed" and the membrane potential at the terminal is zero. If the ratio is of the order of 1, there is sig-nificant termination current and the fiber is in an intermediate state between sealed and unsealed. For a myelinated fiber terminating with nodal membrane, this latter situation appears to apply. An important feature of end-structure stimulation is its time constant.

Analyses of magnetic stimulation of finite-length neuronal structures have been performed using computer simulations [69]. Models of finite neuronal structures in the presence of extrinsically applied electric fields indicate that excitation can be characterized by two driving functions: one due to field gra-dients and the other to fields at the boundaries. It was found that *boundary field-driving functions* play an important role in governing excitation charac-teristics during magnetic stimulation. Simulations have indicated that axons whose length is short compared to the spatial extent of the induced field are easier to excite than longer axons of the same diameter and also that inde-pendent cellular dendritic processes are probably not excited during magnetic stimulation. Analysis of the temporal distribution of induced fields indicated that the temporal shape of the stimulus waveform modulates excitation thresholds and propagation of action potentials. Those results are based on simulations only and need to be confirmed by experimental results.

As mentioned in Section 2.1.2, there are excellent discussions of various models, including Hodgkin–Huxley, Frankenhaeuser–Huxley, Chiu–Ritchie–Rogert–Staff–Stagg, and Schwarz–Eikhof [7]. Large differences in model behavior are obvious for strong signals. The results also depend on fiber parameters, influence of myelin, irregularities, branching, electrode geometry, and so on.

A number of questions remain unanswered. There is a lack of quantitative results illustrating the significant differences, if any, between CW, pulsed, and ELF-modulated waves. Systematic experiments are necessary to illustrate pos-sible significant differences at different frequencies. Little information is avail-able about the quantitative effects of millimeter waves either on the nervous system or on its constituents, although there might be fundamental differences between microwave and millimeter-wave excitation. The wavelength of these may have a size similar to that of some constituents of the nervous system and hence cause resonance effects. The microwave syndrome, if it exists, needs clarification and experimental validation. Some experiments on animals need a more precise comparison between actual and sham exposure. Direct experiments on the brain or the spinal cord should be welcome to characterize thermal, microthermal, and possible nonthermal effects. Microthermal effects should be carefully investigated to avoid any erroneous conclusion about possible nonthermal effects. In a number of calculations related to the link between the nervous system and transmitted fields, one should not only calculate the deposited power but also use the bioheat equation to calculate the temperature elevation. With this in view, low-level exposure should be looked at very carefully. One main conclusion is that the parameters of exposure should be more carefully controlled. As an example, in some experiments, a comparison is made between CW exposure at some power level and pulsed- or modulated-wave exposure at another power level [17]. This should of course be avoided.

3.3   CELLS AND MEMBRANES

The use of weak EM fields to study the sequence and energetics of events that couple humoral stimuli from surface receptor sites to the cell interior has identified cell membranes as a primary site of interaction with low frequency fields in the pericellular fluid. Field modulation of cell surface chemical events indicates a major amplification of initial weak triggers associated with bindings of hormones, antibodies, and neurotransmitters to their specific binding sites. Calcium ions play a key role in this stimulus amplification, probably through highly cooperative alterations in binding to surface glycoproteins, with spreading waves of altered calcium binding across the membrane surface. Protein particles spanning the cell membrane form pathways for signaling and energy transfer. Fields millions of times weaker than the membrane potential gradient of $10^7 \, V \, m^{-1}$ modulate cell responses to surface-stimulating molecules. The evidence supports nonlinear, nonequilibrium processes at critical steps in transmembrane signal coupling. Cancer-promoting phorbol esters act at cell membranes to stimulate ornithine decarboxylase, which is essential for cell growth and DNA synthesis. This response is enhanced by weak microwave fields, also acting at cell membranes. There is strong evidence that cell membranes are powerful amplifiers of weak EM events in their vicinity [70].

Fluxes of ions, in particular calcium ions, have received much interest in recent years. The Ca^{2+} ions relay electrochemical messages to the cell surface and its biochemical mechanisms. Evolution seems to have favored calcium in preference to other neighboring ions (Mg^{2+}, Na^+, K^+, Cl^+) because calcium is able to bond without causing deformation to either membrane proteins or soluble proteins of cytoplasm or organelles. These proteins themselves act as intermediaries. They have several ion-bonding sites, and by changing their configuration as a function of the occupied site, they excite a specific target enzyme. The Ca^{2+} ions use specific channels to cross the membrane along the concentration gradient. These voltage-gated channels are normally closed in excitable cells, but they open in response to the action potential, that is, the transmembrane voltage pulse induced by the arrival of a messenger on the cellular surface. The membrane resting potential is $-90\,mV$, but it may reach $+40\,mV$. The channel opens from $-30\,mV$. The phenomenon lasts for a millisecond and allows the passage of about $3000\,Ca^{2+}$ ions, after which the outward migration of K^+ ions returns the potential to its equilibrium value. The calcium ion flux excites the endoplasmic reticulum, which itself liberates Ca^{2+} ions. The evacuation of the ions must take place against the concentration gradient. It is supported by the enzyme ATPase, or calcium pump, which acquires the necessary energy by dissociating ATP molecules [13].

A new model of the neuronal membrane electrical activity was presented in 1994 [71]. The main differences with previous models consist both in taking into account the temperature dependence of the various parameters and in inserting the synaptic inputs described as ionic channels. The model consists of two interconnected schemes: (1) a circuital model, representing the ionic currents crossing the membranes, and (2) a block model, representing the intracellular calcium concentration dynamics. It provides a *deterministic and stochastic analysis*. In the deterministic model the time course of the membrane voltage and the firing frequency are univocally determined once given the initial conditions. The insertion of the synaptic inputs in the model causes stochastic time fluctuations of the instantaneous membrane voltage and of the interspike interval, without inducing variations in the spike shape. The model yields a good simulation of some known responses of the membrane in terms of its firing frequency and resistance, validated by experimental results. It can be used in a variety of applications and may be particularly promising in the study of the interaction between EM fields and neuronal membrane activity for the comprehension of interaction mechanisms. The field-induced modifications of the membrane stochastic behavior can be simulated by suitable alterations of specific model parameters according to the type of stimulus (frequency, wave shape, incident energy, etc.).

The model has been used for analyzing microscopic effects of signals of GSM on voltage-dependent membrane channels of one single cell by computer simulation [72]. The theory is based on Markov finite-state models. The effect of mobilophony signals has been investigated on calcium, potassium, and sodium channels. Pulsed signals are shown to act more on the reduction of the opening probability than on CWs. The low-frequency components of the signals induce a variation of 30% of the opening probability in Na^+ and K^+ channels and 60% in Ca^{2+} channels. It is interesting to observe that the two low-frequency sine waves of GSM, at 8.3 and at $217\,Hz$, respectively, act much less on the opening probability than the composite pulsed GSM signal, in which the frequencies 8.3 and $217\,Hz$ are present.

3.4 MOLECULAR LEVEL

Effects of microwaves on in vitro V79 and human lymphocyte cell cultures have been followed up, with a special emphasis on DNA synthesis. The cells were exposed to 7.7 GHz at 0.5, 10, and 30 mW cm^{-2} with an exposure time of 10, 20, 30, and 60 min, respectively. When compared to controls, exposed cells exhibit a significant increase in the number of specific chromosome lesions, that is, unstable aberrations such as dicentrics and rings. The micronucleus test shows evidence of changes in the genome [73].

Teratological effects of long-term, *low-level* CW and AM-modulated in utero microwave exposure on mice have been investigated as well as morphological and biochemical evaluations of the sequential generation. Protein synthesis is measured on the level of translation in the brain and liver tissue. The morphological evaluation of development in sequential generations of mice after long-term, low-level exposure is obtained by separating into normal and not-normal (abnormal, malformed, nondeveloped) groups of generations while the amino acylation is measured using C_{14} amino acid in transfer-RNA (tRNA) and amino acid tRNA synthesis in the brain and liver tissues of irradiated mice and rats [74].

In a 4-year study on the *genetic effects* of microwaves and RFs, bacteria were tested for different mutations or prophage induction, with no increase in the mutation frequency and a small but significant increase in the cellular rate growth for all microwave exposures. Embryos of *Drosophila* were tested at 2.45 GHz. A sensitive somatic test system was used in which mutagenicity was measured as the frequency of somatic mutations in a gene controlling eye pigment. No increase in mutation frequency is obtained, when compared with the nonexposed controls [75].

Considering the *genome* role in the physical mechanics of cell response to low-intensity EM fields, a general model has been presented based on internal field sources in a living cell. The generation of collective modes arising from the hydrogen bond system in the DNA is possible at definite frequencies in a wide frequency range. A resonant effect of *Escherichia coli* cells at 51.7 GHz on rat or human leucocytes is observed on the genome conformational state (GCS). The low-intensity millimeter-wave exposure (200 µW cm^{-2}) of rat thymocytes resulted in GCS changes resonant at 41.65 GHz. The time course of Ca^{2+} proved to be most sensitive: There is a 10% reduction in the intracellular Ca^{2+} after 10 min exposure, so that millimeter-wave-induced changes in the GCS may be related to the Ca-dependent processes in the cell. Millimeter-wave exposure (46.35 GHz, 100 µW cm^{-2}, 2 h) of early *Drosophila melanogaster* embryos following an ionizing treatment (1.5 Gy of X-rays) resulted in strong modification of radiation injury. Hence, nonthermal millimeter waves can be used as a radio modificator with a high selective influence. The modification has only "positive" results and can protect biological objects from radiation hazards. *Drosophila melanogaster* under low-intensity millimeter-wave exposure (<100 µW cm^{-2}) exhibited large and significantly nonrandom variations of a survival in control groups, determining two-directional embryo response to millimeter-wave exposure [76].

A theoretical model has been proposed for evaluation of the dielectric properties of the cell nucleus between 0.3 and 3 GHz as a function of its nucleic acids concentration. It is based on literature data on dielectric properties of DNA solutions and nucleoplasm. In skeletal muscle cells, the SAR ratio between nucleoplasm and cytoplasm is found to be larger than 1 for a nucleic acids concentration above $30 \, mg \, mL^{-1}$. A nearly linear relationship has been found between this concentration and this nucleo-cytoplasmic SAR ratio. Considering the nanoscale of the layer of condensed counterions and bound water molecules at the nucleic acid–solution interface, the power absorption per unit volume has been evaluated at this precise location. It has been found to be between one and two orders of magnitude above that in the surrounding solution and in muscle tissue as a whole. Under realistic microwave exposure conditions, however, these SAR inhomogeneities do not generate any significant thermal gradient at the scale considered here. Nevertheless, the question arises of a possible biological relevance of nonnegligible and preferential heat production at the location of the cell nucleus and of the nucleic acid molecules [77].

3.5 LOW-LEVEL EXPOSURE AND ELF COMPONENTS

The effects of low-level RF and microwave exposure of brain tissue and animal behavior, more specifically the effects on the BBB (protein permeability and saccharide permeability), on calcium ion exchange in brain tissue and on animal behavior were extensively reviewed in 1986 [78]. The study offers a comprehensive critical discussion of the relevance of the experimental data derived from a number of studies to the health and safety of exposed people as well as a number of references.

3.5.1 Microwave Syndrome

For humans exposed to very low power densities (between a few microwatts per square centimeter and a few milliwatts per square centimeter), East European epidemiological studies [13, 79, 80] have revealed a variety of reversible asthenic problems that constitute the hypothetical *microwave syndrome* (headache, perspiration, emotional instability, irritability, tiredness, somnolence, sexual problems, loss of memory, concentration and decision difficulties, insomnia, and depressive hypochondriac tendencies). There was, however, no control group. Furthermore, these complaints are very subjective and their evaluation is difficult in the absence of well-established dosimetric data. Indeed, individuals suffering from a variety of chronic diseases may exhibit the same dysfunction of the central nervous and cardiovascular systems. Hence, it is extremely difficult, if not impossible, to rule out other factors in attempting to relate microwave exposure to clinical conditions. These problems may well be due to environmental factors unrelated to microwaves, but a possible nonthermal mechanism cannot be completely ruled out [81]. On the other hand, the controversy regarding whether RF radiation sickness is a medical entity has recently been revisited [82].

3.5.2 Low-Level Pulsed Exposure

Exposure to low-level *pulsed microwaves* has been reported to affect brain neurochemistry in a manner broadly consistent with responses to *stress*. The acute exposure of rats to pulsed 2.45 GHz (2-ms pulses, whole-body average SAR 0.6 W kg^{-1}, specific absorption 1.2 mJ kg^{-1} per pulse) was found to alter cholinergic activity in various regions of the forebrain [50]. Central cholinergic activity was increased after 20 min of exposure but decreased after 45 min, and repeated exposure engendered compensatory changes in the concentration of cholinergic receptors [83]. Stressors such as noise and acute restraint can induce similar changes in central cholinergic responses, suggesting that low-level microwave exposure may be a source of mild stress. Pretreatment with narcotic antagonists blocked the effects, suggesting the involvement of endogenous opioids as a mediating role in some of the neurological effects of microwaves and that parameters of microwave exposure are important determinants of the outcome of the microwave effects. The similarity of the effects of microwaves and those of established sources of stress led to the speculation that microwave irradiation is a "stressor" [43].

The hypothesis that low-intensity pulsed microwave exposure can be a source of stress has recently received an additional support by investigating effects of single and repeated exposure to low-intensity, pulsed microwaves on benzodiazepine receptors in three areas of the brain: cerebral cortex, hippocampus, and cerebellum [46]. The cerebral cortex and hippocampus are known to play significant roles in stress responses and, along with the cerebellum, contain high concentrations of benzodiazepine receptors. The authors of the study investigated the effects of single (45-min) and repeated (10 daily 45-min sessions) microwave exposure (2.45 GHz, 1 mW cm^{-2}, average whole-body SAR 0.6 W kg^{-1}, pulsed 500 pps with pulse width 2 ms) on the concentration and affinity of benzodiazepine receptors in the cerebral cortex, hippocampus, and cerebellum of the rat. The mechanism by which acute stress affects benzodiazepine receptors in the brain is not known.

A better understanding and evaluation of the possible physiological consequences of the effects of microwave exposure on benzodiazepine receptors require further research for the following reasons [47]: (1) the differential effects of microwave exposure parameters, such as power density, duration, and absorption pattern, because different patterns of effects can result from different parameters of exposure; (2) the type and location in the brain of benzodiazepine receptors affected by microwaves, because different populations of central benzodiazepine receptors seem to serve anxiolytic, anticonvulsant, and sedative-hypnotic functions; and (3) the effects of microwaves on other neurotransmitter systems in the brain, because benzodiazepine receptors interact intensively with other neurotransmitters.

An important conclusion of the mentioned research was that the long-term biological consequences of *repeated microwave exposure* also depend on the exposure parameters. Further experiments showed that changes in cholinergic receptors after repeated microwave exposure also depended on endogenous opioids in the brain. The microwave effects could be blocked by pretreatment, before each session of daily exposure, with the narcotic antagonist naltrexone. At present, there is no convincing evidence that repeated exposure to low-level microwaves could lead to irreversible neurological effects.

3.5.3   ELF Components

The development of wireless (mobile) telephony has resulted in the transmission and propagation in the atmosphere of microwaves modulated by very specific signals. Radio waves transmitted by mobile phones of the GSM type (Global System for Mobile telecommunications) present a characteristic pattern that results from the particular time structure of such a signal (time division multiple access, TDMA). It is an ELF-modulated pulsed microwave carrier. This is not the case for analog radio and television. One may say that digital cellular phones using the GSM system transmit information in bursts of microwaves. The presence of ELF components in the signal and the *bursting* activity of these waves raised a new controversial question: Can this signal structure exert a negative influence on human head tissues and more specifically on the brain, possibly inducing nonthermal effects?

In Section 3.3, it was mentioned that a model for the operation of one living cell simulated the effect of the two low-frequency sine waves included in the GSM signal, at 8.3 and at 217 Hz, respectively, as well as that of the composite pulsed GSM signal, in which the frequencies 8.3 and 217 Hz are present. It has been observed that the two low-frequency sine waves of GSM, at 8.3 and at 217 Hz, respectively, act much less on the opening probability than the composite pulsed GSM signal, in which the frequencies 8.3 and 217 Hz are present. This will be investigated in more detail in Section 3.10.

An in vitro experimental investigation has been conducted to verify if pure magnetic fields at 8.3 and 217 Hz could induce any effect on the spontaneous bioelectric activity of the neurons from the brain ganglia of the snail *Helix aspersa*. Changes have indeed been observed, as well as the reversibility of the effects induced under exposure to magnetic fields of low magnetic flux densities in the ranges of 0.37–6.68 and 0.6–3.6 mT for 8.3 and 217 Hz, respectively. The first results indicated that, in most cases, the neurons reacted to the lowest values of applied magnetic flux density, that is, 50 μT, and that some neurons would probably react to a lower exposure level [84]. A second series of results shows some interesting neuron behavior which demonstrates the ability of the neurons to recover their spontaneous activity once it has been modified under exposure to 8.3 and 217 Hz sinusoidal applied magnetic flux density values between 0.6 and 6.68 mT. These results show the *reversibility* of the bioelectric induced alterations on neurons under the specified experimental conditions [85]. This investigation is in progress in view of obtaining practical conclusions about sensitivity threshold and reversibility for actual mobile phone ELF magnetic field exposure.

3.6 EAR, EYE, AND HEART

Some studies have been devoted to the effects of RFs and microwaves on specific organs of the human body, such as the *heart*. The only aspects covered in this section are biological ones, excluding interference with electric and electronic devices, such as cardiac and acoustic devices. These will be covered in Section 3.10.

As for the *ear*, the microwave auditory phenomenon is one of the most interesting biological effects of microwave radiation. Short rectangular microwave pulses impinging on the heads of humans and animals have been shown to produce audible sounds. When microwave radiation impinges on the head, the absorbed energy is converted into heat, which produces a small but rapid rise in temperature. This temperature rise, occurring in a very short time, generates rapid thermoelastic expansion of tissues in the head, which then launches an acoustic wave of pressure that is detected by hair cells in the cochlea. Pulsed microwaves were shown to induce acoustic pressure waves in the cat brain which propagate with an acoustic wave velocity of 1523 m s^{-1} [86]. Cat brains were irradiated with pulsed 2.45-GHz microwaves. Short rectangular pulses (2 ms, 15 kW peak power) were applied singly through a direct-contact applicator located at the occipital pole of a cat's head. Acoustic pressure waves were detected by using a small hydrophone transducer which was inserted stereotaxically into the brain of an anesthetized animal through a matrix of holes drilled on the skull. The mean speed of propagation was based on an ensemble of 64 measurements made at six different sets of distance. The amplitude attenuation experienced by the thermoelastic pressure wave in the brain follows an exponential law and has an attenuation coefficient of 0.56 cm^{-1}.

It should be mentioned that the internal structure of the ear is much more complex than that of the brain, for instance, because it contains both hydrated and nonhydrated materials. This makes the microwave power in the ear rather difficult to determine. This complexity, however, increases undulatory phenomena, which reduces the EM power density. Also, the vascularization of the ear is important, because of the number of vessels and capillary blood vessels. As a consequence, the temperature elevation of the ear can be important when submitted to microwave exposure. As mentioned in Section 3.1.3, a thermographic camera has been used to measure the thermal distribution over the surface of a human face exposed to the emission of a mobile telephone handset through a pattern of 10,000 thermocouples located in the lens. Analyzing separately the ear area, it has been found that the most significant surface temperature occurs on the ear lobe. Its temperature increases with time for about 20 min, after which thermoregulation limits the temperature. For bad conditions—a phone conversation established from the basement of a modern building—the temperature of the ear lobe increases by 1.5 and 2.5°C after a 10- and 15-min conversation, respectively [7].

In recent years, particular interest has been expressed in the biological effects of pulse-modulated RFs and microwaves. It is well established that humans can perceive pulse-modulated radiation between 200 MHz and 6.5 GHz as a buzzing or clicking noise, depending on modulation characteristics. The effect is generally attributed to thermoelastic expansion of brain tissue following the small but rapid increase in temperature on the absorption of the incident energy, generating a sound wave in the head that stimulates the cochlea. Some of the observed effects of pulsed radiation may relate to this phenomenon. Speed of propagation and attenuation are both functions of frequency and temperature [87]. The thermoelastic waves excited by the absorbed energy of pulsed microwaves in a human head have been analyzed numerically [88].

High-power pulsed microwave systems are used in applications such as radar, communications, telemetry, and electronic warfare. Exposure to high-peak-power microwave pulses may cause specific behavioral responses, some of which can be related to microwave hearing. Such pulses have been reported to reduce acoustic startle responses of mice, modify performance cognitive tasks, and evoke body movements. A startle response is a neuromuscular reflex to an intense stimulus, and exposure to a single 1-ms pulse of 1.25-GHz radiation has been shown to suppress the amplitude of these responses in rats and mice to a brief, loud noise presented 50 ms later. The specific absorption was 100 mJ kg^{-1} for rats and 200 mJ kg^{-1} for mice. These values are significantly above the auditory threshold of rodents of about 1 mJ kg^{-1} for pulsed RF [89]. More information on the microwave auditory phenomenon can be found in the literature [90].

The cornea and the crystalline lens of the *eye* are not vascularized and their metabolism is slow. The eye has no thermal sensors and does not produce any arm sensation. Hence, there is no protective reflex with respect to a heating process. Significant damage may appear in the case of imprudence. Several experiments have been led on the various components of the nonhuman primate eye. At 0.9, 1.2, 2.45, and 2.85 GHz, either pulsed or continuous waves, they demonstrated that relatively low level exposure can result in significant ocular changes in the nonhuman primate. These changes range from cellular disruption to altered visual function. The experiences demonstrated (1) the possibility of cataract at power densities higher than 100 mW cm^{-2} at frequencies above 1 GHz; (2) corneal endothelial lesions, increased iris vascular permeability, degenerative cell changes in both the iris and retina, and altered electroretinograms, indicating a significant decrease in visual functions at $10\,mW\,cm^{-2}$; (3) corneal damage at a SAR of $2.6\,W\,kg^{-1}$ at 2.45 GHz; and (4) retinal damage at a SAR of $4\,W\,kg^{-1}$ in the frequency range 1.25–2.45 GHz. Pulsed microwaves with an average power density of $10\,mW\,cm^{-2}$ (SAR $2.6\,W\,kg^{-1}$) produced effects for which levels of $20–30\,mW\,cm^{-2}$ (SAR 5.3–$7.8\,W\,kg^{-1}$) CW exposure were required to produce similar changes [91]. Later, another series of experiments was conducted where the nonhuman primate eye was submitted to an ophthalmic pretreatment. Immediately before microwave exposure, one or both eyes of an anesthetized primate were treated topically with one drop of 0.5% timolol maleate or 2% pilocarpine. The power density threshold was observed to decrease by a factor of about 10 (from 10 to $1\,mW\,cm^{-2}$) for induction of corneal endothelial lesions and for increased vascular permeability of the iris at a SAR of $0.26\,W\,kg^{-1}$ at pulsed 2.45 GHz [92]. In another experimental investigation, however, the same authors reported the absence of any detectable ocular damage after either single or repeated exposure to $10\,mW\,cm^{-2}$ from a 60-GHz CW source [93].

Possible effects of RFs and microwaves on the *heart* have been investigated also. The location the heart, however, placed quite inside the body, together with the small penetration of microwaves due to the skin effect are such that the heart is not submitted to high microwave fields. Hearts of chicken embryos have been isolated. Heartbeat stimulation and the control effect of microwaves on the electrical activity of the heart have been analyzed. The hearts were exposed to low-power, pulse-modulated microwaves at 2.45 GHz, 10 mW peak power, and 10% duty cycle. The estimated incident peak power density was $3\,mW\,cm^{-2}$. The repetition frequency was within normal physiological limits (1–3 Hz). Before being exposed, the heart rhythm was rather irregular. When microwaves with a pulse repetition rate of 2.4 Hz were tuned on, the heartbeat became regular at about the same frequency. By increasing the repetition frequency, the heartbeat increased likewise until, above 2.65 Hz, the heart came back to beat irregularly. Hence, the heartbeat was synchronized with the signal from the source within normal physiological limits [94]. This phenomenon is explained by an effect of the pulsed modulation of the source on currents due to the calcium ions, mentioned in Sections 3.2 and 3.3.

Further experimental results by the same authors showed that the dragging and regularization effects observed when the samples are exposed for short durations also appear during longer exposure durations, lasting through all the exposure time. However, CW exposure at the same peak power as the experiments with pulsed modulation does not show any significant modification of the heartbeat. As the authors have pointed out, this suggests a nonthermal effect induced by the pulse-modulated microwaves. The temperature of the sample is lower than at CW excitation; therefore, no heartbeat increase can be related to temperature variations [95].

3.7   INFLUENCE OF DRUGS

From data obtained at low-level microwave irradiation, it has been concluded that *endogenous opioids* may play a mediating role in some of the neurological effects of microwaves, summarized as follows [45]:

1. Microwaves enhance morphine-induced catalepsy in the rat.
2. Microwaves attenuate the naloxone-induced wet-dog shake, a morphine withdrawal symptom, in morphine-dependent rats.
3. Narcotic antagonist blocks a transient increase in body temperature after microwave exposure.
4. There is an effect of acute microwave exposure on amphetamine-induced hyperthermia.
5. Microwave-induced changes in high-affinity choline uptake in the brain can be blocked by narcotic antagonists.
6. Changes in concentrations of muscarinic cholinergic receptors in the brain after repeated sessions of microwave exposure can be blocked by pretreatment with narcotic antagonists before each session of microwave exposure.
7. The three major subtypes of opioid receptor, μ, δ, and κ, are involved in the effect of microwaves on hippocampal high-affinity choline uptake.

Some research has been led under the assumption that the effects of microwaves can be better understood while an animal is under the influence of a drug. It was found that each brain region responds differently to microwaves depending on exposure parameters. Effects on the frontal cortex are independent of the exposure system and of the use of pulsed or CW microwaves. The hippocampus responds to pulsed but not to CW microwaves. The response of the striatal choligernic system depends on the exposure system used. Also, under the same exposure conditions, different brain regions have different sensitivities to microwaves. Hence the areas of the brain that show changes in cholinergic activity are not correlated with localized SAR in the brain.

An important conclusion of investigations is that the *long-term biological consequences* of repeated microwave exposure also depend on the parameters of irradiation. Furthermore, some experiments showed that changes in cholinergic receptors after repeated microwave exposure also depend on endogenous opioids in the brain, as mentioned in Sections 3.2 and 3.5. The microwave effects could be blocked by pretreatment, before each session of daily exposure, with the narcotic antagonist naltrexone [83]. At present, there is no convincing evidence that repeated exposure to low-level microwaves could lead to irreversible neurological effects.

The action of drugs has already been mentioned in Section 3.6 relating to effects on the eye. When the eye of a nonhuman primate was submitted to an ophthalmic pretreatment (opioids), the sensitivity to microwave radiation was observed to increase by about a factor 10, from 10 to $1\,\mathrm{mW\,cm^{-2}}$ for induction of corneal endothelial lesions [92].

3.8 NONTHERMAL, MICROTHERMAL, AND ISOTHERMAL EFFECTS

Thermal effects have been investigated in Section 3.1. It was said that the possibility of either nonthermal or microthermal effects is not a recent question. It was also mentioned that one needs to be very careful about the conditions of an experimental study when investigating the possibility of nonthermal effects, to be sure to take into account all the power components. It is obviously indispensable to be able to distinguish between thermal and nonthermal effects.

There is quite a controversy about the possibility of nonthermal or microthermal effects. This controversy is not only scientific; it is largely political and commercial. Accepting the idea that RFs may cause nonthermal effects or microwave exposure implies that such an exposure could be of a low or very low level, and this is not well accepted. On the other hand, it is a rather common mistake to consider that biological effects are almost necessarily pathogenic for human beings. This is not true: Biological effects may or may not result in an adverse health effect. In fact, this underlies the whole question about how to establish guidelines for limiting electromagnetic field (EMF) exposure: They must provide protection against known adverse health effects. This subject will be investigated in detail in Section 3.11, devoted to hazards and standards. The question of accepting or rejecting nonthermal effects is not a minor question. In 1971, Michelson and Dodge, comparing Soviet and Western views on the biological effects of microwaves, mentioned: "The importance of the difference between the Soviet and Western views is readily apparent when it is realized that practical consideration of Maximum Permissible Exposure (MPE) is based on the acceptance or rejection of non-thermal effects of microwaves as biologically significant" [96, p. 109].

This section offers a discussion on the scientific reasons for which non-thermal or microthermal effects might be detected as well as the tools to be used for the investigation. One main reference is a monograph by Chukova, essentially based on the Soviet and Russian literature on nonthermal effects [97]. In this literature there is a consensus that such effects do exist and that they can be observed in weak fields, hence at a low level of absorbed energy. Another main reference is the set of contributions by Fröhlich, for instance [98]. Other significant references for this section are Morse [99] and Sommerfeld [100].

In Section 1.4.1, the concepts of EM energy and power were developed. It was said that Poynting's theorem expresses equality between the space variation of EM power and the time variation of EM energy. It is well known that EM theory is connected with the structure of the problem, in particular with boundary conditions. It must be noticed, however, that temperature is not an EM parameter: It is a consequence of energy absorption at RFs and microwave frequencies. This is illustrated for instance by Eqns. (2.19) and (2.21): The SAR is proportional to absorption losses, and there is a temperature elevation when the SAR is positive; if there is absorption, there is a temperature elevation. Hence, from a phenomenological point of view, it should be emphasized that EM theory does not possess the mathematical tool for imposing a constant temperature. As a consequence, it cannot investigate the possibility of nonthermal effects: When using EM theory, only thermal effects can be evaluated. In other words, using exclusively EM tools offers no chance to display nonthermal effects.

If EM energy is not an adequate tool for investigating nonthermal RF and microwave effects, then what is the good tool?

Obviously, other considerations have to be taken into account in which temperature is a parameter. This of course leads to thermodynamics. Contrary to EM theory, thermodynamics has no connection with the structure of the system. It considers the system as a "black box," with four parameters: volume, pressure, temperature, and entropy. Hence, thermodynamics is able to investigate effects at constant volume or constant pressure or also constant temperature. In other words, to investigate the possibility of *isothermal* effects, electromagnetics and thermodynamics have to be used jointly, combining Poynting's theorem with basic thermodynamic equations. Hence, one has to investigate to what extent energy and entropy can be used in combination to evaluate isothermal effects. This of course seriously complicates the study.

On the other hand, investigating the possibility of isothermal effects does not preclude the attention to be paid to "nonthermal" effects, which should probably better be termed *microthermal* effects [11]. The question is: Can extremely weak EM exposure have large biological effects, and how is this possible? One then has to consider the possibility of *trigger action* by microwaves.

The study of nonthermal/microthermal effects, on the one hand, and isothermal effects, on the other, will be separated and handled in Sections 3.8.1 and 3.8.2, respectively. It should be reminded that biomolecules, like proteins for instance, have remarkable dielectric properties, different in many ways from more ordinary, small molecules.

3.8.1 Microwaves as a Trigger

It must be emphasized that an appropriate discussion on experimental results with regard to microthermal effects must frequently consider thermal effects as well. Energy transfer from the radiation to the material produces thermal effects. These vary slowly with frequency and are largely governed by dielectric loss. Within reasonable limits this loss is proportional to the intensity of the radiation. Microthermal effects, on the other hand, occur in certain frequency regions only and usually exhibit saturation at rather low intensity. As a consequence, thermal effects might drown microthermal effects. This is particularly to be expected when the signal of the considered effect in the nonthermal region is opposite to that of the thermal ones.

It might be tempting to link frequency-dependent biological effects of resonance type to absorption bands in certain biomolecules. Fröhlich has demonstrated, however, that such resonances are properties of the *whole system* and may depend on the biological activity. Biological systems have to be considered in terms of their activities, which require a high degree of organization. Such organization may be of a complex nature, but it does require consideration of the system as a whole. This section is largely based on his contributions [98].

It is important to realize that in some instances biosystems can exhibit properties similar to those of the most refined electronic instruments, achieving this in a way that is not well understood. For instance, at low intensities, the human visual system has a *sensitivity* that is close to the theoretical limit. Comparing the energy of a light quantum with that of a nerve impulse, there is a gain of more than 10^6. Clearly the light quantum acts as a trigger for a nerve impulse, the energy of which is provided by the biological system.

Control of activities represents another important set of in vivo biological properties. It must be well understood that, for instance, the absence of control of cell division consists in cancer. Of particular interest among materials are enzymes, which through their catalytic action regulate most biological chemical processes. Of particular interest also is the maintenance of the electric potential difference across biological membranes, with a thickness of about 10 nm, described in Section 3.3. In this section we discuss how this might be involved in the interaction with microwaves.

Macroscopic organization is, of course, uniquely correlated to details of microscopic structure. This does not mean, however, that knowing all microscopic details will reveal the interesting macroscopic properties. The number of microstates is so enormous that it cannot be handled. Furthermore, the relevant macroscopic properties are expressed in terms of concepts that do not exist in microphysics: They are collective properties. In these circumstances, the use of the concept of *information*, which is negative entropy, cannot be useful. For instance, the enormous number of microstates may lead to a corresponding information content of the order of $200 \log 20$.

It is a general feature of active biological systems that energy is always available, through metabolic processes, and that this causes nonlinear changes in molecules or large subsystems. Hence, from the point of view of physics, there are nonlinear effects that change with time and are maintained through constant energy supply. It is dangerous, therefore, to extrapolate from properties of biological molecules obtained by extracting them from the living system to their behavior in vivo, although in some cases this can be done. Still, from the point of view of physics, biological systems are relatively stable from a microscopic point of view: The thermal vibrations of single atoms are practically the same as in a corresponding nonbiological system. In some respect, however, involving relatively few degrees of freedom, they are very far from the thermal equilibrium, and these degrees of freedom dominate the overall behavior of the rest. They may be described in terms of collective properties or organizations that do not exist in individual particle physics. These collective properties evolve as a consequence of supply of energy (metabolism) and usually represent extreme nonlinear displacements.

Two theoretical physical models have been presented in detail [98]. An essential feature in both cases is the basic importance of the *nonlinear characteristics* in conjunction with a supply of energy. One is quasi-static, showing that under very general conditions metastable states with very high electric polarization exist. The second is dynamic in terms of *coherent excitation of electric vibrations*. It shows that under more stringent conditions coherent electric vibrations may be excited by random metabolic energy. This leads to long-range selective forces that supplement short-range forces, including those leading to standard chemical processes. The selectivity of the long-range forces depends on the frequency of the oscillating systems, which in turn depends on their structure.

Biological membranes have been considered as a basic material, showing oscillations with displacement perpendicular to the surface, so that for the longest wavelength the membrane thickness equals half the wavelength. Assuming elastic properties corresponding to a sound velocity of $10^3\,\mathrm{m\,s^{-1}}$ yields a frequency $0.5 \times 10^{11}\,\mathrm{Hz}$, that is, the microwave frequency $50\,\mathrm{GHz}$. The membrane is not homogeneous, however, because a considerable number of proteins are dissolved in it, so that small sections of the membrane may of course vibrate separately from the rest. The membrane is very strongly polarized electrically, with a field of about $10^7\,\mathrm{V\,m^{-1}}$. Hence excitation of vibration of a particular section of the membrane is connected with a vibrating electric dipole. Multipoles may lead to higher frequencies. Of course, molecules outside the membranes may also oscillate as well as other sections of a cell. There may also be other sources of oscillations, such as plasma modes of the unattached electrons. As a consequence, a great variety of possible frequencies will exist.

Quite logically now, the question is: What is the possibility of a trigger action by microwaves? Fröhlich pointed out theoretically the possibility of coherent excitations, showing that when certain conditions are fulfilled, random energy supply to the modes of a band of electric polarization waves may lead to strong, that is, coherent, excitation of a single mode provided the energy supply exceeds a critical value. The trigger action of microwaves is one feature of the model. The other is that it predicts strong excitation of polar modes through biological pumping.

The nonthermal action of microwaves can be evoked at this stage. Microwaves penetrating into a material deliver energy to it as a linear response proportional to the intensity of the radiation and with a magnitude and phase dependence governed by the complex permittivity of the material. There is an action of microwaves on the thermal motion of the electric dipoles of the material. This motion is only very minutely perturbed, but this induces an increase in temperature: Under stationary conditions, the energy removed by heat conduction is equal to the energy supplied by the absorption of the microwaves. Consequently one should not expect too much the detection of nonthermal effects in regions of high dielectric absorption.

Theoretical investigation of the quantitative detailed microscopic requirements would of course be extremely difficult. So the experiment has to decide about the existence of coherent excitation. Such experiments, however, are difficult to set up. There is of course no totally convincing experimental evidence of coherent excitation in biological systems; otherwise there would be no controversy about nonthermal/microthermal effects. A number of experimental results, however, are in favor of this theory. Such results are available essentially in the millimeter-wave region, with for instance the influence of low-intensity radiation on *E. coli* at frequencies of 70–75 and 136 GHz, yeast cells at 41–42 GHz, protein metabolism at 40 GHz, bone marrow cells of mice around 40 GHz, and the induction of colicins near 45 GHz [101, 102]. Colicins are a class of proteins manufactured at certain occasions by many bacteria of the *E. coli* group. They act as antibiotics that kill cells of bacterial strains related to the strains that make them.

There is also one general argument in favor of such a mechanism. One of the most general biological features is indeed that energy supplied in terms of food, or of sunlight, is in part used to build up and maintain a very complex organization. The sun may well warm a plant. Nobody says, however, that the action of the sunlight on a plant is exclusively thermal.

3.8.2 Entropy

The fundamentals of thermodynamics have been summarized in Section 2.3, to which the reader is referred.

The concept of entropy has been shown to be necessary to evaluate correctly some phenomena in electronics engineering. One typical example from literature is luminescence, with the problem of conversion of heat into luminescence radiation. The heat comes from the thermal energy of the crystal lattice. It has been shown that the ratio between the luminescence energy and the absorbed excitation radiation energy—the energy efficiency—can be larger than unity, at the expense of the thermal energy of the lattice [97]. This results in the *cooling* of the lattice, often termed optical cooling. The phenomenon can be explained by analogy with a heat engine. When a motor transforms heat into mechanical work, less ordered energy is transformed into a more ordered one, and the efficiency is limited by thermodynamic considerations. On the other hand, when the motor transforms mechanical work into heat, more ordered energy is transformed into a less ordered one, and the limit efficiencies are larger than 1. Hence, a limit efficiency larger than unity for luminescence implies a process in which the emitted radiation is less ordered than the energy being transformed.

It has been mentioned in Section 2.3 that thermodynamics considers the interaction of the system with the environment, taking into account entropy and energy at the system input and output, dealing only with three different types of systems, however:

1. An *isolated* system has no exchange at all with the environment.
2. A *closed* system can exchange energy with the environment.
3. An *open* system can exchange both energy and mass with the environment.

The luminescence just described is a process in which EM radiation interacts with matter, producing a direct conversion of radiant energy into electrical energy. Luminescent systems are closed systems in the thermodynamic sense of the word: They can exchange energy with the environment. Another example is a photochemical system: Such systems can be open systems, able to exchange both energy and mass with the environment. They can also be closed systems, however.

The ratio of the useful electrical work to the absorbed part of the EM radiation energy is termed the *efficiency* of the converter. On the other hand, the ratio of the useful electrical work to the total EM radiation energy is termed the *effectiveness*. The effectiveness is equal to the efficiency when the EM radiation energy is totally absorbed, that is, when the converter is matched to the incident radiation. When there is a mismatch, part of the incident radiation is reflected, only part of the energy is absorbed, and the effectiveness is smaller than the efficiency. It is rather obvious that effectiveness is used in many experiments for simplicity, because the degree of matching is difficult to evaluate. However, thermodynamics only deals with energy efficiency.

Luminescence has been demonstrated when exposing an air–water interface to millimeter waves. This has been called the Saratov phenomenon and was published (in Russian) in 1999 by Sinitsin et al. [103]. We follow here the description given by Chukova [97], who also published (in Russian) a thermodynamic explanation of the Saratov data in 2001. The conditions of exposure of the water medium to millimeter waves are the following:

1. The power density was less than $10\,\mu W\,cm^{-2}$, usually $1\,\mu W\,cm^{-2}$.
2. The millimeter-wave (MM) excitation was in a sweeping mode.
3. The response of the water medium was observed in the decimeter-wave (DM) region and was measured at 0.4 and 1.0 GHz.
4. The power of this DM radiation was low, of the order of $10^{-16}\,W$.

At higher MM exposure levels, from $10\,\mu W\,cm^{-2}$ to $10\,mW\,cm^{-2}$, the DM radiation was not observed, while at a power level above $10\,mW\,cm^{-2}$ a thermal effect is observed.

A number of water media have been tested: for example, water with NaCl, water with ice, alkaline and acid water, blood, plasma of blood, serum of blood, erythrocytes, water with narcotic, water with stimulant, mouse under narcotic, mouse under stimulant, alcohol (C_2H_5OH), glycerin [$C_3H_5(OH)_3$], water with NH_3 (25%), milk, white of egg, and yolk of egg. All these media have exhibited the DM radiation in three frequency ranges, near 50, 65, and 100 GHz. In each of these ranges there are two peaks, at 50.3 and 51.8, 64.5 and 65.5, and 95 and 105 GHz, respectively. Sinitsin et al. explain the existence of these two peaks on the basis of the water structure. Figure 3.1 shows decimeter luminescence measured on water and on human tissue. The Saratov spectrum measured on human tissue is very similar to that measured on water.

The researchers have noted that the Saratov spectrum changes when the physiological functional condition of the tissue changes. They have deduced a new diagnosis method from this particularity, investigating the DM radiation from a body exposed to millimeter waves. Measuring the spectrum, they have also showed that abnormal tissues such as thyroid gland with diffuse toxic goiter turn into normal ones after a certain number of sessions of millimeter-wave exposure. Hence, they consider the Saratov phenomenon not only as a diagnosis method but also as a measure of a medical treatment due to

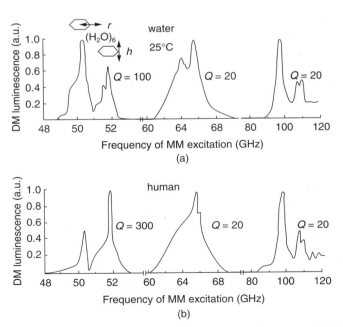

FIGURE 3.1 Decimeter luminescence of water and human tissue. (Adapted from [97].)

millimeter-wave exposure. The thermodynamic explanation published by Chukova in 2001 takes into account luminescence, change of chemical bond energy, and heat.

3.9   EPIDEMIOLOGY STUDIES

The widespread use of hand-held mobile phones has as a consequence that many people routinely place RF/microwave transmitters against their heads. This is a good reason to warrant examination of the safety of this form of radiant energy. There should have been good reasons to be concerned about exposure to television and FM radio transmitters earlier too, but this did not happen. It is the use of mobile phones that has raised concern about microwave exposure as well as, mainly in Europe, exposure to the relay stations of digital mobile telephony. This has led to epidemiology investigations involving statistical analyses of health records and standardized tests on animals. Both types of investigations have been led on cancer and genotoxicity, while brain cancer dominates public discussion. On-going research is reviewed at regular intervals [104, 105].

Up to now, studies do not establish clearly that RFs lead to cancer. To establish such effects, the difficult question is *dosimetry*, as explained in Section 3.1.2: Measuring the exposure to which a human being is submitted is one thing; estimating what it has been in the past is much more difficult. Because of this difficulty, added to the fact that low-level exposure is investigated, it is very difficult to prove either one effect or the opposite. Identifying links between cancer and environmental exposure of any kind is extremely difficult because of the absence of a single cause of cancer and for a variety of other reasons. Even if mobile phones had no connection to cancer, given the hundreds of millions of mobile phone users around the world, thousands of users would develop brain cancer every year.

Brain cancer takes years or decades to develop, and the studies say nothing about future risks. Detecting small or long-term cancer risks through statistical analyses of health records is not an easy task. Detecting small increases in risk would necessitate large studies that are difficult to control and may be controversial in their interpretation. Any valid study would have to assess an individual's use of mobile phones over a decade or more, an assessment complicated by the rapid technological developments in this industry.

Animal studies are the other main source of information used in cancer risk assessment. They are easier to control than epidemiology studies. They have, however, uncertain relevance to human health. It should be remembered that frequency scaling, defined in Section 3.1.2, enables use of results obtained with a given object and to adapt them to predict the results to be obtained with another object, similar in form to the first and differing only by a scale factor. This can be strictly true only in loss-free situations, however, which is not the case in living tissue.

About 10 epidemiology studies on cancer are usually mentioned in the literature, although some others have been led. The results of some studies exhibit an increase of the relative risk of cancer for the population submitted to microwaves. This increase, however, is not statistically significant. On the other hand, some other studies exhibit negative results. In general, these studies are very difficult to lead, because accurate dosimetry is difficult. In such a case, investigations in vitro can be very useful. In this area, results are mostly negative, but not all of them. Hence, although investigations have been made on transgenic mice at a rather high exposure level, there is presently no firm conclusion. The cancers for which further work may be most valuable on the basis of studies to date are leukemia in both adults and children and brain cancer in relation to mobile phone use because of the dose considerations.

Genotoxic effects on biological cells may lead to adverse health effects such as cancers and neurological or other diseases. Hence, investigating *genotoxicity* is important for studies on cancer as well as on possible effects. It can be investigated in humans, animals, and tissue cultures. Many of these investigations are in vitro, which makes experiments easier to control. They consist in integrating the genomic analysis of protein and RNA components of the cell and investigating the properties of molecular, cellular, and organismal function. This is genomic epidemiology. The extent to which cellular molecular research can predict properties of the organism, whether it is a simple form of life or a human being, is not completely clear yet, but the advancement in the understanding of molecular mechanisms has been spectacular at the end of the twentieth century. The majority of the studies led until 2002 have not demonstrated the genotoxicity of RFs. There are, however, some positive results, showing among others an increase of DNA alteration in brains of rats and mice exposed to RFs. This remains a controversial question.

3.10 INTERFERENCES

Although interferences produced by RF/microwave devices on electric and especially electronic equipment are not a biological effect, they affect the health of patients. It seems appropriate to devote a small section to this subject. The potential for equipment problems induced by electromagnetic interference (EMI) has increased significantly as the use of electronic systems has proliferated in the hospital setting, on the one hand, and as personal health devices such as pacemakers or comfort devices such as hearing aids, on the other hand. Unfortunately there is little quantitative information available on these possible EMI threats.

On-site EMI tests have been conducted in hospitals, however [106, 107]. Data have been collected for radiated EM fields and conducted power line disturbances in several different hospital environments, ranging from typical patient rooms to X-ray, magnetic resonance imaging (MRI), and surgical suites. These devices often produce broadband energy at frequencies below 10 MHz but may use higher frequency narrow-band energy as part of the operation. For example, an MRI system uses a 64-MHz source, and lasers often use 13- or 27-MHz sources. If well shielded, these systems do not usually pose a threat to nearby electronic equipment; if unshielded, however, devices such as electrosurgical units can cause serious interference.

As expected, the principal sources of radiated EMI are *intentional radiators*, usually situated within the hospital, such as telemetry and paging transmitters, hand-held radios, and cellular phones. Electrosurgical units are also a major source of radiated energy. In all cases, the proximity to the source is the key factor affecting the radiated EMI levels. Such sources can cause monitoring, therapeutic, and diagnostic medical devices to malfunction. Rather surprising, however, are the very low levels of radiated EMI from equipment such as lasers, X-ray and MRI equipment, and computed tomography scanners. While these systems use high levels of energy to perform their functions, the resulting radiated EMI levels are well below the levels from nearby transmitters.

There is a concern about EMI induced by cellular phones, due in part to the rapid expansion in use and proliferation of these devices. There is also another reason. A hand-held two-way radio emits RF energy only when the user intentionally transmits; hence, the user can use discretion about operating the radio in a clinical area. However, the cellular telephone can transmit without the user's knowledge when placed in the standby mode because the phone periodically transmits to the system receiver to indicate that it is in the cell and ready to receive calls. Equipment responses have resulted when a cellular phone was operated within a distance of 1 m [106]. It should be noted that a directive from the European Union about EMI imposes electric and electronic equipment to be exposed to an electric field of $3 \, \text{V m}^{-1}$ up to 1 GHz without dysfunction to obtain a quality label. This value is obtained at a distance of about 2.5 m from a transmitting cellular phone of the GSM type.

GSM cell phones can also interfere with ionizing radiation dose-monitoring equipment. A total of 13 personal electronic dosimeters, portable dose monitors, and contamination monitors were assessed. Six of the personal dosimeters showed abnormal responses when exposed to mobile phone transmission. This should be taken into account when distributing these devices and when assessing results generated by them. Electromagnetic compatibility testing should form part of the commissioning and specification protocol for new dose-monitoring equipment [107].

Electromagnetic interference of pacemakers by mobile phones has also been investigated, in particular for European GSM mobile cellular phones. Two studies investigated about 250 different models of pacemakers from more than 20 manufacturers [108, 109]. The result was that about 30% of patients might have problems with a cellular phone at a distance of 10–15 cm from the pacemaker. A distance of 20 cm is sufficient to guarantee integrity of the pacemaker with respect to hand-held phones. Recent pacemaker models are better protected against interference risk. The conclusion is that it is desirable for implanting physicians to use only pacemakers with immunity against mobile phones as guaranteed by the manufacturers.

Electromagnetic interference has also been evaluated from a global system for mobile communication telephones with a model of a hearing aid used in the ear canal [110]. The electric and magnetic fields were calculated in the ear canal using a finite-difference method for two models of the ear and three positions of the telephone. One result was that, for a given model of telephone, the level of EMI-induced sound pressure levels vary by more than 40 dB, depending on the design of the hearing aid.

3.11   RADIATION HAZARDS AND EXPOSURE STANDARDS

Although equipment that utilizes or emits EM energy provides benefits to mankind, it also constitutes hazards to the individual through uncontrolled and excessive emissions. There is a need to set limits on the amount of exposure to radiant energies individuals can accept with safety. These limits are still subject to change. It can be expected that adequate limits should be frequency dependent. At higher frequencies, indeed, the depth of penetration in biological tissue is limited to the superficial layers. Hence, any concern about potential hazards should focus on the tissues that are both superficial and biologically sensitive. Ocular tissue meets both of these criteria because of its unique structure, location, biochemistry, physiology, and sensitivity to various physical agents. Limits can be based on body-averaged SAR, specifying mostly SAR in 1 or 10 g tissue. Of course, local SAR cannot be measured, in the human brain for instance, and there is a need for reliable evaluation of exposure risks. Investigation of human exposure is difficult because the human body possesses a complex geometry and heterogeneous tissues.

3.11.1   Standards and Recommendations

The limits have been quite different from one country to another in the past. In 1975 already, there was a ratio of 1000 between the U.S. standard ($10\,mW\,cm^{-2}$) and the USSR standard ($10\,\mu W\,cm^{-2}$ for an exposure time beyond 2 h a day). This has been the subject of a number of investigations [111, 112]. Such a huge difference in representing unhealthy effects comes from the fact that the USSR standards took into account not only thermal but also isothermal effects. These have been described in Section 3.8. The controversy has not ceased because, as said above, accepting the possibility of nonthermal effects introduces an extra safety factor of at least 100. It is sometimes said that the Soviet/Russian standard for microwaves is the single standard that does not require a modification, because it already accounts for the existence of isothermal effects, as a result of a large experimental program made in the middle of the twentieth century [95]. Hence, the picture about protection is complicated, especially when remembering that guidelines for limiting EM field exposure provide protection against known adverse health effects while biological effects may or may not result in a health effect.

In the present recommendations, two kinds of limitations are considered:

- Basic restrictions that should be always respected.
- Reference levels that could be exceeded when the basic limitations are not exceeded.

The reason is simple. The basic restrictions are expressed in quantities that are internal to the body and are not measured, such as SAR. On the other hand, the reference levels are expressed in quantities that are measured *in the absence of human beings*, such as an electric field. There are theories and estimations relating these two sets of quantities.

Only one biological effect of microwaves is well known: *heating*. Hence, the present recommendations, being based only on scientific evidence, are limited to heating processes. As an example, the Scientific Steering Committee of the European Commission stated in June 1998 that as regards nonthermal exposure to EMF, the available literature does not provide sufficient evidence to conclude that long-term effects occur as a consequence of EMF exposure. The conclusion was therefore that any recommendation for exposure limits regarding nonthermal long-term effects cannot be made at this stage on a scientific basis. Arguments other than scientific might however be considered, for instance, observations made by medical doctors on public health, published in the 2002 Freibuerger Appeal [113].

The recommendations are based on one single source: They originally come from the World Health Organization (WHO), 1993 [2]. Today, they are essentially based on documents produced by the International Commission on Non-Ionizing Radiation Protection (ICNIRP), with a main document establishing in 1998 guidelines for limiting exposure to EM fields up to 300 GHz [114]. There are, however, ambiguities in the basic texts. The 1993 WHO document states that a biological effect is produced from 1 to $4 \, \mathrm{W \, kg^{-1}}$ while calculating the safety factor from 4 and not from $1 \, \mathrm{W \, kg^{-1}}$. A further factor of 5 is recommended for the public at large, yielding safety factors of 50 and 12.5, when starting from 4 and $1 \, \mathrm{W \, kg^{-1}}$, respectively. Most documents, however, refer to a safety factor of 50. The same discrepancy is found in the 1998 ICNIRP document.

The text of the WHO 1993 document [2, p. 21] is based on the known effect of "increasing the body central temperature by less than 1°C when exposing healthy adults for 30 minutes to a microwave exposure of 1–4 W/kg." The safety factor has to take into account several elements: The temperature increase should be much less than 1°C; the exposure may be 24 h a day and not 30 min; adults are not healthy; there are nonadults (children); all children are not healthy; and there are "unfavorable, thermal, environmental, and possible long-term effects." Health epidemiologists have to evaluate if the safety factor is large enough.

One may compare a temperature elevation of "less than 1°C" by microwave heating with a temperature elevation of less than 1°C due to physical efforts. In an experimental study, 59 healthy young men have been submitted to a maximal exercise test on a treadmill (stress test) for 18 min, during which the limits of the cardiorespiratory physiology are reached. During the last minute of the test, the energetic expense of the adult (20 kcal) is 20 times higher than that of the same adult at rest. Considering that approximately 22.5% of the calories generated during work in aerobiose are transformed into useful external work, the mean energy density was found $102 \, \text{W min kg}^{-1}$. The temperature elevations were measured, yielding an average temperature elevation of 0.736°C. On the other hand, the total energy density due to 30 min microwave exposure with an SAR of $4 \, \text{W kg}^{-1}$ is equal to $120 \, \text{W min kg}^{-1}$. The comparison of the two values shows that, if the two temperature elevations are proportional, the temperature due to this microwave exposure would produce a temperature elevation equal to 0.866°C. It is interesting to observe that this external microwave heating and the maximal exercise testing produce about the same temperature elevation [115].

As can be seen, establishing standards is not an easy thing to do. As an example, it is worth considering the values used in Europe for the standard related to European GSM mobile telephony, at 900 MHz, for the general public, expressed in volts per meter: (1) The WHO, ICNIRP, and European Union recommend not to exceed $41.2 \, \text{V m}^{-1}$; (2) several European governments have adopted lower values, such as Belgium ($20.6 \, \text{V m}^{-1}$), Italy ($20 \, \text{V m}^{-1}$, $6 \, \text{V m}^{-1}$ for an exposure of 4 h or more), Switzerland (4 or $6 \, \text{V m}^{-1}$), and Luxembourg ($3 \, \text{V m}^{-1}$); (3) effects on BBB permeability have been observed at $0.016 \, \text{W kg}^{-1}$, that is, $18 \, \text{V m}^{-1}$; (4) considering the possibility of isothermal or microthermal effects implies an extra factor of about 100 in power, yielding $4 \, \text{V m}^{-1}$; (5) two epidemiology studies out of four on TV/FM exposure evidenced a twofold increase of leukemia under 2 to $4 \, \text{V m}^{-1}$ exposures; (6) the Belgian High Council for Health has recommended an extra safetyfactor of 100–200, yielding 4 to $3 \, \text{V m}^{-1}$; and (7) in February 2003, the City of Paris, France, obtained from the operators not to exceed a value between 1 and $2 \, \text{V m}^{-1}$, depending on the power transmitted at 1800 MHz. Hence, the picture is complex because a number of arguments can be used which do not lead to the same conclusions.

3.11.2 Tissue Phantoms and SAR Measurements

The increase in public concern about possible health risks from microwave energy has stimulated a number of investigations in vivo and in vitro. A step necessary for assessing the potential risks is to analyze and quantify the EM energy induced in the human body, in particular the human head, by external sources, such as cellular telephones and the associated base stations.

Studies carried out by numerical simulation will be reviewed in Section 3.11.3. Another procedure consists of producing a phantom: a biological tissue equivalent with the same permittivity and conductivity as brain and skull tissue. Several papers in [116] describe the procedure and mention a number of references. These phantoms make it possible to accomplish highly reliable and precise estimation of the SAR in biological tissue. They are most frequently human head models, because of both the main concern, about the possible effects into the head, and the practical realizability of the model. Phantom models of cube, sphere, and realistic human heads have been fabricated. Measurements are then performed to estimate the SAR in the human head models exposed to microwave sources by using a thermographic method. More specifically, such phantoms offer the possibility of measuring the local peak SAR. This is extremely interesting, in particular in the case of local external microwave source, such as a cellular telephone.

A typical composition of a brain-equivalent phantom is given in Table 3.2. It is in accordance with recommendations from the Action of the European Cooperation in the field of Scientific and Technical Research (COST)244 [117]. The agar is used for maintaining the shape of the phantom by itself. The relative permittivity is controlled with additional rate of polyethylene powder. In order to mix water with the polyethylene, TX-151 is selected for stickiness. The sodium azide is a preservative. In addition, the loss factor depends on the concentration of sodium chloride. The relative permittivity and conductivity can be controlled by the quantity of polyethylene powder and sodium chloride.

The composition of a skull-equivalent solid phantom is shown in Table 3.3. The glycerol is used as a solvent of the skull phantom. Skull tissue is a low-loss medium; therefore, the solvent should be low loss and hydrophilic. However, the relative permittivity would be too small with only glycerol, and silicone emulsion is added. The constant can be controlled by the mixture rate of the glycerol and silicone emulsion.

These phantoms do not need rigid shells because they contain and shape the jelly material. Therefore it is easy to directly measure temperature rise on

TABLE 3.2   Composition of Brain-Equivalent Phantom

Material	Amount (g)
Deionized water	3375.0
Agar	104.6
Sodium chloride	21.5
Sodium azide	2.0
TX-151	57.1
Polyethylene powder	548.1

Source: From [118].

Note: Batch is approximately 4000 cm^3.

TABLE 3.3 Composition of Skull-Equivalent Phantom

Material	Amount (g)
Silicone emulsion	250.0
Glycerol	200.0
TX-151	12.0
Agar	20.0
Polyethylene powder	130.0

Source: From [118].

Note: Batch is approximately $500\,cm^3$.

TABLE 3.4 Average Parameter Values of Phantom

Parameter	Brain	Skull
Relative permittivity	43.1	17.3
Conductivity ($S\,m^{-1}$)	0.83	0.25
Density ($kg\,m^{-3}$)	880	910

Source: From [118].

Note: Conductivity and relative permittivity are values measured at 900 MHz.

a section or surface of the phantom. The SAR is measured either by using an E-field probe or by thermographic methods. Equation (3.6) gives the relation between the SAR and temperature.

The relative permittivity and conductivity of these phantoms have been measured from 200 to 3000 MHz [118]. The measured values are in very good agreement with the COST244 brain and skull tissue. The average values of the parameters are given in Table 3.4.

The thermal characteristic of the phantom is very important and the precautions to be taken for thermographic analysis of phantoms are the same as for actual human heads. Several methods are based on thermal measurements: (1) *calorimetric methods*, particularly suited for in vitro measurements, in which heating and cooling data can be analyzed to estimate the energy absorbed by an irradiated sample; (2) *thermometric methods*, used to measure the temperature due to microwaves with particular types of nonperturbing thermometer, but only a few are commercially available; and (3) *thermographic techniques*, used to measure the temperature with particular thermographic cameras. In thermography, when using a thermographic camera, emissivity of the object is an important parameter. It is defined as the ratio of the radiation flux per unit area of the emitter to that of a blackbody radiator at the same temperature and under the same conditions. In short, it measures how much radiation is emitted by the object compared to that emitted by a perfect blackbody. Normally, emissivities range from approximately 0.1 to 0.95.

The human skin exhibits an emissivity close to 1.0. Determination of surface temperature as well as temperature variations over the surface depends directly upon the surface emissivity. The following parameters have to be accurately determined prior to starting the measurements: emissivity, ambient temperature, atmospheric temperature, relative humidity of the air, and distance [7].

Calibration is of course important. At the beginning of a set of measurements, the temperature indicated by the IR camera has to be calibrated in several points of the image using data read by thermometers. The temperature of a reference point is measured using a thermometer, for instance a thermocouple. The emissivity is then altered until the temperature calculated by the IR camera agrees with the thermometer reading. This is the emissivity value of the reference object. The temperature of the reference object must, however, not be too close to the ambient temperature for this to work. The procedure for measuring a phantom of course offers more flexibility than that for a human head. Measurements can, for instance, be calibrated by painting part of the phantom surface in black and comparing results obtained from the bare part of the phantom to that of the black part of the phantom [118].

It is rather usual—and recommended—to first operate with canonical models: spheres, cylinders, and homogeneous, single-layer phantoms. This permits the comparison between calculated and measured values.

Operating with phantoms has demerits. One is that high power is in general necessary for the experiment. Hence, actual telecommunication mobile equipment cannot be tested. Another is that three-dimensional measurements cannot be done. Therefore, it is not efficient to measure the 1- or 10-g average SAR by thermographic experiments. On the other hand, the use of thermographic experiments is efficient in the SAR evaluations for a medium of complicated shape such as the inner ear and earlobe.

3.11.3   Computational Methods for SAR Evaluation

There are a number of studies carried out by numerical simulation, in particular about SAR values in the human head when exposed to microwave radiation from hand-held devices [18]. Several papers in [116] cover this subject while mentioning a number of references.

Exposures to radiation from various types of antennas, effect of frequency, and separation distances, for example, have been examined. The effect of the human body on the absorption by the human head has also been considered, in particular the conditions under which it is necessary to take the effect of the human body into account in the calculations [119].

The frequency-domain method of moments (MoM) and FDTD method are widely used for analyzing complex EM problems such as biological structures. Treating electrically large and/or penetrable structures with the MoM requires extensive computational resources. On the other hand, accurate models of wire and curved structure are difficult to implement in the FDTD method.

The FDTD method is usually preferred for cellular telephone simulations because it allows the inclusion of arbitrarily heterogeneous objects in the region to be simulated. Highly realistic heterogeneous human head models have been developed and successfully used. These models are characterized by resolutions as low as 1 mm, and as many as 32 tissue types have been identified.

The major drawbacks of these simulations are the execution time and computer memory requirements, and methods have been proposed to reduce both. Subgridding has been introduced for representing with high resolution the regions of maximum absorption and lower resolution the regions where the EM coupling is weak. In this approach, the major drawback is the need to implement the subgridding scheme in the FDTD code. Methods have been developed for accurate modeling of the region of interest (e.g., ear) and increased coarser modeling of the weakly exposed regions of the human head. The use of the absorbing perfectly matched layer (PML) for the purpose of truncating the head model has been investigated, with the possibility of truncating the head model in all the directions [120]. This method allows truncation along all three axes. It has been shown that by using only a half-truncated head model it is possible to achieve accurate radiation patterns at both 835 and 1900 MHz. Therefore, when complete characterization of the handset performance is needed, it is possible to use the proposed approach with a considerable saving for SAR evaluation and a more limited saving for radiation pattern calculation in both execution time and memory requirements. Reductions of the head volume down to 4% of the original volume have been achieved, and memory savings of up to 82% have been obtained. Execution times were no more than 7% of the original values.

A hybrid method between the MoM and the FDTD method has been developed [121]. It is capable of analyzing a system of multiple discrete regions by employing the principle of equivalent sources to excite their coupling surfaces. It has been shown that this theory has advantage for accurately modeling complex and arbitrarily oriented mobile telephone handset antennas.

3.11.4 Exposure of Body to Cell Phone and Base Station

Guidelines for limiting EM exposure provide protection against known adverse health effects. Biological effects, on the other hand, may or may not result in an adverse health effect. There is a serious concern among the population, in particular in Europe, about possible adverse biological effects due to cellular telephone base stations. The question of the "microwave syndrome" rises again, the possible effects of low-intensity exposure are discussed, recommendations are analyzed in detail, and health issues are reviewed. Surprisingly, however, there is no concern about TV and FM exposure. Main conclusions are at the present: There are still a number of uncertainties; a substantial database is available on thermal biological effects; most countries have protection standards; and there is a variety of these standards.

Biological effects due to microwave exposure depend upon the electric field inside the tissues. As has been seen earlier, thermal effects depend on the SAR distribution (Section 3.1). Are there nonthermal effects? Some say yes, many say no. The so-called nonthermal effects might be microthermal effects, while there might also be isothermal effects (Section 3.8). It should be well remembered that microwaves as a whole form a "family" of frequencies, traditionally from 100 MHz to 1 THz, inducing essentially the same effects throughout the whole frequency range. From a biological point of view, it is even wise to consider microwaves from 50 MHz up, because of the TV emissions at those frequencies.

Absorbed power density is the main cause of effects: $1 \, W \, kg^{-1}$ may yield an increase of (less than) $1°C$ in human body, taking thermal regulation into account [2]. Measurements on the face have shown that the temperature increase is obtained after 10–12 min [7]. It should be well noted that the SAR is the ratio of absorbed power to absorbing mass: The skin effect has to be taken into account. At 900 MHz, the lowest GSM cellular telephone frequency, the skin depth is approximately 1.5 cm. As an indication, using a simple cylindrical approximation, the absorbing mass of a human body weighting 65 kg and 1.70 m high is approximately 8.5 kg at 900 MHz. Pulsed waves produce a detectable effect at power levels smaller than at CW (Section 3.5.2). For mobile telephony, the interaction between a handset antenna and the human body is obviously of special interest [18].

The question of the "microwave syndrome" (Section 3.5.1) was raised several decades ago in Eastern Europe, related to a number of manifestations, such as headache, perspiration, emotional instability, irritability, tiredness, somnolence, sexual problems, loss of memory, concentration and decision difficulties, insomnia, and depressive hypochondriac tendencies. The evaluation is difficult because of the absence of a control group and well-established dosimetric data. A recent paper supports the RF sickness syndrome as a possible medical entity [83].

Experiments concerning thermal effects due to GSM exposure from handheld mobile telephones have been made [7]. The time evolution of different parts of the head has been measured when exposed to microwaves from GSM at 900 GHz. Three different phones have been used from three different manufacturers located in Europe, the United States, and Japan, respectively. An IR camera measured the temperature evolution by taking a thermographic image every 20 s using a pattern of 10,000 thermocouples located in the lens. Measurements were made during the different phases of the phone operation: stand-by, ringing, talking, and switching off. Mainly two different locations were used: favorable reception–transmission conditions on the fourth floor of a building and unfavorable conditions in the basement, 4 m below ground level. Total temperature was measured, taking into account microwave heating, possible heating due to the electronic part of the device, and possible reduction in ventilation because of the presence of the device near the face. The most significant temperature increase has been observed on the ear lobe, at the end of the talking phase, for the highest duration (20 min.). This temperature increases from 1.0 to 2.4°C, depending on location and phone type. The temperature trend for cheek and neck areas follows a similar trend. The temperature increases in the exposure the first 5–7 min. A value as high as 0.7°C is reached in basement conditions.

A good question is obviously: Is protection effective? Section 3.11.1 shows that there are a variety of arguments when trying to answer the question. The present recommendations, based only on "scientific evidence," recommend limits related only to heating processes, which is the only well-known microwave effect. As stated by the Scientific Steering Committee of the European Commission in June 1998, as regards nonthermal exposure to EMFs, the available literature does not provide sufficient evidence to conclude that long-term effects occur as a consequence of EMF exposure. The conclusion being therefore that any recommendation for exposure limits regarding nonthermal long-term effects cannot be made at this stage on a scientific basis. It can of course be argued whether this conclusion is valid [122, 123].

Part of the difficulty is that some people describe themselves as hypersensitive to EM fields. Hypersensitivity is a syndrome defined essentially by the patient complaining about effects that he or she attributes to the use or the proximity of equipment emitting fields, electric, magnetic, and/or EM. The causes mainly cited are electric distribution at 50/60 Hz and base stations for cellular telephony. This hypersensitivity is mentioned at levels largely below those of international recommendations. In 2004, the relationship between electric and magnetic fields and associated effects was far from being demonstrated and a dose–effect relationship at such low levels had not been established yet.

REFERENCES

[1] M. A. Stuchly, S. S. Stuchly, "Experimental radio and microwave dosimetry," in C. Polk and E. Postow (Eds.), *Handbook of Biological Effects of Electromagnetic Fields*, Boca Raton, FL: CRC Press, 1996.

[2] World Health Organization, *Electromagnetic Fields (300 Hz to 300 GHz)*, Geneva: WHO, 1993.

[3] NCRP, "Radiofrequency electromagnetic fields: Properties, quantities and units, biophysical interaction, and measurements," NCRP Report No. 67, Washington, DC: National Council on Radiation and Measurements, 1981.

[4] A. Vander Vorst, *Electromagnétisme. Champs et Circuits*, Brussels: De Boeck, 1994.

[5] J. A. Stratton, *Electromagnetic Theory*, New York: McGraw-Hill, 1941.

[6] J. P. Reilly, *Electrical Stimulation and Electropathology*, New York: Cambridge University Press, 1992.

[7] M. D. Taurisano, A. Vander Vorst, "Experimental thermographic analysis of thermal effects induced on a human head exposed to 900-MHz fields of mobile phones," *IEEE Trans. Microwave Theory Tech.*, Vol. 48, No. 11, pp. 2022–2032, Nov. 2000.

[8] J. Teng, D. Carton de Tournai, F. Duhamel, A. Vander Vorst, "No nonthermal effect observed under microwave irradiation of spinal cord," in A. Rosen and A. Vander Vorst (Eds.), Special Issue on Medical Applications and Biological Effects of RF/Microwaves, *IEEE Trans. Microwave Theory Tech.*, Vol. 44, No. 10, pp. 1942–1948, Oct. 1996.

[9] S. Baranski, "Histological and histochemical effect of microwave irradiation of the central nervous system of rabbits and guinea pigs," *Am. J. Phys. Med.*, Vol. 51, pp. 182–191, 1972.

[10] A. Vander Vorst, "Microwave bioelectromagnetics in Europe," *IEEE MTT-S Microwave Int. Symp. Dig.*, Atlanta, 1993, pp. 1137–1140.

[11] A. Vander Vorst, "Biological effects, introduction to workshop," *Proc. Workshop Biological Effects Medical Applications*, 27th Eur. Microwave Conf., Jerusalem, Sept. 1997, pp. 1–3.

[12] S. M. Michaelson, J. C. Lin, *Biological Effects and Health Implications of Radiofrequency Radiation*, New York: Plenum, 1987.

[13] J. Thuery, *Microwaves: Industrial, Scientific and Medical Applications*, Boston, MA: Artech House, 1992.

[14] C. Polk, E. Postow, *Handbook of Biological Effects of Electromagnetic Fields*, Boca Raton, FL: CRC Press, 1996.

[15] P. R. Riu, K. R. Foster, D. W. Blick, E. R. Adler, "A thermal model for human thresholds of microwave-evoked warmth sensations," *Bioelectromagnetics*, Vol. 18, pp. 578–583, 1997.

[16] N. Kuster, Q. Balzano, "Energy absorption mechanism by biological bodies in the near field of dipole antennas above 300 MHz," *IEEE Trans. Vehicular Tech.*, Vol. 41, pp. 17–23, 1992.

[17] A. Vander Vorst, F. Duhamel, "1990–1995 Advances in investigating the interaction of microwave fields with the nervous system," in A. Rosen and A. Vander Vorst (Eds.), Special Issue on Medical Applications and Biological Effects of RF/Microwaves, *IEEE Trans. Microwave Theory Tech.*, Vol. 44, No. 10, pp. 1898–1909, Oct. 1996.

[18] M. Okoniewski, M. A. Stuchly, "A study of the handset antenna and human body interaction," in A. Rosen and A. Vander Vorst (Eds.), Special Issue on Medical Applications and Biological Effects of RF/Microwaves, *IEEE Trans. Microwave Theory Tech.*, Vol. 44, No. 10, pp. 1855–1864, Oct. 1996.

[19] L. Dubois, J. P. Sozanski, V. Tessier, J.-C. Camart, J.-J. Fabre, J. Pribetich, M. Chivé, "Temperature control and thermal dosimetry by microwave radiometry in hyperthermia," in A. Rosen and A. Vander Vorst (Eds.), Special Issue on Medical Applications and Biological Effects of RF/Microwaves, *IEEE Trans. Microwave Theory Tech.*, Vol. 44, No. 10, pp. 1755–1761, Oct. 1996.

[20] D. Sullivan, "Three-dimensional computer simulation in deep regional hyperthermia using the finite-difference time-domain method," *IEEE Trans. Microwave Theory Tech.*, Vol. 38, pp. 204–211, 1990.

[21] C. Rappaport, F. Morgenthaler, "Optimal source distribution for hyperthermia at the center of a sphere of muscle tissue," *IEEE Trans. Microwave Theory Tech.*, Vol. 35, pp. 1322–1327, 1987.

[22] C. Rappaport, J. Pereira, "Optimal microwave source distributions for heating off-centers tumors in spheres of high water content tissue," *IEEE Trans. Microwave Theory Tech.*, Vol. 40, pp. 1979–1982, 1992.

[23] D. Dunn, C. Rappaport, A. Terzuali, "Verification of deep-set brain tumor hyperthermia using a spherical microwave source distribution," in A. Rosen and A. Vander Vorst (Eds.), Special Issue on Medical Applications and Biological Effects of RF/Microwaves, *IEEE Trans. Microwave Theory Tech.*, Vol. 44, No. 10, pp. 1769–1777, Oct. 1996.

[24] V. Hombach, K. Meier, M. Burkhardt, E. Kühn, N. Kuster, "The dependence of EM energy absorption upon human head modeling at 900 MHz," in A. Rosen and A. Vander Vorst (Eds.), Special Issue on Medical Applications and Biological Effects of RF/Microwaves, *IEEE Trans. Microwave Theory Tech.*, Vol. 44, No. 10, pp. 1865–1873, Oct. 1996.

[25] S. Watanabe, M. Taki, T. Nojima, O. Fujiwara, "Characteristics of the SAR distributions in a head exposed to electromagnetic fields radiated by a hand-held portable radio," in A. Rosen and A. Vander Vorst (Eds.), Special Issue on Medical Applications and Biological Effects of RF/Microwaves, *IEEE Trans. Microwave Theory Tech.*, Vol. 44, No. 10, pp. 1874–1883, Oct. 1996.

[26] O. P. Gandhi, G. Lazzi, C. M. Furse, "Electromagnetic absorption in the human head and neck for mobile telephones at 835 and 1900 MHz," in A. Rosen and A. Vander Vorst (Eds.), Special Issue on Medical Applications and Biological Effects of RF/Microwaves, *IEEE Trans. Microwave Theory Tech.*, Vol. 44, No. 10, pp. 1884–1897, Oct. 1996.

[27] K. Ito, "Hyperthermia: Review and present developments," *Proc. Workshop Biological Effects and Medical Applications*, 27th Eur. Microwave Conf., Jerusalem, Sept. 1997.

[28] O. P. Gandhi, "Conditions of strongest electromagnetic power deposition in man and animals," *IEEE Trans. Microwave Theory Tech.,* Vol. 23, pp. 1021–1029, 1975.

[29] O. P. Gandhi, "Biological effects and medical applications of RF electromagnetic fields," *IEEE Trans. Microwave Theory Tech.*, Vol. 30, pp. 1831–1847, 1982.

[30] D. I. McRee, "Cooperative program between the United States and Soviet Union in environmental health," *Proc. US-USSR Workshop Phys. Factors-Microwaves and Low-Frequency Fields*, Res. Park, 1985, pp. 3–10.

[31] C. L. Mitchell, D. I. McRee, N. J. Peterson, H. A. Tilson, M. G. Shandala, M. K. Rudnev, V. V. Varetskii, M. I. Navakatikyan, "Results of a United States and Soviet Union joint project on nervous system effects of microwave radiation," *Environ. Health Persp.*, Vol. 81, pp. 201–209, 1989.

[32] M. I. Rudnev, M. I. Navakatikyan, "Dynamics of changes in behavioral reactions induced by microwave radiation," *Proc. US-USSR Workshop Phys. Factors-Microwaves and Low-Frequency Fields*, Res. Triangle Park, 1981, pp. 113–132.

[33] J. M. Zhou, W. X. Lou, J. W. Jiang, X. D. Cao, "The role of NE in the preoptic area of rabbit's brain under the acupuncture analgesic effect," *Acta Phys. Sinica*, Vol. 38, No. 4, pp. 415–421, 1986.

[34] J. Teng, D. Vanhoenacker, A. Vander Vorst, "Biological effects of microwaves in acupuncture," *Proc. Eur. Microwave Conf.*, London, 1989, pp. 918–923.

[35] A. Vander Vorst, J. Teng, D. Vanhoenacker, "The action of microwave electromagnetic fields on the nervous system," *Proc. J. Int. Nice Antennes*, Nice, 1992, pp. 111–119.

[36] A. L. Lehninger, "Components of the respiratory chain and approaches to its reconstruction," in *The Mitochodrion*, New York: W. A. Benjamin, 1965.

[37] A. P. Sanders, W. T. Joines, J. W. Allis, "The differential effects of 200, 591, and 2,450 MHz radiation on rat brain energy metabolism," *Bioelectromagnetics*, Vol. 5, pp. 419–433, 1984.

[38] R. Pethig, *Dielectric and Electronic Properties of Biological Materials*, New York: Wiley, 1979.

[39] E. H. Grant, R. F. Sheppard, G. P. South, *Dielectric Behavior of Biological Molecules in Solution*, Oxford: Oxford University Press, 1978.

[40] J. Teng, H. Yan, D. Vanhoenacker, A. Vander Vorst, "Variations of pain threshold and norepinephrine release in rabbits due to microwave stimulation," *Proc. MTT Symp., Boston*, 1991, pp. 801–804.

[41] J. C. Lin, "The blood-brain barrier, cancer, cell phones, and microwave radiation," *IEEE Microwave Mag.*, Vol. 2, No. 4, pp. 26–30, Dec. 2001.

[42] H. Goldman, J. C. Lin, S. Murphy, M. F. Lin, "Cerebral permeability to ^{86}Rb in the rat after exposure to pulsed microwaves," *Bioelectromagnetics*, Vol. 5, pp. 323–330, 1984.

[43] L. G. Salford, A. Brun, F. Sturesson, J. L. Eberhart, B. R. Persson, "Permeability of the blood-brain barrier induced by 915 MHz electromagnetic radiation, continuous wave and modulated at 8, 16, 50 and 200 Hz," *Microscopy Res. Tech.*, Vol. 27, pp. 535–542, 1994.

[44] J. C. Lin, "Microwave radiation and leakage of albumin from blood to brain," *IEEE Microwave Mag.*, Vol. 5, No. 3, pp. 22–28, Sept. 2004.

[45] H. Lai, "Research on the neurological effects of nonionizing radiation at the University of Washington," *Bioelectromagnetics*, Vol. 13, pp. 513–526, 1992.

[46] H. Lai, M. A. Carino, A. Horita, A. W. Guy, "Single vs. repeated microwave exposure: Effects on benzodiazepine receptors in the brain on the rat," *Bioelectromagnetics*, Vol. 13, pp. 57–66, 1992.

[47] A. P. Sanders, W. T. Joines, J. W. Allis, "Effects of continuous-wave, pulsed, and sinusoidal-amplitude-modulated microwaves on brain energy metabolism," *Bioelectromagnetics*, Vol. 6, pp. 89–97, 1985.

[48] D. O. Brown, S. T. Lu, E. C. Elson, "Characteristics of microwave evoked movements in mice," *Bioelectromagnetics*, Vol. 15, pp. 143–161, 1994.

[49] H. Lai, A. Carino, A. Horita, A. W. Guy, "Low-level microwave irradiation and central cholinergic activity: A dose response study," *Bioelectromagnetics*, Vol. 10, pp. 203–208, 1989.

[50] H. Lai, A. Horita, A. W. Guy, "Microwave irradiation affects radial-arm maze performance in the rat," *Bioelectromagnetics*, Vol. 15, pp. 95–104, 1994.

[51] S. Takashima, "Effects of modulated RF energy on the EEG of mammalian brains," in S. S. Stuchly (Ed.), *Electromagnetic Fields in Biological Systems*, Ottawa: Springer-Verlag, 1978.

[52] G. Thuroczy, G. Kubinyi, M. Bodo, L. D. Szabo, "Simultaneous response of brain electrical activity (EEG) and cerebral circulation (REG) to microwave exposure in rats," *Rev. Environ. Health*, Vol. 10, No. 2, pp. 135–148, 1994.

[53] Y. E. Moskalenko, *Biophysical Aspects of Cerebral Circulation*, Oxford: Pergamon, 1980.

[54] S. Bawin, W. Adey, I. Sabbot, "Ionic factors in release of 45Ca^{++} from chicken cerebral tissues by electromagnetic fields," *Proc. Natl. Acad. Sci. U.S.A.*, Vol. 75, No. 12, pp. 6314–6318, 1978.

[55] C. Blackman, J. Elder, C. Weil, S. Benane, D. Eichinger, D. House, "Modulation-frequency and field-strength dependent induction of calcium-ion efflux from brain tissue by radio-frequency radiation," *Radio Sci.*, Vol. 14, No. 6 (S), pp. 93–98, 1979.

[56] R. MacGregor, "A possible mechanism for the influence of electromagnetic radiation on neuroelectric potentials," *IEEE Trans. Microwave Theory Tech.*, Vol. 27, pp. 914–921, Nov. 1979.

[57] B. Nilsson, L. Petersson, "A mechanism for high-frequency electromagnetical field-induced biological damage?" *IEEE Trans. Microwave Theory Tech.*, Vol. 27, pp. 616–618, June 1979.

[58] R. W. Adey, "Electromagnetic fields and the essence of living systems," in J. B. Andersen (Ed.), *Modern Radio Science*, Oxford: Oxford University Press, 1990.

[59] Z. Somossy, G. Thuroczy, T. Kubasova, J. Kovacs, L. D. Szabo, "Effects of modulated and continuous microwave irradiation on the morphology and cell surfaces of 3T3 fibroblasts," *Scan. Microsc.*, Vol. 5, pp. 1145–1155, 1991.

[60] D. R. McNeal, "Analysis of a model for excitation of myelinated nerve," *IEEE Trans. Biomed. Eng.*, Vol. 23, No. 4, pp. 329–337, Apr. 1976.

[61] F. Rattay, M. Aberham, "Modeling axon membranes for functional electrical stimulation," *IEEE Trans. Biomed. Eng.*, Vol. 40, No. 12, pp. 1201–1209, Dec. 1993.

[62] A. T. Barker, B. H. Brown, I. L. Freston, "Modeling of an active nerve fiber in a finite volume conductor and its application to the calculation of surface action potential," *IEEE Trans. Biomed. Eng.*, Vol. 26, No. 1, pp. 53–56, 1979.

[63] J. J. Struijk, J. Holsheimer, B. K. van Veen, H. B. K. Boom, "Epidural spinal cord stimulation: Calculation of field potentials with special reference to dorsal column nerve fibers," *IEEE Trans. Biomed. Eng.*, Vol. 38, No. 1, pp. 104–110, Jan. 1991.

[64] W. K. Sin, B. Coburn, "Electric stimulation of the spinal cord: A further analysis relating to anatomical factors and tissue properties," *Med. Biol. Eng. Comp.*, Vol. 21, pp. 264–269, 1983.

[65] P. S. Tofts, "The distribution of induced currents in magnetic stimulation of the nervous system," *Phys. Med. Biol.*, Vol. 35, No. 8, pp. 1119–1128, 1990.

[66] B. J. Roth, P. J. Basser, "A model of the stimulation of a nerve fiber by electromagnetic induction," *IEEE Trans. Biomed. Eng.*, Vol. 37, No. 5, pp. 588–597, June 1990.

[67] P. J. Basser, R. S. Wijesinghe, B. J. Roth, "The activated function for magnetic stimulation derived from a three-dimensional volume conductor model," *IEEE Trans. Biomed. Eng.*, Vol. 39, No. 11, pp. 1207–1210, Nov. 1992.

[68] J. T. Rubinstein, "Axon termination conditions for electrical stimulation," *IEEE Trans. Biomed. Eng.*, Vol. 40, No. 7, pp. 654–663, July 1993.

[69] S. S. Nagarajan, D. M. Durand, E. N. Warman, "Effects of induced electric fields on finite neuronal structures: A simulation study," *IEEE Trans. Biomed. Eng.*, Vol. 40, No. 11, pp. 1175–1187, Nov. 1993.

[70] W. R. Adey, "Cell membranes: The electromagnetic environment and cancer promotion," *Neurochem. Res.*, Vol. 13, No. 7, pp. 671–677, 1988.

[71] P. Bernardi, G. d'Inzeo, S. Pisa, "A generalized ionic model of the neuronal membrane electrical activity," *IEEE Trans. Biomed. Eng.*, Vol. 41, No. 2, pp. 125–133, Feb. 1994.

[72] F. Apollonio, G. D'Inzeo, L. Tarricone, "Theoretical analysis of voltage-gated membrane channels under GSM and DECT exposure," *IEEE MTT-S Microwave Int. Symp. Dig.*, Denver, 1997, pp. 103–106.

[73] V. Garaj-Vrhovac, A. Fucic, D. Horvat, "The correlation between the frequency of micronuclei and specific chromosome aberrations in human lymphocytes exposed to microwave radiation in vitro," *Mutation Res.*, Vol. 281, pp. 181–186, 1992.

[74] G. Thuroczy, private communication, National Research Institute, Budapest, 1992.

[75] P. Hojevik, P. J. Sandblom, S. Galt, Y. Hamnerius, "Ca^{2+} ion transport through patchclamped cells exposed to magnetic fields," *Bioelectromagnetics*, Vol. 16, pp. 33–40, 1995.

[76] I. Y. Belyaev, Y. D. Alipov, V. S. Shcheglov, "Chromosome DNA as a target of resonant interaction between *Escherichia coli* cells and low-intensity millimeter waves," *Electro- and Magnetobiology*, Vol. 11, No. 2, pp. 97–108, 1992.

[77] J. Vanderstraeten, A. Vander Vorst, "Theoretical evaluation of dielectric absorption of microwave energy at the scale of nuclear acids," *Bioelectromagnetics*, Vol. 25, pp. 380–3890, 2004.

[78] R. P. Blackwell, R. D. Saunders, "The effects of low-level radiofrequency and microwave radiation on brain tissue and animal behaviour," *Int. J. Radiat. Biol.*, Vol. 50, No. 5, pp. 761–787, 1986.

[79] A. Presman, *Electromagnetic Fields and Life*, English translation, New York: Plenum, 1970.

[80] K. Marha, J. Musil, H. Tuha, *Electromagnetic Fields and the Living Environment*, San Francisco: San Francisco Press, 1971.

[81] C. Silverman, "Epidemiological studies of microwave effects," *Proc. IEEE*, Vol. 68, pp. 78–84, Jan. 1980.

[82] A. G. Johnson Liakouris, "Radiofrequency (RF) sickness in the Lilienfeld study: An effect of modulated microwaves?" *Arch. Environ. Health*, Vol. 53, No. 3, pp. 236–238, May–June 1998.

[83] H. Lai, A. Carino, A. Horita, A. W. Guy, "Naltrexone pretreatment blocks microwave-induced decreases in central cholinergic receptors," *Bioelectromagnetics*, Vol. 12. pp. 27–33, 1991.

[84] D. Lederer, M. J. Azanza, A. C. Calvo, R. N. Pérez Bruzón, A. del Moral, A. Vander Vorst, "Effects associated with the ELF of GSM signals on the spontaneous bioelectricity activity of neurons," *Proc. 5th Int. Cong. Eur. BioElectromagnetics Assoc.*, Helsinki, Sept. 2001, pp. 194–195.

[85] M. J. Azanza, R. N. Pérez Bruzón, D. Lederer, A. C. Calvo, A. Vander Vorst, A. del Moral, "Reversibility of the effects induced on the spontaneous bioelectric activity of neurons under exposure to 8.3 and 217.0 Hz low intensity magnetic fields," *2nd Int. Workshop Biol. Effects of EMFs*, Rhodes, Oct. 2002, pp. 651–659.

[86] J. C. Lin, J.-L. Su, Y. Wang, "Microwave-induced thermoelastic pressure wave propagation in the cat brain," *Bioelectromagnetics*, Vol. 9, pp. 141–147, 1988.

[87] Z. J. Sienkiewicz, N. A. Cridland, C. I. Kowalczuk, R. D. Saunders, "Biological effects of electromagnetic fields and radiation," in W. Ross Stone (Ed.), *Review of Radio Science 1990–1992*, Oxford: Oxford University Press, 1993.

[88] Y. Watanabe, T. Tanaka, M. Taki, S. Watanabe, "FDTD analysis of microwave hearing effects," in A. Rosen and A. Vander Vorst (Eds.), Special Issue on Medical Applications and Biological Effects of RF/Microwaves, *IEEE Trans. Microwave Theory Tech.*, Vol. 44, No. 10, pp. 2126–2132, Oct. 1996.

[89] R. L. Seaman, D. A. Beblo, T. G. Raslear, "Modification of tactile startle amplitude by microwave pulses," *Abstracts 1st World Congress for Electricity and Magnetism in Biology and Medicine*, Orlando, June 1992, p. 60.

[90] J. C. Lin, "Hearing microwaves: The microwave auditory phenomenon," *IEEE Microwave Mag.*, Vol. 3, No. 2, pp. 30–34, June 2002.

[91] H. A. Kues, L. A. Hirst, G. A. Lutty, S. A. D'Anna, G. R. Dunkelberger, "Effects of 2.45-GHz microwaves on primate corneal endothelium," *Bioelectromagnetics*, Vol. 6, pp. 177–188, 1985.

[92] H. A. Kues, J. C. Monahan, S. A. D'Anna, D. S. McLeod, G. A. Lutty, S. Koslov, "Increased sensitivity of the non-human primate eye to microwave radiation following ophthalmic drug pretreatment," *Bioelectromagnetics*, Vol. 13, pp. 379–393, 1992.

[93] H. A. Kues, S. A. D'Anna, R. Oslander, W. R. Green, J. C. Monahan, "Absence of ocular effects after either single or repeated exposure to 10 mW/cm² from a 60 GHz CW source," *Bioelectromagnetics*, Vol. 20, No. 8, pp. 463–473, 1999.

[94] C. C. Tamburello, L. Zanforlin, G. Tiné, A. E. Tamburello, "Analysis of microwave effects on isolated hearts," *IEEE MTT-S Microwave Int. Symp. Dig.*, Boston, 1991, pp. 804–808.

[95] M. Abbate, G. Tiné, L. Zanforlin, "Evaluation of pulsed microwave influence on isolated hearts," in A. Rosen and A. Vander Vorst (Eds.), Special Issue on Medical Applications and Biological Effects of RF/Microwaves, *IEEE Trans. Microwave Theory Tech.*, Vol. 44, No. 10, pp. 1935–1941, Oct. 1996.

[96] S. M. Michelson, C. H. Dodge, "Soviet views on the biological effects of microwaves—An analysis," *Health Phys.*, Vol. 21, pp. 108–111, July 1971.

[97] Y. P. Chukova, *Advances in Nonequilibrium Thermodynamics of the Systems under Electromagnetic Radiation*, Moscow: Academy of Sciences, 2001.

[98] H. Fröhlich, "The biological effects of microwaves and related questions," in *Advances in Electronics and Electron Physics*, New York: Academic, 1980, pp. 85–152.

[99] P. M. Morse, *Thermal Physics*, New York: Benjamin, 1964.

[100] A. Sommerfeld, *Thermodynamics and Statistical Mechanics*, New York: Academic, 1964.

[101] H. Fröhlich, "Further evidence for coherent excitations in biological systems," *Phys. Lett.*, Vol. 110A, No. 9, pp. 80–81, Aug. 1985.

[102] S. Baranski, P. Czerski, *Biological Effects of Microwaves*, Stroudsburg, PA: Dowden, Hutchinson & Ross, 1976.

[103] N. I. Sinitsyn, V. I. Petrosyan, V. A. Yolkin, N. D. Devyatkov, Yu. V. Gulaev, O. V. Betskii, "Particular role of a MM wave – water medium system in nature," *Biomedical Radioelectronics* (in Russian), No. 1, pp. 3–21, 1999.

[104] S. Johnston, "Review of on going research on radio frequencies and health 2002–05," *Proc. 2nd. Intern. Workshop, Biological Effects of EMFs*, Rhodes, Oct. 2002, pp. 1004–1013.

[105] W. D. Kimmel, D. D. Gerke, "Electromagnetic interference in the hospital environment," *Med. Dev. Diag. Indust.*, Vol. 17, No. 5, pp. 97–101, May 1995.

[106] G. Knickerbocker, "Guidance article: Cellular telephones and radio transmitters—Interference with clinical equipment," *Health Dev.*, Vol. 22, Nos. 8–9, pp. 416–418, Aug.–Sep. 1993.

[107] P. Gilligan, S. Somerville, J. T. Ennis, "GSM cell phones can interfere with ionizing radiation dose monitoring equipment," *Brit. J Radiol.*, Vol. 73, No. 873, pp. 994–998, Sept. 2000.

[108] V. Barbaro, P. Bartolini, A. Donato, C. Militello, G. Altamura, F. Ammirati, M. Santini, "Do European GSM mobile cellular phones pose a potential risk to pacemaker patients?" *Pac. Clin. Electrophysiol.*, Vol. 18, No. 6, Pt. 1, pp. 1218–1224, June 1995.

[109] W. Irnich, L. Batz, R. Müller, R. Tobisch, "Electromagnetic interference of pacemakers by mobile phones," *PACE*, Vol. 19, pp. 1431–1450, Oct. 1996.

[110] K. Caputa, M. A. Stuchly, M. Skopec, H. I. Bassen, P. Ruggera, M. Kanda, "Evaluation of electromagnetic interference from a cellular telephone with a hearing aid," in A. Rosen, A. Vander Vorst, and Y. Kotsuka (Eds.), Special Issue on Medical Applications and Biological Effects of RF/Microwaves, *IEEE Trans. Microwave Theory Tech.*, Vol. 48, No. 11, pp. 2148–2154, Nov. 2000.

[111] S. M. Michaelson, "Protection guides and standards for microwave exposure," NATO, *Lecture Series No. 78 Radiation Hazards*, Sept. 1975, pp. 12.1–6.

[112] A. Vander Vorst, A. Laloux, *Les Hyperfréquences*, Louvain-la-Neuve: Cabay, 1982.

[113] Interdisciplinäre Gesellschaft für Umweltmedizin e.V., *Freibuerger Appeal*, Bad Säckingen, Octt. 2002.

[114] ICNIRP Guidelines, "Guidelines for limiting exposure to time-varying electric, magnetic, and electromagnetic fields (up to 300 GHz)," *Health Phys.*, Vol. 74, pp. 494–522, Apr. 1998.

[115] G. Pirquin, L. Stevens, A. Vander Vorst, "Body temperature elevation during maximal exercise testing compared with microwave heating," *Microwaves UCL Rep. 03.02.11*, Louvain-la-Neuve: Microwaves UCL, Feb. 2003.

[116] A. Rosen, A. Vander Vorst, Y. Kotsuka (Eds.), Special Issue on Medical Applications and Biological Effects of RF/Microwaves, *IEEE Trans. Microwave Theory Tech.*, Vol. 48, No. 11, Pt. II, pp. 1977–2198, Nov. 2000.

[117] COST244, "Proposal for numerical canonical models in mobile communications," *Proc. COST244 Symp.*, Europ. Union, Rome, Nov. 1994, pp. 1–7.

[118] Y. Okano, K. Ito, I. Ida, M. Takahashi, "The SAR evaluation method by a combination of thermographic experiments and biological tissue-equivalent phantoms," in A. Rosen, A. Vander Vorst, and Y. Kotsuka (Eds.), Special Issue on Medical Applications and Biological Effects of RF/Microwaves, *IEEE Trans. Microwave Theory Tech.*, Vol. 48, No. 11, Pt. II, pp. 2094–2103, Nov. 2000.

[119] M. F. Iskander, Z. Yun, R. Quintero-Illera, "Polarization and human body effects on the microwave absorption in a human head exposed to radiation from hand-held devices," in A. Rosen, A. Vander Vorst, and Y. Kotsuka (Eds.), Special Issue on Medical Applications and Biological Effects of RF/Microwaves, *IEEE Trans. Microwave Theory Tech.*, Vol. 48, No. 11, Pt. II, pp. 1979–1987, Nov. 2000.

[120] G. Lazzi, O. P. Gandhi, D. M. Sullivan, "Use of PML absorbing layers for then truncation of the head model in cellular telephone simulations," in A. Rosen, A. Vander Vorst, amd Y. Kotsuka (Eds.), Special Issue on Medical Applications and Biological Effects of RF/Microwaves, *IEEE Trans. Microwave Theory Tech.*, Vol. 48, No. 11, Pt. II, pp. 2033–2039, Nov. 2000.

[121] M. A. Mangoud, R. A. Abd-Alhameed, P. S. Excell, "Simulation of human interaction with mobile telephones using hybrid techniques over coupled domains," in A. Rosen, A. Vander Vorst, and Y. Kotsuka (Eds.), Special Issue on Medical Applications and Biological Effects of RF/Microwaves, *IEEE Trans. Microwave Theory Tech.*, Vol. 48, No. 11, Pt. II, pp. 2014–2021, Nov. 2000.

[122] A. Vander Vorst, M. D. Taurisano, B. Stockbroeckx, "Cellular telephones: Hazards or not?" *IEEE MTT-S Microwave Int. Symp. Dig.*, Boston, June 2000, pp. 937–940.

[123] K. R. Foster, J. E. Moulder, "Are mobile phones safe?" *IEEE Spectrum*, Vol. 37, No. 8, pp. 23–28, Aug. 2000.

PROBLEMS

3.1. A plane wave at 30 GHz is incident at a right angle upon the skin ($\varepsilon = 18 - j19$).

(a) Find the reflection coefficient Γ.

(b) Show that the SAR is given by $SAR(0)e^{-2x/\delta}$, where

$$SAR(0) = \frac{2p_{inc}\left(1-|\Gamma|^2\right)}{\delta}$$

is the SAR at the interface, p_{inc} is the incident power per unit area, ρ is the mass density of skin, and δ is the penetration depth in skin. Assume that the skin has a density of 1000 kg/m³.

(c) Plot the SAR as a function of depth for $p_{inc} = 10$ mW cm⁻².

3.2. Assume that the monopole of problem 1.13 is fed by a 33 Ω coaxial cable.

(a) Find the available power of a source feeding the cable if the cable is lossless, and the monopole terminal voltage is 0.5 V.

(b) A numerical analysis of the equivalent dipole antenna with $V_0^e = 1$ V has provided an SAR profile in the medium as given by the following table[1].

If 2 W of power is available through the coaxial cable, calculate the temperature rise at $r = 5$ mm, for axial distances $z = 5, 15,$ and 20 mm, when the power

[1] J. P. Casey, R. Bansal, "The near field of an insulated dipole in a dissipative dielectric medium," *IEEE Trans. Microwave Theory Tech.*, Vol. MTT-34, No. 4, pp. 459–463, Apr. 1986.

Absorbed Power (W/kg)	$z = 5\,\text{mm}$	$z = 15\,\text{mm}$	$z = 20\,\text{mm}$
$r = 0.8\,\text{mm}$	1.82	2.48	2.95
$r = 1.0\,\text{mm}$	1.42	1.74	2.00
$r = 2.0\,\text{mm}$	0.75	0.66	0.65
$r = 5.0\,\text{mm}$	0.30	0.21	0.17
$r = 10.0\,\text{mm}$	0.12	0.08	0.06

is applied for 100 s. Assume that tissue has a specific heat $c = 3500\,\text{J}\,\text{kg}^{-1}\text{C}^{-1}$. Ignore the thermal conduction in the tissue.

3.3. A monopole antenna with metal tip is fabricated from a 2.2-mm coaxial cable, where $l = 9\,\text{mm}$ and $t = 2\,\text{mm}$ and is used for microwave catheter ablation of heart at 2.45 GHz. Numerical simulation of the structure with 1 V voltage at the input to the antenna yields a radial SAR (SAR_1) profile at $z = 10\,\text{mm}$ as given in Figure P3.3.

   (a) How much is the input voltage if a temperature rise of 10°C is required at $z = 10$, and $r = 3\,\text{mm}$ after 1 min of applying the signal. Assume that tissue has a specific heat of $c = 3500\,\text{J}\,\text{kg}^{-1\circ}\text{C}^{-1}$ and $\rho = 1000\,\text{kg/m}^3$.

   (b) If an input impedance of Z_{in} is measured for the antenna, and the cable is lossless with a characteristic impedance of $50\,\Omega$, what is the relation between the available power and Z_{in}?

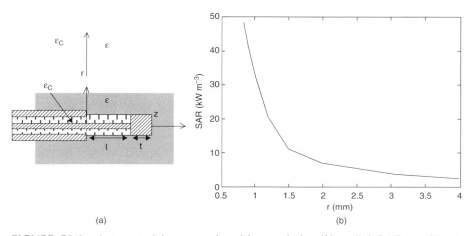

(a)   (b)

FIGURE P3.3   Antenna: (*a*) monopole with metal tip; (*b*) radial SAR profile at $z = 10\,\text{mm}$ (after Labonte et al.[1]).

[1] Labonte et al, "Monopole antennas for microwave catheter ablation" *IEEE Trans. Microwave Theory Tech.,* Vol. MTT-44, No. 10, pp. 1832–1840, Oct. 1996.

3.4. (This problem addresses the reader familiar with Maxwell's equations.) Frequency scaling (Section 2.5.2) is based on what is called the *electrodynamic similitude theorem*. The purpose of this problem is to establish the theorem.

   (a) Write Maxwell's equations for the curl of electric and magnetic fields in the time domain for a ε, μ, σ medium.

   (b) Express all the quantities involved in the equations as a product of a dimensional and a dimensionless quantity, such as, for the electric field,

$$\overline{E} = e\overline{e}$$

   where $\overline{E}$ is the actual electric field and e the scale factor in volts per meter while $\overline{e}$ is dimensionless. Such separation is to be done for both the electric and magnetic fields, permittivity, permeability, conductivity (writing permittivity for instance as $\varepsilon_0\varepsilon$), and length (included in the partial derivatives of the curls to be written, e.g., $\partial l_0 x$) and time (included in the time derivatives).

   (c) Derive the two dimensionless curl equations for $\overline{e}$ and $\overline{h}$. In these equations, it is then possible to group all dimensionless quantities in three factors. You now have the similitude theorem.

   (d) Divide these factors by each other to eliminate e and h and obtain two new dimensionless quantities: $\mu_0\varepsilon_0(l_0/t_0)^2$ and $\mu_0\varepsilon_0 l_0^2 t_0$. These must remain constant for the same dimensionless curl equations to characterize identical problems, for instance at different frequencies. Observe what happens to the other parameters, in particular the characteristic length, when frequency is multiplied, for instance, by a factor 2. Observe that the simple rule for scaling frequency [Eqn. (2.20)] is valid only when the conductivity is zero.

3.5. The pattern of energy absorption in the body contributes to the microwave effect, while inhomogeneities affect the local power absorption. This raises the question of whether the whole-body average SAR can be used as the only determining factor in evaluating biological effects of low-level microwaves. Discuss the advantages and disadvantages of using a local SAR, for instance averaged over 10-g masses instead of over the whole body.

3.6. The parameters of microwave exposure are important to consider when investigating biological effects. For instance, pulsed-wave exposure with the influence of duration, waveform, and peak versus average power can lead to different biological effects than CW exposure. Discuss the influence of some parameters. More specifically, consider radar-type pulsed microwaves, observe the relations between peak power and repetition rate, and discuss possible consequences of a high peak power.

CHAPTER FOUR

Thermal Therapy

4.1 INTRODUCTION TO THERMOTHERAPY

Historically, using heat as a medical treatment is nothing new. Heat utilized to treat lesions can be traced to the age of Hippocrates in 4 B.C. [1, 2]. In general terms, thermotherapy as medical treatment has been widely used, for example, in rheumatism and joints and muscle diseases. In these cases, EM waves, IR rays, ultrasonic waves [3], warm water, and so on, have been used as heating energy sources. These heating methods are seldom different from the ones presently used in thermotherapy. Therapies using physical energy have been recognized as a kind of physiotherapy. More specifically, these sorts of therapy utilizing EM waves at wavelengths from several hundred to several tens of meters have been called *diathermy*. They have been called short-wave or microwave diathermy, ultrasonic wave diathermy, and so on, depending on the difference in radiation form or wavelength. Such thermotherapies have been put into practice in a number of frequency regions along a frequency axis as shown in Figure 4.1. Among these thermotherapies, a hyperthermic method for the malignant tumor has been called *hyperthermia*. In this chapter, we describe the physical facets of thermotherapy, principally referring to hyperthermia (see Section 2.3).

4.2 HEATING PRINCIPLE

In the medical field, radiators using, for example, an EM wave or an ultrasound wave are called *applicators*. The applicators using EM waves are principally divided into two types based on heating principle: the *dielectric heating*

RF/Microwave Interaction with Biological Tissues, By André Vander Vorst, Arye Rosen, and Youji Kotsuka

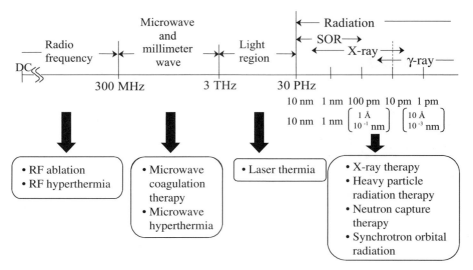

FIGURE 4.1   Thermal therapy and frequency.

applicator and the *inductive heating applicator*. In this chapter, the heating principles of each applicator are described theoretically.

4.2.1   Foundations of Dielectric Heating Principle

Dielectric heating is based on the principle of generating heat for dielectrics (insulators) such as plastic or rubber by applying an alternating electric field. Generally, the particles such as the atoms or the molecules which construct the dielectric are mutually bound by intermolecular force. When the alternating electric field is applied to the dielectric, these particles exhibit a kind of resistance due to intermolecular force. Accordingly, when the alternating electric field is applied to the dielectric, it causes energy loss. As a result, the dielectric is causing heat generation. This phenomenon is related to the polarization in the dielectric material. To explain it in more detail, let us first consider the case of one atom when an electrostatic field is applied to it, as shown in Figure 4.2. Usually, protons in this case are surrounded by electrons in the form of central-point symmetry, as shown in Figure 4.2*a*.

If an electrostatic field is applied to the atom, the electrons slightly shift in the direction opposite to the electrostatic field, as shown in Figure 4.2*b*. Consequently, the atom forms a kind of electric dipole, as shown in Figure 4.2*c*. This is called electronic polarization because the electrons themselves are shifted, and this movement creates an electric dipole. As a special case, there is a state where the electric dipole is naturally formed without applying an electrostatic field, as seen in hydrogen chloride (HCl) and water (H_2O), as shown in Figure 4.3. This is because the molecular structure does not have a

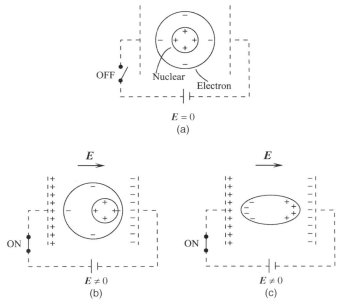

FIGURE 4.2 Polarization of one atom: (*a*) original state when $E = 0$; (*b*) electronic polarization; (*c*) formation of electric dipole (equivalent to polarized) in part (*b*).

FIGURE 4.3 Example of permanent dipole.

center of symmetry. These molecules are called *polar*, forming a permanent dipole. Furthermore, another case is atomic polarization, in which the atom itself is displaced in a molecule, since the applied field modifies the molecular distance between electrons and the bond angle.

Now, let us consider the case of a liquid including a permanent dipole, typically that of water, and suddenly put in an electrostatic field. Then, the permanent dipole begins to move in the electric field direction. However, this

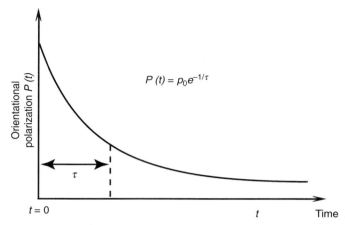

FIGURE 4.4   Return of polarization to zero. Electric field is cut when $t = 0$.

movement is interrupted by an intermolecular binding force and is submitted to resistance from the thermal motion. In this case, there is a time delay associated with the motion of the permanent dipole toward the electric field direction. Accordingly, under the condition which has produced a uniform polarization P (generally a vector representation) by applying the electrostatic field on this liquid beforehand, if the electrostatic field is suddenly shut off in time ($t = 0$), it takes some time until the polarization returns to zero, as shown in Figure 4.4. The parameter τ in this figure is called *relaxation time*: It is the time used by the polarization to decrease in $1/e$ of its steady-state value. Based on these phenomena, the relationship between polarization P and electrostatic field E is shown by the following equation:

$$p = \varepsilon_0 E(\varepsilon_S - 1) \qquad \text{C m}^{-2} \tag{4.1}$$

As mentioned above, when an alternating electric field $E = E_0 e^{jwt}$ is applied to dielectric material exhibition relaxation, the polarization P has a phase delay with respect to the applied electric field. Therefore, *complex permittivity* has been introduced as the way for expressing the delay of this polarization. Therefore, the dielectric constant is expressed as a complex permittivity. By denoting the complex permittivity $\dot{\varepsilon}_s = \varepsilon_s' - j\varepsilon_s''$, the expression related to Eqn. (4.1) when applying alternating electric field $E = E_0 e^{j\omega t}$ is described by the equation

$$\dot{p}(t) = \varepsilon_0 (\dot{\varepsilon}_S - 1) E_0 e^{j\omega t} \tag{4.2}$$

When the alternating voltage $V = V_0 e^{j\omega t}$ is applied to the dielectric material by a parallel-plate applicator, as shown in Figure 4.5, the electric power loss W absorbed in the dielectric is calculated using the complex permittivity. As

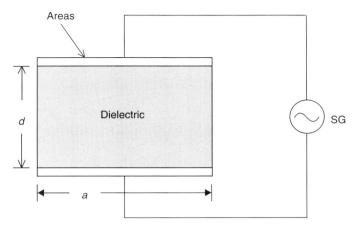

FIGURE 4.5   Parallel-plate applicator.

a result, the power loss per unit volume in the dielectric is given by the following equation using the relation $E = V/d$, expressed by the root-mean-square value of the electric field:

$$W_0 = \omega \varepsilon'' E^2 = 2\pi f \varepsilon'' E^2 \qquad \text{W m}^{-3} \qquad (4.3)$$

where $\omega = 2\pi f$ is the angular frequency and f the frequency.

This power loss means that energy is absorbed by the dielectric material from the electric field, measured per unit volume and unit time, which generates heat for the dielectric. Consequently, it is proven that the dielectric material can be heated proportionally to the imaginary part of the complex permittivity of the dielectric material, frequency f, and the square of the electric field E.

4.2.2   RF Dielectric Heating Applicator

Theory   As a dielectric-heating-type applicator driven by an RF wave, the applicator system called the *capacitive coupling* type, already shown in Figure 4.5, has been developed [4]. The operating frequency is 8 or 13.56 MHz. As shown in the same figure, the heating process is conducted by holding the human body lesion position between both conducting plates.

The heating principle in the present case is complicated because a human body is not pure dielectric material. Beside, a human body is inhomogeneous. From the electric constant point of view, *conductivity* should be taken into consideration in addition to permittivity to explain the capacitive heating principle. To directly state the conclusion of this heating principle, Joule's heat and

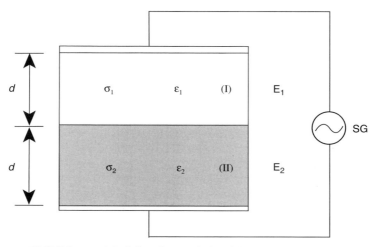

FIGURE 4.6   Modeling for analysis with two-layer medium.

dielectric heating are generated, based both on an ionic current and the dielectric constant.

Using the analytical model of a two-layer medium shown in Figure 4.6, heating is described as follows. (A more detailed investigation will be given in Section 4.2.6). The complex dielectric constant of this two-layer medium, to compare with a single-layer medium, is shown by the equation

$$\dot{\varepsilon}_S = \varepsilon'_S - j\varepsilon''_S = \varepsilon_e \left(1 + \frac{k}{1 + \omega^2\tau^2} \right) - j\left(\frac{\sigma}{\omega\varepsilon_0} + \frac{\omega k \varepsilon_e \tau}{1 + \omega^2\tau^2} \right) \qquad (4.4)$$

where ε_0 is the permittivity in free space, $\varepsilon_e = \varepsilon_1\varepsilon_2/(\varepsilon_1 + \varepsilon_2)$, $\tau = (\varepsilon_1 + \varepsilon_2)\varepsilon_0/(\sigma_1 + \sigma_2)$, and $k = (\varepsilon_1\sigma_2 + \varepsilon_2\sigma_1)^2/[\varepsilon_1\varepsilon_2(\sigma_1 + \sigma_2)^2]$, $\sigma = \sigma_1\sigma_2/(\sigma_1 + \sigma_2)$.

Therefore, the electric power loss per unit volume is obtained as follows by substituting the imaginary part of the complex dielectric constant of Eqn. (4.4) in Eqn. (4.3):

$$W_0 = \omega\varepsilon_0 \left(\frac{\sigma}{\omega\varepsilon_0} + \frac{\omega k \varepsilon_e \tau}{1 + \omega^2\tau^2} \right) E^2 = \left(\sigma + \frac{\omega k \varepsilon_0 \varepsilon_e \tau}{1 + \omega^2\tau^2} \right) E^2 \qquad (4.5)$$

Under the ideal material condition, when there is no energy absorption and exothermic reaction accompanying the chemical or physical change in the dielectric, heat is generated according to the above expressions. The first term in Eqn. (4.5) shows Joule's heat, while the second is loss caused by the RF. As suggested by this simple model, for a medium in which the dielectric constant, permittivity, and conductivity heterogeneously intermingle, Joule's heat contributes to heat generation when the RF electric field is applied to the parallel-plate electrodes. It means that, in the applicator structure shown in Figure 4.6 and at relatively low RF power, Joule's heat seems to greatly contribute, although both heat generations originating from the polarization and Joule's heat intermingle in principle.

Pure dielectric heating is small even in the usual dielectric heating, and dielectric heating is applied to materials in which the dielectric constant and conductivity are present in many industrial cases. It is for this reason that we classify this applicator as a dielectric heating type, although in general the applicator exemplified here is called the *capacitive coupling type*.

Actual Dielectric Heating Applicator Systems   As an example of the RF dielectric heating applicator, we introduce an 8-MHz RF capacitive heating device [4]. The device is schematically illustrated in Figure 4.7. This applicator system has a self-excited oscillation circuit at 8 MHz and 1.5 kW maximum output power. A generator provides the RF power to two disc electrodes via two coaxial cables. The RF is applied through a pair of electrodes placed on opposite sides of the body. The RF power is distributed locally or regionally based on electric fields produced between the parallel-opposed electrodes. To control the heating position of the body, the gantry with the electrodes can be rotated by 180°. Since rotating the gantry can change the electrode positions, this applicator system can easily change treatment sites at different angles. Further, the treatment couch is motorized for vertical and horizontal movement. A pair of electrodes are connected to the pillars of the gantry. The metal plate of the electrodes is attached to a flexible water pad. Temperature-controlled water is circulated in the water pad so that excessive heating of the skin and subcutaneous fat can be avoided. This pad plays an important role for smoothly attaching the electrodes to the body surface. The water temper-

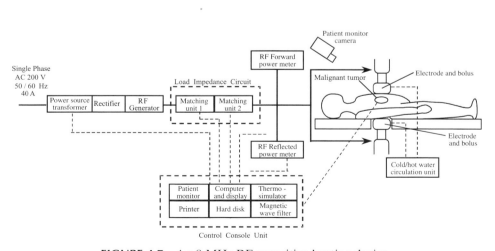

FIGURE 4.7   An 8-MHz RF capacitive heating device.

ature is maintained at 30–40°C for superficial tumors and at approximately 10°C for subsurface or deep tumors. The electrodes can be changed from 4 to 30 cm in diameter. Optimum electrodes are selected depending on the size and location of the tumors. The present applicator system has a thermometry system with four Teflon-coated probes of copper–constantan microthermo-couples. The thermometry system with the microthermocouples is connected to an automatic temperature–power feedback controller with an accuracy of 0.2°C. To protect the thermometry system from RF interference, a high RF wave filter is inserted in the thermometry system and makes it possible to measure temperature even during heating. The temperatures measured at four points in the heated tissue are continuously displayed both graphically and digitally on the computer screen. These data are also continuously recorded on a floppy disc, and a hard copy can be obtained on the internal printer. The power absorbed by the heated site is also continuously displayed graphically and digitally and is recorded.

In the RF capacitive heating, the excessive heating of subcutaneous fat or head skull should be noticed because an electric field becomes perpendicular to these high-impedance layers of human body. It has been shown that a patient with a subcutaneous fat layer of more than 1.5–2 cm thickness is diffi-cult to be heated with this heating modality [5]. The advantage of this kind of capacitive applicator system, however, is that it has wide applicability to various anatomical sites, inducing relatively small systemic stress.

To more widely apply this applicator system to a portion of the thin sub-cutaneous fat such as the neck, a localized heating technique is needed. Gen-erally, in a dielectric applicator using a pair of circular conductive electrodes in RF, a large-size electrode is needed to heat a deep-site tumor uniformly. This means that, in the case of a capacitive applicator with a pair of circular electrodes, a diameter *a* of applicator more than 1.5 times the space *d* between both electrodes is needed to achieve uniform heating inside the human body. If the height of the heating region is 15 cm, 22.5 cm is needed as the diameter of the electrode to uniformly heat the human body in this area. If the diame-ter is smaller while the volume remains constant, hot spots will arise under the electrodes and heating will not be deep. To illustrate this, the electric field dis-tributions using a cubic agar phantom (a medium considered equivalent to the muscle) in the case of heating by the capacitive-coupling-type applicator are shown in Figures 4.8*a,b* [1]. It is found that the electric field concentrates in the area under the pair of electrodes when the electrode diameter is small, as in Figure 4.8*b*. Actually, even in experiment, these parts are observed as hot spots (local heat spots), as shown in Figure 4.9.

To solve this problem and achieve regional heating, a capacitive applicator with a double electrode has been developed, as shown in Figure 4.10 [6]. By introducing a subelectrode consisting of a ferrodielectric material under the main electrode, the electric field can concentrate between a pair of subelec-trodes of ferrodielectric materials. The size of the beam spot is proportional to the diameter of the semicircular subelectrode. If the semicircular subelec-

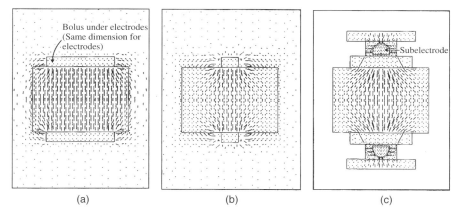

(a)   (b)   (c)

FIGURE 4.8   Electric field distribution by capacitive heating applicator.

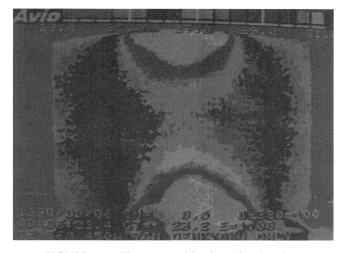

FIGURE 4.9   Thermographic view after heating.

trode of 5 cm in diameter is used, a 5-cm beam of electric field to a distance of 20 cm between a pair of electrodes can be obtained, for example.

Based on the same principle, another type of capacitive heating applicator has been developed. Figure 4.11 shows a heating system for treatment of a body cavity tumor. One electrode is placed on the surface of the body and the other electrode is inserted into the site of the tumor in the body cavity. This electrode is covered by balloon made of silicon rubber and mounted thermosensor. Maximum power is 250 W and the frequency is 13.56 MHz.

As a very similar RF dielectric heating applicator, a small electrode consisting of needles (inner hooks) has been developed as the treatment for

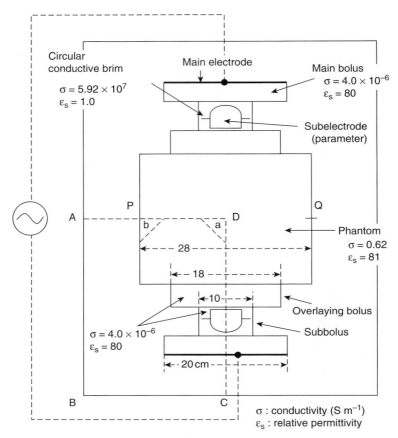

FIGURE 4.10 Double-electrode applicator.

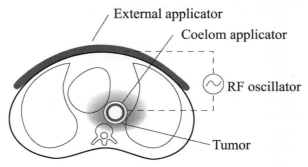

FIGURE 4.11 Applicator for body cavity.

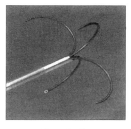

Cannula tip

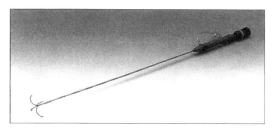

Disposable handpiece with temperature sensor

(electrode, effective length 25 cm)

FIGURE 4.12 Radio-frequency heating applicator used for interstitial tissue ablation.

interstitial tissue ablation. This system is comprised of needle electrodes and an outer electrode attached to the body surface. The RF current concentrates on the outer electrode. This heating principle is based on both Joule's heat and dielectric heating. The electrode is housing inside a fine pipe before treatment. When the tip of the needle reaches a treatment region, multineedles expand in the shape of a hanging bell, as shown in Figure 4.12. A thermosensor is attached to the tip of each needle. The frequency of 460 or 480 kHz is used and the maximum input power extends from 50 to 100 W.

4.2.3 Microwave Dielectric Heating

A microwave dielectric heating applicator is based on the same principle as the microwave oven. Most of the organism is composed of water molecule, and from the viewpoint of electronic polarization described in Section 4.2.1, the permanent dipole greatly contributes to heat generation. That is, in the microwave band, the permanent dipole cannot follow the quick fluctuation of a high-frequency electric field and large loss arises. The power loss in this case generates heat. The complex relative permittivity in the high-frequency region, based on this orientation polarization, is expressed by

$$\varepsilon_s' = \varepsilon_\infty + \frac{\varepsilon_1 - \varepsilon_\infty}{1 + \omega^2 \tau_0^2} - j \frac{(\varepsilon_1 - \varepsilon_\infty)\omega \tau_0}{1 + \omega^2 \tau_0^2} \tag{4.6}$$

where ε_∞ is the value of relative permittivity at extremely high frequency, $\tau_0 = (\varepsilon_1 + 2)/(\varepsilon_\infty + 2)\tau$ and $\tau = \xi/(2k_0 T)$, ζ is the internal friction force, T is the absolute temperature, and k_0 is the Boltzmann constant.

Therefore, electric power loss per unit volume in the heating model of Figure 4.5 is expressed by substituting the imaginary part of Eqn. (4.2.6) in Eqn. (4.2.3):

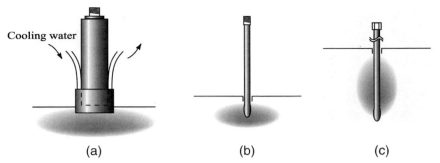

Cooling water

(a)   (b)   (c)

FIGURE 4.13 Classification of microwave applicators: (*a*) body surface adhesive applicator; (*b, c*) coelom interpolation applicatiors.

$$W_0 = \omega \varepsilon_0 \varepsilon_s'' E^2 = \varepsilon_0 (\varepsilon_1 - \varepsilon_\infty) \frac{\omega^2 \tau_0^2}{1 + \omega^2 \tau_0^2} E^2 \qquad (4.7)$$

Under the ideal material condition, this electric power generates heat.

Since the principle of microwave dielectric heating is that of the microwave oven, various kinds of applicators are developed as microwave dielectric heating applicators. As shown in Figure 4.13, applicators are mainly classified into contact or noncontact applicators to the body surface and applicators using a small-diameter coaxial cable which is inserted in the coelom or is stabbed into the lesion position to radiate microwaves [7, 8]. These applicators operate in many cases at a frequency of 2.45 GHz. Therefore, achieving deep heating is limited by skin effect (see Chapter 1). The skin depth is determined by

$$d = \sqrt{\frac{\rho}{\pi f \mu}}$$

where ρ is the resistivity and μ the permeability. This is why, recently, applicators of the invasive type shown in Figures 4.13*b,c* have become popular for microwave coagulation therapy. An applicator system aimed at local heating has been developed at 430 MHz. It has adopted the EM wave lens [9].

4.2.4 Foundation of Inductive Heating Principle

Besides dielectric heating described in the previous section, based on an alternating electric field, there is another type of heating called *inductive heating*. It uses an alternating magnetic field. This inductive heating is based on Joule's heat arising when eddy currents flow in a resistor. This heating method is also used for hyperthermia treatment.

When the coil is wound around the cylindrical conductor and an alternating magnetic flux is applied to it, as shown in Figure 4.14*a*, an electromotive force is produced and the current flows in the vortex state. This current is called

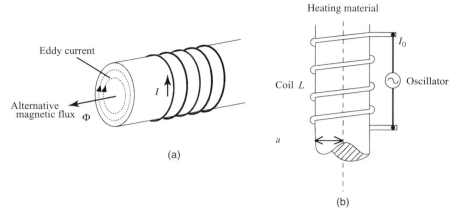

FIGURE 4.14 Principle of inductive heating.

an eddy current; it flows in the vortex state. Its calculation derives from the electromagnetic induction law of Faraday, as shown in the following equations. The derivative type of the EM induction law is

$$\nabla \times \boldsymbol{E} = \frac{\partial \boldsymbol{B}}{dt} \qquad (4.8)$$

where $\boldsymbol{E}$ is the electric field and $\boldsymbol{B}$ the magnetic flux density. The relationship between the electric field and the current density $\boldsymbol{J}$ is expressed as

$$\boldsymbol{J} = \kappa \boldsymbol{E} \qquad \text{A m}^{-2} \qquad (4.9)$$

where κ is the conductivity. From Eqns. (4.8) and (4.9), the following equation is obtained:

$$\nabla \times \boldsymbol{J} = -\kappa \frac{\partial \boldsymbol{B}}{\partial t} \qquad (4.10)$$

This equation suggests that there is current rotation when a magnetic flux changes in the conductor, which means the generation of current in the vortex state.

Now, let us consider the case in which the coil is wound around the heating material of the cylindrical conductor, as shown in Figure 4.14*b*, with high-frequency current flowing in this coil to generate an alternating magnetic field in the form of a sine wave. The result is that the heating material is heated by the eddy current generation.

In the analysis of the power loss (absorbed electric power) in this case, modified Bessel functions appear (ber and bei functions). The analysis becomes complicated, which is why only the result is shown here: The power loss over an axial length *l* (in centimers) for a cross section *A* (in square centimeters) of the cylindrical conductor is given by the expression

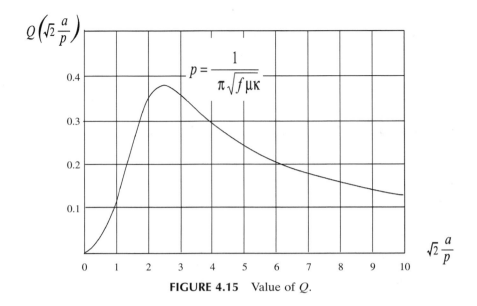

FIGURE 4.15   Value of Q.

$$W_0 = \tfrac{1}{2} f \mu H_a^2 A l Q \left(\sqrt{2}\, \frac{a}{p} \right) \times 10^{-7} \qquad \overline{W} \tag{4.11}$$

where f is the frequency (in hertz), μ the permeability, a the cylinder radius, k the conductivity, H the magnetic field at the surface of cylinder, and p the skin depth of the current near the heating body surface. The value of Q is obtained from Figure 4.15. Since the size of the heating material chosen in the analytical model is small, centimeter-gram-second (cgs) units are used in this analysis. The characteristics of the maximum power loss appear clearly in Figure 4.15: The electric power absorption goes through a maximum value when $\sqrt{2}\, a/p$ is equal to 2.5, that is, when a/p nearly equals 1.75. Expression (4.11) is related to the skin depth p of the current, which becomes small when the frequency becomes high. Therefore, when higher frequency is used, only the surface vicinity can be heated.

4.2.5   Actual Inductive Heating Applicator

To easily understand the difference between dielectric heating and inductive heating, let us first introduce experimental heating characteristics [10, 11] before describing actual inductive heating applicators. Figure 4.16 shows a thermographic view of an experimental result of RF capacitive heating of a rectangular agar phantom which is sandwiched between two layers of pig fat perpendicularly to the RF electric field at 3 MHz. A conductive sphere is buried in the center of the phantom. White areas indicate high temperature.

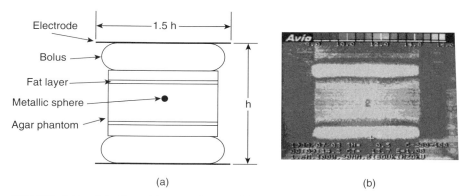

FIGURE 4.16 (*a*) Capacitive applicator. (*b*) Thermographic view of experimental result.

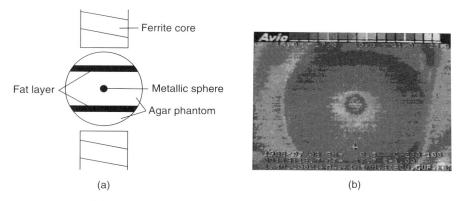

FIGURE 4.17 (*a*) Toroidal ferrite core with two poles. (*b*) Thermographic view of experimental results.

The figure shows that the fat layers are well heated compared to the conductive sphere. As a difference with the capacitive heating systems, inductive heating has the advantage of being able to heat a human body without generating a hot spot in the fat layers. Figure 4.17 shows experimental results from an inductive heating system with a toroidal ferrite core with two poles at 1.5 MHz. A cylindrical agar phantom with two fat layers and a metallic sphere is placed in the central position between two pig fat layers. It is clear that the conductive sphere is well heated without heating the fat layers. Remembering that every tissue of a human body is covered with biomembrane, with high insulation and high impedance, this experimental result confirms the fact that inductive heating is effective when every tissue of a human body is covered with biomembrane. This is because biomembrane is composed of lipid and protein. It is considered to have high insulation or high impedance. A magnetic field can penetrate these kinds of insulating materials or biomembranes. This experimental result confirms that a human body is covered with the bio-membrane, with the high insulation or high impedance from the organization level to the cellular level. Subcutaneous adipose layer also has high insulation. The magnetic field can penetrate these insulation materials on principle. Therefore, the eddy current is generated on the inside of the insulation materials of conductivity exits inside of insulation material. Consequently, inductive heating has a special feature, which is that the lesion position is heated without being affected by high-impedance tissues such as the subcutaneous adipose layer and the skull of the human body.

Hence inductive heating seems to be superior to dielectric heating. However, as described in the previous section, the eddy current distribution in inductive heating has a tendency to generate on a human body surface. Therefore, actually, it becomes difficult to heat positions deep inside the human body.

Many methods for inductive heating have so far been proposed [12–16]. As an example on how to control the eddy current, let us introduce an applicator system using ferrite cores [10]. Figures 4.18a and b show a deformed ferrite core and an applicator for an elliptic shape cross section of phantom. If the magnetic field is strongly coupled between all the magnetic poles A, B, C and D, as shown in Figure 4.19a, eddy currents are distributed as shown in Figure 4.19b: There are no eddy currents in the center portion of the phantom. This has been experimentally verified using an elliptic shape agar phantom with pig fat layer as the subcutaneous adipose layer. Figure 4.20 shows the actual situation after heating an agar phantom. It is found that maximum heating regions represented by the white area are located in four places that correspond to the regions where eddy currents are generated in Figure 4.19b. To improve this heating characteristic, conductive plates acting as electrodes are introduced, as shown in Figure 4.21a.

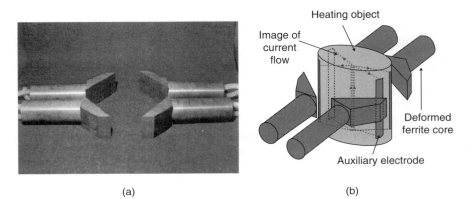

(a)   (b)

FIGURE 4.18   (a) Deformed ferrite core and (b) heating principle.

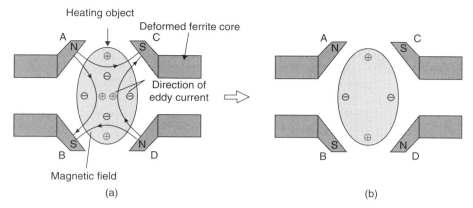

FIGURE 4.19 Heating principle.

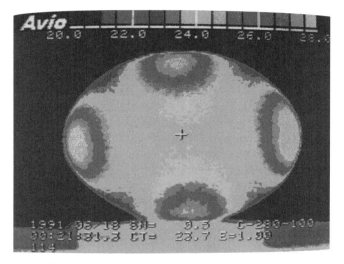

FIGURE 4.20 Thermographic view after heating by deformed core applicator.

This conductive plate also operates as a magnetic shield plate. It is called an auxiliary electrode. By attaching the auxiliary electrode to the surface of the phantom, magnetic couplings between the poles A and C and B and D are reduced. So, it is considered that only the magnetic field coupling between the poles A and B and D and C remains. As a result, eddy currents distribute as shown in Figure 4.21*b*. We should notice here that these eddy currents flow into the auxiliary electrode because it has a conductivity higher than the phantom. Consequently, the total current flow forms a closed loop, as shown

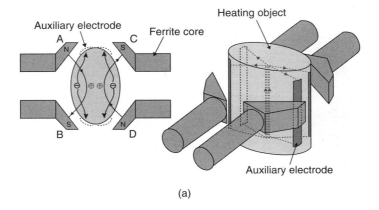

(a)

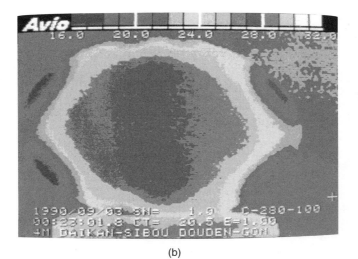

(b)

FIGURE 4.21 (*a*) Heating principle. (*b*) Thermographic view of phantom after heating.

in Figure 4.21*a*. The actual auxiliary electrode is made of silicon rubber containing silver powder to make it conducting, while it can circulate cooling water inside the electrode, as shown in Figure 4.22. Figure 4.23 shows a thermographic view of an experimental result after heating an elliptic shaped cylindrical phantom. The thick black area at the center of the figure is that of highest temperature.

Other examples of how to control the eddy current, are the applicator systems developed for breast hyperthermia. Figure 4.24 shows a system with a pair of ferrite cores. The magnetic field couples mainly between a pair of

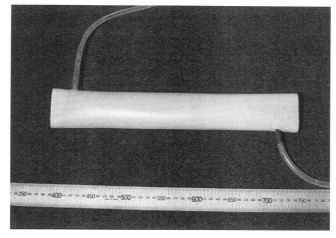

FIGURE 4.22   Auxiliary electrode.

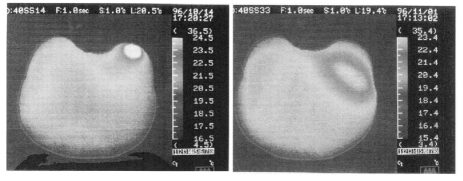

FIGURE 4.23   Thermographic view after heating (white portion shows highest temperature).

magnetic poles. The breast is put between a pair of ferrite cores. Thin shield plates are used to control the heating position in depth. A thin nonmagnetic conductive plate is also effective to control the magnetic field distribution. This is because the edge of a thin nonmagnetic conducting plate generates a new vertical magnetic field in the same direction of the magnetic field being caused by a pair of ferrite cores. As a result, all the magnetic fields enforce each other and come to make a detour. Based on this principle, heating position can be controlled deeply by moving down a single ferrite core arranged outside the phantom. Figure 4.25 shows a thermographic view of agar phantom after heating using shield plates. The white portion in the thermographic view

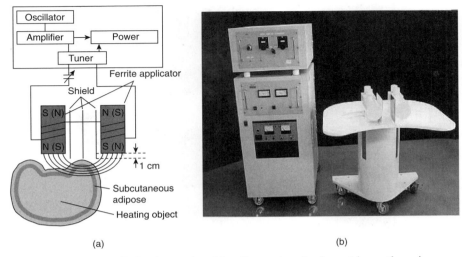

(a)

(b)

FIGURE 4.24   Inductive regional heating system for breast hyperthermia.

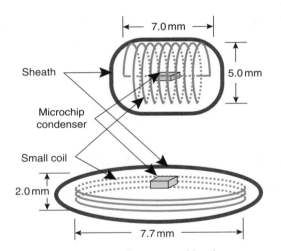

FIGURE 4.25   Structure of implant.

corresponds to the highest temperature. We can verify that the heating position is well controlled by moving the outside ferrite core vertically.

Another special feature of inductive heating is that it is able to cause local heating. It is well known that, by applying an alternating magnetic field to some materials that can be heated by eddy currents and magnetic loss, local heating becomes possible by implanting these materials in the lesion position. Various kinds of implant have thus far been proposed using magnetic materials [17–24]

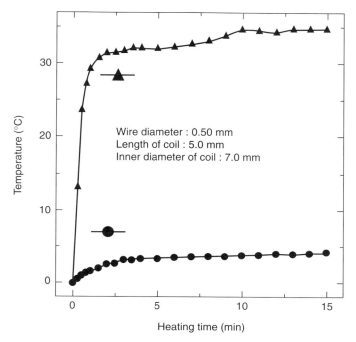

FIGURE 4.26   Rise time of (▲) resonant and (●) nonresonant implants.

and fine-granular material [25–29]. However, a small implant with a millimeter size is very difficult to use, because implant material cannot be heated efficiently by inductive heating when the implant size is less than 5 mm or so [30].
To solve this problem of inductive deep local heating, a small high-efficiency
implant has been proposed, not from conventional material compositions but
from the point of view of electrical circuit theory [31, 32]. This new implant is
simply composed of a small coil and a microchip condenser. It is heated efficiently because of resonance, when RF magnetic field is applied. Figure 4.26
shows the fundamental construction of this implant, with an example of a
prolate shape (▲) and an oblate shape (●). The implant is covered with insulating material. The coil and the microchip condenser constitute a series resonant circuit together with the coil resistance. When an alternating magnetic
field is applied to the coil, a large coil current flows due to resonance. As a
result, the coil generates Joule's heat with high efficiency. The introduction of
the microchip condenser is a key to make the implant small.

Figure 4.27 shows the temperature difference between the initial temperature and the elevated temperature, both when taking resonant phenomena
into consideration (curve A) or not (curve B) The prolate-type coil is used
with a length of 7.0 mm and a diameter of 5.0 mm in this heating experiment.

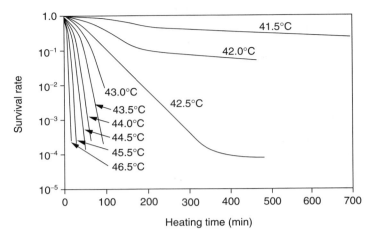

FIGURE 4.27 Survival rate curve of biological cell.

It is clear that the temperature rises are strongly improved by introducing the resonant circuit for implant coil design. In this experiment, a ferrite core applicator is used to investigate the heating characteristic at 4 MHz. The output power is normally 500 W. The heating tests have been conducted using a 20-cm agar phantom cube that conforms to the guideline assigned by the QAC in the *Journal of the American Society of Hyperthermia Oncology*. The implant is put in the center of the phantom.

4.2.6 Detailed Theory of RF Dielectric Heating

In Section 4.2.2, an outline of the heating principle of RF capacitive coupling applicator has been given. The human body, which consists of inhomogeneous medium, was simplified by introducing a two-layer medium differing in dielectric constant and conductivity. In addition, the permittivities of the two-layer medium have been converted into the equivalent permittivity of a single-layer medium. The electric power loss per unit volume, that is, the heating capacity, is calculated using this equivalent permittivity. This approach is based on the heterogeneous theory of Maxwell and Wagner (see Section 2.2).

In this section, the method for deriving expressions (4.4) and (4.5) is described in detail. The beginner who reads this chapter for the first time may skip this section. The derivation procedure consists of replacing each permittivity of the two-layer medium by the permittivity of the equivalent mono-layer medium. To do this, we derive the equivalent admittance Y between parallel-plates and express the current I_d flowing between the parallel-plate electrodes considered as a static capacitance. This yields the equivalent permittivity (4.4).

Going back to Figure 4.6, the sinusoidal alternating voltage V applied between the parallel-plate electrodes is expressed as

$$V = V_0 e^{j\omega t} \tag{4.12}$$

The field intensities in the two-layer medium, E_1 and E_2, and the electric displacements D_1 and D_2 are given by the equations

$$E_1 = E_2 = \frac{100V_0}{d} e^{j\omega t}$$

$$D_1 = \varepsilon_0 \varepsilon_1 E_1 = \frac{100 \varepsilon_0 \varepsilon_1 V_0}{d} e^{j\omega t}$$

$$D_2 = \varepsilon_0 \varepsilon_2 E_2 = \frac{100 \varepsilon_0 \varepsilon_2 V_0}{d} e^{j\omega t} \tag{4.13}$$

where ε_0 is the vacuum permittivity. Further, the time variation of the electric displacements D_1 and D_2 are as follows:

$$\frac{dD_1}{dt} = j\omega \varepsilon_0 \varepsilon_1 \frac{100V_0}{d} e^{j\omega t} = j\omega \varepsilon_0 \varepsilon_1 E_1$$

$$\frac{dD_2}{dt} = j\omega \varepsilon_0 \varepsilon_2 \frac{100V_0}{d} e^{j\omega t} = j\omega \varepsilon_0 \varepsilon_2 E_2 \tag{4.14}$$

Further, each conducting current density in medium 1 and 2, respectively, is expressed as

$$i_1 = \sigma_1 E_1 \qquad i_2 = \sigma_2 E_2$$

The current which flows between the parallel-plate electrodes is the sum of the conductive current and the displacement current when the alternating voltage is applied. Accordingly, the total current density when applying alternating voltage $V = V_0 e^{j\omega t}$ to the electrodes is given by the expression

$$i = (\sigma_1 + j\omega \varepsilon_0 \varepsilon_1)E_1 = (\sigma_2 + j\omega \varepsilon_0 \varepsilon_2)E_2 \tag{4.15}$$

while the relation between voltage and electric field is

$$V = \frac{d}{100}(E_1 + E_2) \tag{4.16}$$

As a consequence, the following equations are obtained when solving Eqns. (4.15) and (4.16) in E_1 and E_2:

$$E_1 = \frac{\sigma_2 + j\omega \varepsilon_0 \varepsilon_2}{\sigma_1 + \sigma_2 + j\omega \varepsilon_0 (\varepsilon_1 + \varepsilon_2)} \frac{100V_0}{d}$$

$$E_2 = \frac{\sigma_1 + j\omega \varepsilon_0 \varepsilon_1}{\sigma_1 + \sigma_2 + j\omega \varepsilon_0 (\varepsilon_1 + \varepsilon_2)} \frac{100V_0}{d} \tag{4.17}$$

Hence, the total current density (4.2.15) is expressed as

$$i = \frac{(\sigma_1 + j\omega\varepsilon_0\varepsilon_1)(\sigma_2 + j\omega\varepsilon_0\varepsilon_2)}{\sigma_1 + \sigma_2 + j\omega\varepsilon_0(\varepsilon_1 + \varepsilon_2)} \frac{100V_0}{d} \tag{4.18}$$

This shows that the two-layer medium between the parallel plates is replaced by a single dielectric medium with the following admittance between these parallel electrode plates:

$$Y = \frac{(\sigma_1 + j\omega\varepsilon_0\varepsilon_1)(\sigma_2 + j\omega\varepsilon_0\varepsilon_2)}{\sigma_1 + \sigma_2 + j\omega\varepsilon_0(\varepsilon_1 + \varepsilon_2)} \tag{4.19}$$

Two special cases are then obtained, when DC voltage ($\omega = 0$) and extremely high frequency ($\omega = \infty$) are applied, respectively:

$$Y_{\omega=0} = \frac{\sigma_1\sigma_2}{\sigma_1 + \sigma_2} = \sigma \tag{4.20}$$

$$Y_{\omega=\infty} = j\omega\varepsilon_0 \frac{\varepsilon_1\varepsilon_2}{\varepsilon_1 + \varepsilon_2} = j\omega\varepsilon_0\varepsilon_e \tag{4.21}$$

where $\varepsilon_e = \varepsilon_1\varepsilon_2/(\varepsilon_1 + \varepsilon_2)$.

Using these expressions, the equivalent admittance Y in the middle frequency region between DC and the extremely high frequency region is determined as follows by introducing a correction factor Y_n:

$$\begin{aligned} Y &= \frac{\sigma_1\sigma_2}{\sigma_1 + \sigma_2} + j\omega\varepsilon_0 \frac{\varepsilon_1\varepsilon_2}{\varepsilon_1 + \varepsilon_2} + Y_n \\ &= \sigma + j\omega\varepsilon_0\varepsilon_e + Y_n \end{aligned} \tag{4.22}$$

The correction factor is easily calculated using expressions (4.19) and (4.22):

$$Y_n = \frac{1}{1 + j\omega\varepsilon_0[(\varepsilon_1 + \varepsilon_2)/(\sigma_1 + \sigma_2)]} j\omega\varepsilon_0 \frac{(\varepsilon_1\sigma_2 - \varepsilon_2\sigma_1)^2}{\varepsilon_1\varepsilon_2(\sigma_1 + \sigma_2)^2} \frac{\varepsilon_1\varepsilon_2}{\varepsilon_1 + \varepsilon_2} \tag{4.23}$$

Introducing $\tau = (\varepsilon_1 + \varepsilon_2)\varepsilon_0/(\sigma_1 + \sigma_2)$ and $k = (\varepsilon_1\sigma_2 - \varepsilon_2\sigma_1)^2/[\varepsilon_1\varepsilon_2(\sigma_1 + \sigma_2)^2]$, expression (4.23) becomes

$$Y_n = \frac{j\omega k\varepsilon_0\varepsilon_e}{1 + j\omega\tau} = \frac{\omega^2 k\varepsilon_0\varepsilon_e\tau}{1 + \omega^2\tau^2} + \frac{\omega k\varepsilon_0\varepsilon_e}{1 + \omega^2\tau^2} \tag{4.24}$$

From Eqns. (4.22) and (4.24), the following equivalent admittance Y expresses the single-layer dielectric medium equivalent to the two-layer dielectric medium:

$$Y = \sigma + j\omega\varepsilon_0\varepsilon_e + \frac{\omega^2 k\varepsilon_0\varepsilon_e\tau}{1+\omega^2\tau^2} + j\frac{\omega k\varepsilon_0\varepsilon_e}{1+\omega^2\tau^2}$$

$$= \sigma + \frac{\omega^2 k\varepsilon_0\varepsilon_e\tau}{1+\omega^2\tau^2} + j\omega\varepsilon_0\varepsilon_e\left(1 + \frac{k}{1+\omega^2\tau^2}\right) \quad (4.25)$$

The admittance (4.19) was defined by Eqn. (4.18). The total current density i between the electrodes can be derived by replacing Y in Eqn. (4.18) by expression (4.25). Accordingly, the final total current flowing between the parallel plates is given by

$$I_d = Ai \times 10^{-4}$$

$$= j\omega\varepsilon_0\left[\varepsilon_e\left(1 + \frac{k}{1+\omega^2\tau^2}\right) - j\left(\frac{\sigma}{\omega\varepsilon_0} + \frac{\omega k\varepsilon_e\tau}{1+\omega^2\tau^2}\right)\right]\frac{AV}{100d} \quad (4.26)$$

where A is the area of one electrode. In addition, the static capacitance C_0, assuming that the medium between the parallel-plate electrodes is a vacuum, is obtained as

$$C_0 = \varepsilon\frac{A}{d} \times 10^{-2}$$

Using this relation, the current I_d in (4.26) can then be transformed as

$$I_d = j\omega C_0\left[\varepsilon_e\left(1 + \frac{k}{1+\omega^2\tau^2}\right) - j\left(\frac{\sigma}{\omega\varepsilon_0} + \frac{\omega k\varepsilon_e\tau}{1+\omega^2\tau^2}\right)\right]V \quad (4.27)$$

Finally, from this equation, the equivalent complex permittivity $\dot{\varepsilon}_e$ is obtained as follows:

$$\dot{\varepsilon}_e = \varepsilon'_e - j\varepsilon''_e$$

$$= \varepsilon_e\left(1 + \frac{k}{1+\omega^2\tau^2}\right) - j\left(\frac{\sigma}{\omega\varepsilon_0} + \frac{\omega k\varepsilon_e\tau}{1+\omega^2\tau^2}\right) \quad (4.28)$$

Accordingly, substituting the imaginary part of this equivalent complex permittivity into expression (4.3) yields the electric power loss per unit volume (4.5).

4.2.7   Detailed Theory of Microwave Dielectric Heating

Microwave dielectric heating is concerned with the permanent dipole of a water molecule, as described in Section 4.2.3. The heat quantity in this case is obtained from the complex permittivity based on the orientation polarization described in expression (4.6). The basis of this derivation is in the dipole theory of Debye (see Section 1.3): The polarization P_m obtained when applying an electrostatic field from the outside to a polar molecule which has a permanent dipole is known as Debye's polarization, shown by

$$P_m(\omega) = \frac{\varepsilon_r - 1}{\varepsilon_r + 2} \frac{M}{\rho} = \frac{N}{3\varepsilon_0}\left(\alpha + \frac{\mu^2}{3k_0 T}\right) \tag{4.29}$$

On the other hand, when applying an alternating electric field and provided expression (4.29) holds, the polarization is given by what is called a Debye dispersion expression:

$$P_m(\omega) = \frac{\varepsilon_r - 1}{\varepsilon_r + 2} \frac{M}{\rho} = \frac{N}{3\varepsilon_0}\left(\alpha + \frac{\mu^2}{3k_0 T}\frac{1}{1 + j\omega\tau}\right) \tag{4.30}$$

Defining ε_0 as the value of ε_r when $\omega = 0$ in this expression yields

$$\frac{\varepsilon_r - 1}{\varepsilon_r + 2} \frac{M}{\rho} = \frac{N}{3\varepsilon_0}\left(\alpha + \frac{\mu^2}{3k_0 T}\right) \tag{4.31}$$

Similarly, defining ε_∞ as the value of ε_r when ω is very high yields

$$\frac{\varepsilon_\infty - 1}{\varepsilon_\infty + 1} \frac{M}{\rho} = \frac{N\alpha}{3\varepsilon_0} \tag{4.32}$$

The permittivity ε_r between the extreme frequency regions defined by (4.31) and (4.32) can be expressed as follows using (4.30), (4.31), and (4.32):

$$\varepsilon_r = \frac{\varepsilon_0/(\varepsilon_0 + 2) + j\omega\tau[\varepsilon_\infty/(\varepsilon_\infty + 2)]}{1/(\varepsilon_0 + 2) + j\omega\tau[1/(\varepsilon_\infty + 2)]} = \frac{\varepsilon_0 + j\omega\tau[(\varepsilon_0 + 2)/(\varepsilon_\infty + 2)]\varepsilon_\infty}{1 + j\omega\tau[(\varepsilon_0 + 2)/(\varepsilon_\infty + 2)]} \tag{4.33}$$

Defining x as

$$x = \omega\tau\frac{\varepsilon_0 + 2}{\varepsilon_\infty + 2} \tag{4.34}$$

the expression (4.34) can be written as:

$$\varepsilon_r = \frac{\varepsilon_0 + jx\varepsilon_\infty}{1 + jx} = \varepsilon_\infty + \frac{\varepsilon_0 - \varepsilon_\infty}{1 + jx}$$

$$= \varepsilon_\infty + \frac{\varepsilon_0 - \varepsilon_\infty}{1 + x^2} - j\frac{(\varepsilon_0 - \varepsilon_\infty)x}{1 + x^2} \tag{4.35}$$

Finally, defining a generalized relaxation time $\tau_0 = \tau(\varepsilon_0 + 2)/(\varepsilon_\infty + 2)$ yields the following equation when using permittivityε_r from Eqn. (4.33):

$$\varepsilon_r = \frac{\varepsilon_0 + j\omega\tau_0\varepsilon_\infty}{1 + j\omega\tau_0} = \varepsilon_\infty + \frac{\varepsilon_0 - \varepsilon_\infty}{1 + j\omega\tau_0}$$

$$= \varepsilon_\infty + \frac{\varepsilon_0 - \varepsilon_\infty}{1 + \omega^2\tau_0^2} - j\frac{(\varepsilon_0 - \varepsilon_\infty)\omega\tau_0}{1 + \omega^2\tau_0^2} \tag{4.36}$$

With this, we have derived the final expression for the complex permittivity based on the orientation polarization. As a consequence, the microwave dielectric heat quantity per unit time and unit volume is obtained when substituting the imaginary part of Eqn. (4.36) into Eqn. (4.3).

4.2.8 Detailed Theory of Inductive Heating

In the previous section, electric power loss in inductive heating has been explained on the basis of the ohmic loss produced by eddy currents that arise in a cylindrical conductor surrounded by a coil. In this section, we shall describe in detail the method for deriving the electric power loss in the form of Eqn. (4.11). We consider the case where a sinusoidal alternating magnetic field is generated by a coil wound around the cylindrical conductor, as in Figure 4.14b, while a high-frequency current I_0 flows through this coil. In the present analysis, we consider the ideal case where there is no leakage magnetic field, assuming that the coil is closely wound around the conductor. It is assumed that only the axial z component H_z of the magnetic field arises in this case and that H_z is distributed circumferentially and symmetrically with respect to the conductor.

We first solve a magnetic field equation derived from Maxwell's equation for the magnetic field. Second, the eddy current density of the cylindrical conductor is calculated using the electric field derived from the magnetic field. Finally, the total electric power loss based on the ohmic loss produced by the eddy current is obtained. We use cgs units such as f (Hz), A (cm^2), and H_a (Oe) because of the small size of the conducting cylinder.

The alternating magnetic field is expressed as

$$H_Z = H_0 e^{j\omega t} \tag{4.37}$$

where ω is the angular frequency. This magnetic field must satisfy the following equation derived from Maxwell's equations:

$$\nabla^2 \boldsymbol{H} = j\omega 4\pi\sigma\mu\boldsymbol{H} \tag{4.38}$$

expressed in cylindrical coordinates (r, φ, z). When taking the symmetrical distribution of the EM field in the circumferential direction of the cylindrical conductor into consideration, the following expression is obtained:

$$\frac{\partial^2 H_0}{\partial r} + \frac{1}{r}\frac{\partial H_0}{\partial r} - j\omega 4\pi\sigma\mu H_0 = 0 \tag{4.39}$$

This equation appears as a Bessel differential equation with respect to zero order when defining $\gamma^2 = -j\pi\mu\omega\sigma$:

$$\frac{d^2 H_0}{dr} + \frac{1}{r}\frac{dH_0}{dr} + \gamma^2 H_0 = 0$$

$$\frac{d^2 H_0}{d(\gamma r)^2} + \frac{1}{\gamma r}\frac{dH_0}{d(\gamma r)} + H_0 = 0 \tag{4.40}$$

Therefore, the solution is the linear combination of Bessel functions of the first and second kinds:

$$H_0 = C_1 J_0(\gamma r) + C_2 Y_0(\gamma r) \tag{4.41}$$

where C_1, C_2 are constants, $J_0(\gamma r)$ is the Bessel function of the first kind of order zero, and $Y_0(\gamma r)$ is the Bessel function of the second kind of order zero.

Applying the boundary conditions to the above expression, the right-hand-side term takes the value of infinity at $r = 0$. However, the actual magnetic field cannot be infinite at the center of the conductive cylinder. Therefore, the constant C_2 must be zero. Another boundary condition is given by the value of the magnetic field H_a on the surface of the conductive cylinder, expressed as

$$H_a = (H_0)_{r=a} = C_1 J_0(\gamma a) \tag{4.42}$$

which yields

$$C_1 = \frac{H_a}{J_0(\gamma a)} \tag{4.43}$$

This determines the magnetic field amplitude,

$$H_0 = H_a \frac{J_0(\gamma r)}{J_0(\gamma a)} \tag{4.44}$$

as well as the magnetic field (4.37),

$$H_z = H_a \frac{J_0(\gamma r)}{J_0(\gamma a)} e^{j\omega t} \tag{4.45}$$

Now, the expression γ^2 related to Eqn. (4.39) is transformed into the following:

$$\gamma = \sqrt{-j4\pi\omega\mu\sigma} = \sqrt{-j}\sqrt{4\pi\omega\mu\sigma} = j\sqrt{j}\gamma_0$$

with $\gamma_0 = \sqrt{4\pi\omega\mu\sigma}$. Introducing this expression in the zero-order Bessel function of the first kind, it becomes possible to express the result using the ber and bei functions, which yields the final form of the magnetic field:

$$H_z = H_a \frac{\mathrm{ber}(\gamma_0 r) + j\,\mathrm{bei}(\gamma_0 r)}{\mathrm{ber}(\gamma_0 a) + j\,\mathrm{bei}(\gamma_0 a)} e^{j\omega t} \tag{4.46}$$

Let us now derive the electric field. It is determined by substituting the magnetic field obtained above into the expression derived from Maxwell's equation:

$$\boldsymbol{E} = -\frac{1}{4\pi\sigma + j\omega\varepsilon} \frac{\partial H_z}{\partial r} \boldsymbol{a}_\varphi \tag{4.47}$$

where $\boldsymbol{a}_\varphi$ is the unit vector in the direction φ. Differentiating the magnetic field in (4.46) with respect to r yields the electric field:

$$E = -\frac{\gamma_0 H_a}{4\pi\sigma + j\omega\varepsilon} \frac{\text{ber}'(\gamma_0 r) + j\,\text{bei}'(\gamma_0 r)}{\text{ber}(\gamma_0 a) + j\,\text{bei}(\gamma_0 a)} e^{j\omega t} \boldsymbol{a}_\varphi \qquad (4.48)$$

Then, the eddy current at distance r from the center of the conducting cylinder is based on the expression

$$\boldsymbol{I} = \sigma\boldsymbol{E} \qquad (4.49)$$

Substituting expression (4.48) into (4.49) yields

$$I = -\frac{\gamma_0}{4\pi} H_a \frac{\text{ber}'(\gamma_0 r) + j\,\text{bei}'(\gamma_0 r)}{\text{ber}(\gamma_0 a) + j\,\text{bei}(\gamma_0 a)} e^{j\omega t} \boldsymbol{a}_\varphi \qquad (4.50)$$

Since the eddy current flows in the tangential direction of the circumference of the conducting cylinder, its amplitude at r is given by the expression

$$I_r = -\frac{\gamma_0}{4\pi} H_a \frac{\text{ber}'(\gamma_0 r) + j\,\text{bei}'(\gamma_0 r)}{\text{ber}(\gamma_0 a) + j\,\text{bei}(\gamma_0 a)} e^{j\omega t} \qquad (4.51)$$

With this, we can derive the expression of the electric power loss based on the ohmic term of the above expression. Let us consider a small area formed by a width dr at a radius r from the center of the cylinder and by a unit length in this position in the axial z direction. The resistance dR of this area is then calculated using the total current through the area, which is expressed by

$$dI_r = -\frac{\gamma_0}{4\pi} H_a \frac{\text{ber}'(\gamma_0 r) + j\,\text{bei}'(\gamma_0 r)}{\text{ber}(\gamma_0 a) + j\,\text{bei}(\gamma_0 a)} dr\, e^{j\omega t} \qquad (4.52)$$

Therefore, the ohmic loss associated with the elementary area with small width dr and unit length in the z direction is

$$\begin{aligned} dP_r = dR|dI_r|^2 &= \rho \frac{2\pi r}{dr} |dI_r|^2 \\ &= \frac{\gamma_0 \rho H_a^2}{8\pi} \frac{\text{ber}'^2(\gamma_0 r) + j\,\text{bei}'^2(\gamma_0 r)}{\text{ber}^2(\gamma_0 a) + j\,\text{bei}^2(\gamma_0 a)} r\, dr \end{aligned} \qquad (4.53)$$

Then, the total power loss is obtained by integrating the expression in dP_r for the whole region of the conducting cylinder:

$$\begin{aligned} P_a &= \int_0^a \frac{\gamma_0^2 \rho H_a^2}{8\pi} \frac{\text{ber}'^2(\gamma_0 r) + j\,\text{bei}'^2(\gamma_0 r)}{\text{ber}^2(\gamma_0 a) + j\,\text{bei}^2(\gamma_0 a)} r\, dr \\ &= \frac{\gamma_0 \rho a}{8\pi} H_a^2 \frac{\text{ber}(\gamma_0 a)\,\text{ber}'(\gamma_0 a) + \text{bei}(\gamma_0 a)\,\text{bei}'(\gamma_0 a)}{\text{ber}^2(\gamma_0 a) + \text{bei}^2(\gamma_0 a)} \end{aligned} \qquad (4.54)$$

using the relations

$$\frac{d}{dr}[r\,\text{ber}'(\gamma_0 r)] = -\gamma_0 r\,\text{bei}(\gamma_0 r) \qquad \frac{d}{dr}[r\,\text{bei}'(\gamma_0 r)] = -\gamma_0 r\,\text{ber}(\gamma_0 r)$$

Expression (4.54) is the total power loss per unit length in the direction of the z axis. Accordingly, the power loss for the actual conducting cylinder with length l is obtained by multiplying the expression in (4.54) by l:

$$P = \frac{\gamma_0^2 \rho a A l}{16\pi^2} H_a^2 \frac{2}{\gamma_0 a} \frac{\text{ber}(\gamma_0 r)\,\text{ber}'(\gamma_0 a) + \text{bei}(\gamma_0 a)\,\text{bei}'(\gamma_0 a)}{\text{ber}^2(\gamma_0 a) + \text{bei}^2(\gamma_0 a)} \qquad (4.55)$$

where $A = \pi a^2$. This expression is then transformed by defining

$$Q(\gamma_0 a) = \frac{2}{\gamma_0 a} \frac{\text{ber}(\gamma_0 a)\text{ber}'(\gamma_0 a) + \text{bei}(\gamma_0 a)\,\text{bei}'(\gamma_0 a)}{\text{ber}^2(\gamma_0 a) + \text{bei}^2(\gamma_0 a)} \qquad \gamma_0 = 8\pi^2 \mu f \sigma$$

Finally, we obtain

$$P = \tfrac{1}{2} f \mu H_a^2 Q(\gamma_0 a) A l \qquad (4.56)$$

Now, using the relation on the skin depth of current, the definition of $Q(\gamma_0 a)$ can be transformed into

$$Q(\gamma_0 a) = Q\left(\sqrt{2}\,\frac{a}{p}\right)$$

$$= \sqrt{2}\,\frac{p}{a} \frac{\text{ber}\left(\sqrt{2}\,\dfrac{a}{p}\right)\text{ber}'\left(\sqrt{2}\,\dfrac{a}{p}\right) + \text{bei}\left(\sqrt{2}\,\dfrac{a}{p}\right)\text{bei}'\left(\sqrt{2}\,\dfrac{a}{p}\right)}{\text{ber}^2\left(\sqrt{2}\,\dfrac{a}{p}\right) + \text{bei}^2\left(\sqrt{2}\,\dfrac{a}{p}\right)} \qquad (4.57)$$

where the skin depth of current is $p = \sqrt{2}/\gamma_0$. Expression (4.56) can then be transformed into the simple form

$$P = \tfrac{1}{2} f \mu H_a^2 Q\left(\sqrt{2}\,\frac{a}{p}\right) A l \qquad (4.58)$$

In all these expressions, cgs units such as $f(\text{Hz})$, $A(\text{cm}^2)$, and $H_a(\text{Oe})$ have been used because of the small size of the conducting cylinder. Therefore, the present power loss is expressed in units of (ergs per second).

Accordingly, when watts are used, the power loss in (4.56) is finally given by the expression

$$P = \tfrac{1}{2}\, f \mu H_a^2 Q\left(\sqrt{2}\,\frac{a}{p}\right) A l \times 10^{-7} \qquad \overline{\text{W}} \qquad (4.59)$$

This is the expression (4.11) obtained in the previous section.

4.3 HYPERTHERMIA

The first paper on hyperthermia was published by W. Bush in 1886 [33]. According to Bush, the sarcoma that occurred on the face of a 43-year-old lady was cured when fever was caused by erysipelas. F. Westermark in Sweden tried to circulate high-temperature water for the treatment of an inoperable cancer of uterine cervix in 1898 and the effectiveness was confirmed [34]. In the early twentieth century, applied research was carried out together with basic research. However, since the heating method and temperature-measuring technology, for example, did not develop sufficiently at that time, the positive clinical application of hyperthermia treatment was not carried out. Therefore, surgery, radiotherapy, chemotherapy, and so on, were dominant as therapy of tumors. Under such situations, a selective antitumor effect for a tumor treated at a temperature of 40–45°C was reported by Westermark. Later, hyperthermia attracted attention and its effect was supported by Westra, Overgaard, and others [35, 36]. Especially after the report of an enhanced effect when hyperthermia and radiotherapy are combined by Crile [37] and Ben-Hur [38], the interest in hyperthermia heightened, and research also rapidly developed. In the 1970s, hyperthermia research activity started at a hyperthermia institute. A symposium organized by the cancer center in the United Sates was held in Washington in 1975 and the activity has been pursued. In 1977, the application to cancer treatment opened a session of the International Microwave Symposium of the Microwave Theory and Techniques (MTT) Society of the IEEE, and a special issue of the *MTT Transactions* was published on the subject the next year [39]. In the United Sates, a hyperthermia group was formed in 1981 while the European Hyperthermia Institute was formed in Europe in 1983. In Japan, hyperthermia research started in 1978 and *The Japanese Society of Hyperthermia Oncology* was established in 1984. Research on multidisciplinary therapy in which hyperthermia is used jointly with radiotherapy, chemotherapy, surgical treatments, immunotherapy, and so on, has become popular recently.

4.3.1   Biological Background of Hyperthermia

This section gives some reasons for the effectiveness of hyperthermia in treating tumors based on biological results on culture cells and tumors of laboratory animals.

Survival Rate and Hyperthermia Sensitivity of Cell   Generally, living cells are submitted to an effect of temperature rise: When the temperature increases, the survival rate of the cell becomes lower. The survival rate of the cell is defined as the ratio of the cell population submitted to heating to the cell population before heating. The result of measuring this survival rate as a function of time is called the *survival rate curve* [40].

Figure 4.27 is a survival rate curve for the Chinese hamster ovarian cell. It is observed that the survival rate rapidly decreases above 42.5°C and decreases when heating time increases. In short, the cell population begins to decrease depending on these factors. The effect due to heating, that is, this hyperthermia sensitivity, depends on various kinds of factors. It has been said that there is no difference in hyperthermia sensity between a normal cell and a cancer cell in every phase of cell growth. However, one feature of the normal cell is that it reaches the fixed size and there is a period of time when cell breeding stops. During this period, hyperthermia sensitivity becomes low, compared to a period of normal cell growth. On the other hand, cancer cell growth does not stop, and it tends to keep increasing indefinitely. Accordingly, the hyperthermia treatment is considered to affect more the cancer cell.

Oxygen Partial Pressure, pH, and Hyperthermia Sensitivity of Cell   The hyperthermia sensitivity of the cell is influenced by oxygen partial pressure and pH, which are environmental factors of the cell. Generally, in the tumor, it is considered that the blood vessel growth cannot follow the tumor growth, and the tumor becomes a no-development condition. Therefore, it is in the anaerobic condition where the oxygen from blood flow is not sufficiently supplied. In this condition, hyperthermia sensitivity is high, as shown in Figure 4.28. It is also observed in the same figure that the sensitivity becomes lower when pH is higher. Generally, when we consider the gross tumor, it tends to

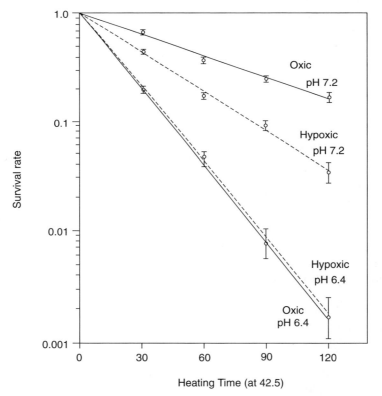

FIGURE 4.28   Survival rate depending on cell environmental factor (oxygen partial pressure and pH).

the anaerobic condition, the anaerobic glycolysis becomes active, and also pH becomes acidic, being lower. Accordingly, this means that hyperthermia is effective when the conditions of hypoxia (low oxygen partial pressure) and low pH hold. The efficiency of hyperthermia has been estimated in view of these environmental factors. The effect of irradiation is considered to be small for the hypoxic cell.

Period and Hyperthermia Sensitivity of Biological Cell Generally, the biological cell, which constitutes the organism, repeats the cell division with a fixed period of time called a *cell cycle*. Cell division does not repeat forever. This is because there are times called the *M phase (mitotic phase)* during which cells divide and the *interphase*, when there is no cell division. This one-cell cycle is separated into four phases: *pre-DNA synthesis phase* (G_1), *DNA synthesis phase* (S), *division preparatory phase* (G_2), and *cell-dividing phase* (M). After all this, the division repeats in terms of initial-phase G_1 phase, S phase, G_2 phase, and M phase, as shown in Figure 4.29. In each phase, it is in the last half of the S phase that hyperthermia sensitivity appears the highest. The next closest is the M phase. On the other hand, the sensitivity to irradiation is high in the M stage, but it is low in the S phase. Therefore, it is possible to promote a decrease in the survival rate of the cell in the S phase by hyperthermia while a large effect is not obtained by irradiation exposure. This is why combining irradiation and hyperthermia is an effective therapy.

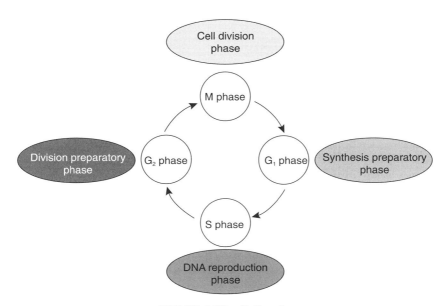

FIGURE 4.29 Cell cycle.

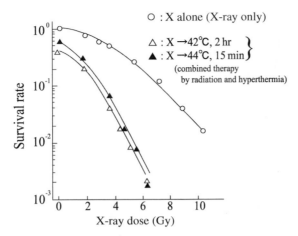

FIGURE 4.30 Survival rate of combined therapy by radiation and hyperthermia.

Hyperthermia and Combined Effect of Irradiation Ben-Hur [38] reported that the sensitization effect of irradiation is obtained by combining radiation and hyperthermia, as shown in Figure 4.30. It was shown that lowering the equivalent survival rate of cancer cells with irradiation only was compensated by using hyperthermia: Figure 4.30 shows that there is a large effect and that the survival rate of cancer cell is greatly reduced even at 42°C when both hyperthermia and irradiation are used versus the hyperthermia-only therapy.

4.4 METHOD OF THERMOMETRY

Generally, in thermal therapy, temperature-monitoring means different from the heating means play an important role. Especially in hyperthermia, the temperature at the treatment location must be controlled around 42.5°. Accordingly, accurate thermometry technology is an important problem. Temperature measurement methods in the intraorganism are generally classified into *invasive methods* and *noninvasive method*. As a thermometry method in thermal therapies such as hyperthermia, the noninvasive method for temperature measurement without stabbing the sensor into the human body is desirable. However, presently, technological means measuring temperature accurately by noninvasive methods have not been established. The following sections describe practical thermometry methods.

4.4.1 Invasive Thermometry

Measurement by Thermocouple Sensor As a thermometric sensor, the thermocouple has been used widely and generally. This thermometry principle is

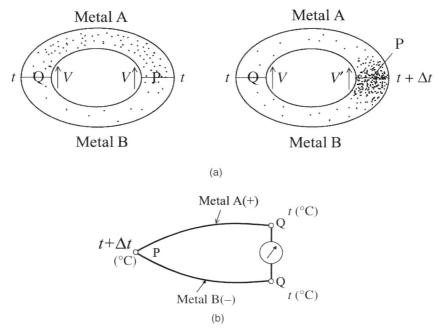

FIGURE 4.31   Explanation of Seebeck effect.

based on the Seebeck effect, discovered by Thomas Johann Seebeck in 1821.

Principle   Seebeck formed a closed loop using two different metal conductors A and B by joining each conductor end together, as shown in Figure 4.31*a*. He found a thermoelectric current flow in the circuit due to the electromotive force generation when he kept these contact parts *P* and *Q* at different temperatures. This phenomenon is called the *Seebeck effect*. When the tips of *P* and *Q* are joined together as shown in Figure 4.31*b*, the temperature at junction *P* varies while keeping the temperature at junction *Q* constant as a reference temperature. The following relation describes the Seebeck effect: When $t + \Delta t$ is the temperature measured at junction *P* while *t* is the reference temperature at *Q*, the thermoelectromotive force *V* generated in this circuit is proportional to Δt. Then, *V* is expressed as

$$V = \alpha \, \Delta t \tag{4.60}$$

where α is the Seebeck coefficient. When the reference temperature at junction *Q* is 0°C, the relationships between the thermoelectromotive force and the temperature of various kinds of thermocouples are shown in Figure 4.32

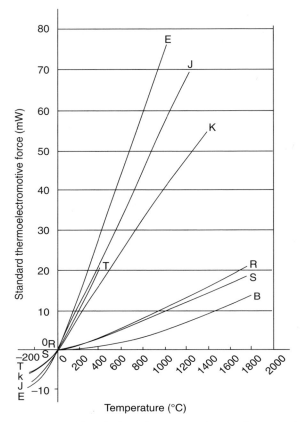

FIGURE 4.32 Example of relation between thermoelectromotive force and temperature.

TABLE 4.1   Examples of Thermocouple Materials

Type of Thermocouple	Materials	
	Positive	Negative
B	Rhodium 30%; the rest platinum	Rhodium 6%; the rest platinum
R	Rhodium 13%; the rest platinum	Platinum
(PR)	Rhodium 12.8%; the rest platinum	Platinum
S	Rhodium 10%; the rest platinum	Platinum
K(CA)	Chromium 10%; the rest nickel	Aluminum, manganese, silicon, etc.; the rest a little nickel
E(CRC)	Chromium 10%; the rest nickel	Nickel 45%; the rest copper (constantan)
J(IC)	Iron	Nickel 45%; the rest copper (constantan)
T(CC)	Copper	Nickel 45%; the rest copper (constantan)

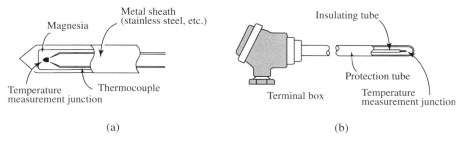

FIGURE 4.33   Construction method of thermocouple: (*a*) structure of temperature measurement junction of sheath-type thermocouple; (*b*) structure of thermocouple with protected tube.

as an example. The materials of the thermocouple in this case are tabulated in Table 4.1. As is clear form curve *T* in Figure 4.32, for example, this thermocouple can be used for low-temperature measurement, and a thermocouple of this kind is mainly used in thermal therapy. This thermocouple is composed of copper and a copper and nickel alloy as tabulated in Table 4.1.

There are two types of thermocouples: a sheath type and a protected-tube type, as shown in Figure 4.33. The sheath type stores the thermocouple wire in a metal sheath (metal pipe) which is filled with inorganic insulation material (MgO or Al_2O_3) to keep mutual insulation and an airtight condition and to prevent corrosion and degradation of the wire. The outside dimension of this thermocouple is of the order of 0.2–8 mm. The thermocouple of the protected-tube type consists of a protected tube, a terminal box, and glass that insulates the thermocouple wire. As such, protected tubes, a metallic type using stainless steel, and a nonmetallic type using alumina magnetism have been produced. The outside dimension is of the order of 3–30 mm. The sheath type has interesting features: It has a rapid response with respect to temperature change because the sheath diameter is relatively narrow and the inside of the sheath is filled with inorganic insulation material when compared to the protected-tube type. Further, the sheath type is excellent even for heat resistance and vibration resistance characteristics, and it is flexible when using a narrow-diameter sheath. This is why the sheath type is used mainly in thermal therapy.

Circuit Structure   In the construction of actual thermocouples, a lead wire in the interval between the temperature-measuring junction *P* and the reference junction *Q* is required, as shown in Figure 4.34.

On the other hand, as mentioned above, the thermoelectromotive force of the thermocouple is determined by the temperature difference between the measuring junction *P* and the reference junction *Q*. Therefore, if the spacing from the thermocouple terminal (compensation junction) to the reference

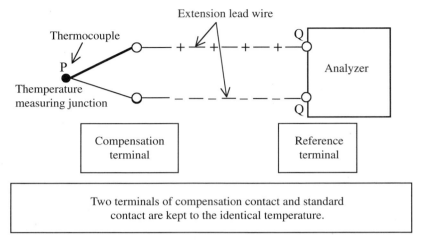

FIGURE 4.34 Wiring technique of thermocouple.

junction Q is combined with a copper wire, for instance, and if there is a temperature difference between both contact points, this position becomes a new thermocouple and an electromotive force arises. As a result, an error will be produced. To avoid this problem, usually, a copper wire with electromotive force characteristics almost equal to that of the thermocouple below temperatures of 90–150°C or less is introduced to compensate for the temperature fluctuation between the compensation terminal and the reference terminal. In short, a relation equivalent to extend the thermocouple from the thermocouple position to the reference terminal is realized by introducing the extension lead wire with in view compensation. However, the electromotive force produced by the thermocouple is generally very small and the input signal is of the order of millivolts, even as a maximum value. Therefore, induction noise from the outside is easily captured when this extension lead wire is used. As a consequence, a countermeasure for this is required. The noise countermeasure is required especially in thermometers used in the environment of hyperthermia because this radiates a strong EM field. Therefore, shielding for the compensation wire is needed as shown in Figure 4.35. To avoid effects due to this kind of noise, the following steps must be taken:

1. Use a pair of twisted extension lead wires.
2. Insert only the extension lead wire in the tube made of iron.
3. Design a measurement environment that keeps the extension lead wire sufficiently away from the induction source.

The fundamental circuit structure of the thermometer is shown in Figure 4.36. In thermometry using thermocouples, the thermoelectromotive force is the

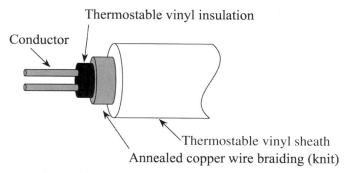

FIGURE 4.35   Compensation copper wire with shield.

temperature difference between the measuring junction and the reference terminal. Accordingly, the temperature at the reference terminal should be kept at 0°, but this is generally difficult. Hence, the temperature of the reference junction is usually measured by the thermosensor, and the electromotive force is compensated for in the electronic circuit. For this purpose, there is a compensation circuit at the reference terminal. In this case, it is necessary that the temperatures of the reference terminal junction and the thermosensor be equal. The input terminal mount in Figure 4.36 is introduced for this purpose, so that the reference junction terminal and the thermosensor are submitted to a homogeneous temperature distribution. The burnout circuit in Figure 4.36 is introduced to not exceed the thermometer scale indication when the thermocouple sensor wire is suspended. Especially, in thermocouples used to measure the temperature of a furnace, for example, in a high-temperature condition, the thermocouple gradually oxidizes and causes a disconnection problem. When this happens, the input signal before disconnection has been charged in an input filter condenser and the data (as for normal temperature) are displayed. As a result, overheating inside the furnace may occur because the measured temperature is not normal when the thermocouple sensor wire is suspended. This can cause a serious accident. To prevent breakdown of a thermocouple wire, a burnout circuit is introduced. In addition, the electromotive force of a thermocouple may exhibit a nonlinear characteristic without being in direct proportion to the temperature. Therefore, a linearization circuit is introduced to correct the nonlinearity. Finally, as described previously, the thermoelectromotive force of the thermocouple is very small, and the induction noise from the outside becomes a problem. As a countermeasure, an isolation amplifier circuit for insulating the input of the thermometry circuit against an output is introduced.

Thermometer Using Thermistor   A temperature sensor called a *thermistor* uses the general semiconductor property that electrical resistance decreases

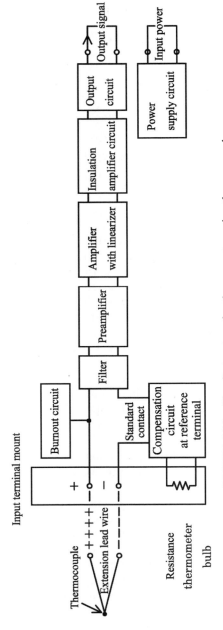

FIGURE 4.36 Circuit structure for thermometer using thermocouple.

192

TABLE 4.2   Classification of Thermistors

Clarification of Thermistor	Characteristics	Application Temperature
NTC thermistor (negative-temperature-coefficient thermistor)	NTC thermistor has the negative temperature coefficient with semilog characteristic. R vs T Log R vs $1/T$	Medium temperature use −50–300°C Low-temperature use −100–0°C High-temperature use 200–700°C
PTC thermistor (positive-temperature-coefficient thermistor)	PTC thermistor has positive temperature coefficient with the switching behaviors. R vs T	−50–150°C
CTR (critical-temperature resistor)	CTR has the negative temperature coefficient with the switching behaviors. R vs T	0–150°C

when temperature rises. Among the various kinds of semiconductors, a thermistor has the appropriate resistivity, temperature coefficient, and stability. The origin of the name thermistor stems from the fact that the oxidic compound sintered material was developed at Bell Laboratories in the United States and was called a *thermal-sensitive resistor*, announced as a thermistor. Then, Philips, in the Netherlands, developed a positive-temperature-coefficient thermistor, with its electric resistance suddenly increasing at some temperature. Further, in Japan, the *critical-temperature resistor* (CTR) thermistor, with its electric resistance suddenly decreasing, was developed. Therefore, the thermistor with an electrical resistance increase is called a *positive-temperature-coefficient* thermistor (PTC), while the conventional-type thermistor with an electrical resistance decrease is called a *negative-temperature-coefficient* thermistor (NTC). The characteristics of these thermistors are given in Table 4.2.

Principle of NTC Thermistor   In this section, we describe in detail the thermistor thermometer, which has been mainly used for clinical thermometry. The relation between temperature T and resistance R in the thermistor is obtained similarly to the case of the intrinsic semiconductor:

$$R = R_0 \exp B\left(\frac{1}{T + 273.15} - \frac{1}{T_0 + 273.15}\right) \quad \Omega \qquad (4.61)$$

where T_0 (K) is the reference temperature, T_0 a reference temperature which has often the value of 0°C (273.15 K) or 20°C (293.15 K), R_0 the value of resistance when temperature is T_0, and B a thermistor constant given by the expression

$$B = \frac{\ln(R/R_0)}{1/(T + 273.15) - 1/(T_0 + 273.15)} \quad °C \qquad (4.62)$$

where ln is the natural log. From these expressions, temperature T can be expressed as

$$T = \left(\frac{1}{T_0} - \frac{\ln(R/R_0)}{B}\right)^{-1} - 273.15 \quad °C \qquad (4.63)$$

and the sensitivity of the thermistor is given by

$$\alpha = -\frac{B}{T^2} \times 100 \quad \%/\text{deg} \qquad (4.64)$$

Thermistor sensors come in various shapes (Fig. 4.37). The *bead-type* thermistor is covered with glass, making it possible to measure a high temperature with reliability and stability. Since the *tipped thermistor* can be mass produced at low cost and with high precision, it has been used as the electronic thermometer. The *disc-type* thermistor has been utilized for consumer products such as air conditioners and refrigerators, because it cannot make a small-size thermometer. Since the thermistor is a resistive element, it cannot convert temperature into an electric signal directly. Therefore, it converts temperature into the electric signal with the aid of the external power source to feed the sensor. The circuit structure is based on the following principles:

1. *Temperature Measurement Method based on Relation of Temperature–Resistance Characteristics after Having Measured Thermistor Resistance.* This method yields temperature T from the measurement of R/R_0 by making the graphic relation between T and R/R_0 on the basis of the expression (1).
2. *Method of Converting Temperature Change into Voltage or Current Using Wheatstone Bridge.* This incorporates the thermistor into the Wheatstone bridge to detect the temperature change as a change of voltage or current using the detection method based on the zero method or the fixing bridge. This bridge circuit method is mainly used for actual thermometers.

- ***Bead type***   Due to excellent reliability and stability it is possible to carry out the high-temperature measurement

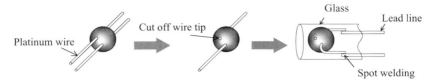

- ***Chip type***   Due to low cost, high precision, it can be mass produced. It is used for electronic thermometerr, etc.

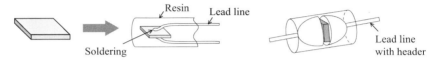

- ***Disk type***   Being capable of mass-production, it is inferior with respect to response speed, since it cannot be miniaturized. It is used for consumer products such as air conditioners

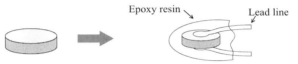

FIGURE 4.37   Various kinds of thermistor sensors.

Optical Fiber Thermometer   It is preferable to use noninductive optical fiber to measure temperature in a strong EM field environment such as for hyperthermia treatment. The optical fiber thermometer is described below. The *optical fiber thermometer* has a simple thermosensor, as shown in Figure 4.38. A sensor is attached to the tip of an optical fiber which is composed of a phosphor capable of excitation by a light-emitting diode (LED). As example, phosphor (YF_3: Yb, E_r) which is excited by the LED of silicon GaAs consists of three fluoridations of yttrium doped by ytterbium. The measuring principle is as follows. The tip of the optical fiber is attached to the measured object and the pulse wave of IR excitation light at a wavelength of 940 nm is applied. This applied pulse wave is converted into visible light, at a wavelength of 550 nm, while at the same time it is modulated by the temperature. At this moment, there is an afterglow for a while, even if the exciting light is cut off. There is a temperature dependency of this afterglow quantity. Therefore, the temperature is measured by the variation of afterglow. In short, by carrying out the sequential sampling of this afterglow quantity in a time series and summing it, after the search of the afterglow integral luminance, the temperature is calculated as shown in Figure 4.39. Actually, the temperature

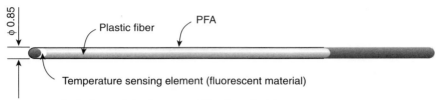

Optical part of the fluorescent fiber in optical thermometer

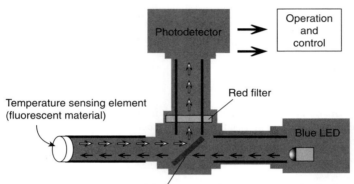

FIGURE 4.38 Construction of optical fiber thermometer.

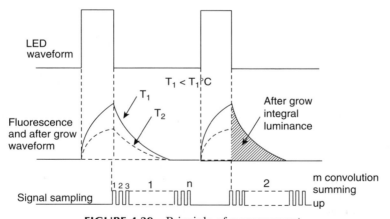

FIGURE 4.39 Principle of measurement.

is calculated as a result of comparing the mean value of the each afterglow integral luminance with the known data (reference temperature) by repeated sampling. The characteristic of the afterglow integral luminance is shown in Figure 4.39. Specifications of the present sensor are given in Table 4.3.

TABLE 4.3   Specifications of Core/Clad Diameter: Sensor

Fiber	Fiber length	2, 4, 6, 10, and 30 m
	Core/clad diameter	305 and 362 μm
	Fiber covering	Diameter 0.80 ± 0.05 mm, material PFA
	Protection tube	Diameter 2.8 ± 0.2 mm, material PVC
Tip	Length	1 m (sensor 8 mm)
	Sensor covering	Diameter 0.80 ± 0.05 mm, material PFA
	Sensor material	YF_3 (Doped material, Yb, Er)

TABLE 4.4   Specifications of Thermometer: Main Body

Thermometry range	30–80°C
Accuracy	±0.2°C
Measurring time	Switching of 1 or 2 s
Display	Four-degit LED display
Output form	Digital/analog output, intermittent control output
Temperature calibration	Automatic calibration function
Power souce	100 V AC (50/60 Hz)
External form	One channel, 325 (W) × 280 (D) × 125 (H)
	Four channels, 430 (W) × 450 (D) × 150 (H)

It is possible to have a length of optical fiber from 2 to 30 m. Usually, the multicomponent system optical fiber is used. Part of the probe is coated by Teflon. The range of measuring temperature is extended from 30 to 80°C. This thermometer may have multiple channels in order to enable the simultaneous multiple-point measurement. The specification for this thermometer is given in Table 4.4.

4.4.2   Noninvasive Thermometry

Thermometer Using IR Photodetector   Noninvasive thermometry using IR photodetectors has been put into practice, as illustrated in Figure 4.40. Such a thermometer can measure only the surface temperature of an object by detecting IR energy that is radiated by the object. The IR ray is an EM wave. Usually, EM waves with wavelengths larger than 0.7 μm and smaller than 1 mm are defined as IR rays. The IR range is specific regarding the corresponding fields: Infrared irradiation is principally based on the vibration of the atom that constitutes the molecule, that is, molecular vibration. Another specificity is lattice vibration, which depends on the molecular structure. Generally, since the IR ray image pickup is often used outdoors, one has to use a wavelength at which the transparency of the atmosphere is excellent. Frequency bands where the

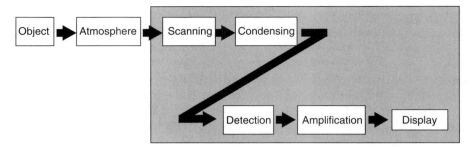

FIGURE 4.40   Construction of thermometer using IR photodetector.

transparency of the atmosphere is good are 3–5 and 8–14 μm wavelength, respectively. At wavelengths below 4 mm, the effect of solar light reflection becomes serious. Therefore, the 8–14 μm range is often used for such a thermometer.

The measurement principle is based on the following physical phenomena.

1. All objects at an absolute temperature above 0 K radiate IR energy.
2. There is a correlation between the IR energy and the temperature of the object, expressed by the *Stefan–Boltzmann law* (see Section 1.4.3) as follows:

$$W(T) = \sigma T^4 \tag{4.65}$$

where $\sigma = 5.637 \times 10^{-12}$ (W cm^{-2} K^{-4}) is the Stefan–Boltzmann constant.

The IR ray detectors are classified depending on the associated physical phenomena, as shown in Table 4.5. The two-dimensional IR detector is called an *infrared focal plane array* (IRFPA). Recently, the *bolometer*-type noncooled detector, also an IRFPA, is used as a photodetector to obtain a good resolution in the temperature measurement. The characteristics of the bolometer-type noncooled detector are as follows: For the typical photodetector vanadium oxide (VO$_x$) is used, with a number of pixels equal to 240 V × 320 H, a pixel pitch of 50 × 50 μm, detection wavelength of 8–12 μm, and a noise equivalent temperature difference (resolution of temperature) below 0.05°C. The temperature-measuring range extends from 0 to 50°C for medical use.

Noninvasive Thermometry Using NMR Technique   A promising method for noninvasive thermometry, application of *computed tomography using nuclear magnetic resonance* (NMR-CT) to a noninvasive temperature monitor for hyperthermia therapy has been proposed and studied [41]. In NMR-CT, the

TABLE 4.5   Classification of Infrared Ray Detectors

Quantum type (cooling)	Photoconductive type	Intrinsic semiconductor: HgCdTe, InSb
		Extrinsic semiconductor: Si–Ga, Si–In
		QWIP: GaAs/AlGaAs
	Photoelectromotive force type	p–n junction: HgCdTe, InSb
		Photoelectric emission: PtSi, GeSi
Thermo type (uncooled)	Pyroelectricity type	BST (barium strontium titanate)
	Thermopile type	Poly-Si p–n junction
	Bolometer type	VO_x

main parameters needed for NMR imaging are considered to be the thermal equilibrium magnetization (M_0) and the relaxation times (T_1, T_2). In 1980, Lewa and Majewska found experimentally the temperature dependence of T_1 for animal and human tissue using an NMR spectrometer in vitro [42]. In 1981, Kato et al. also measured the temperature dependence of NMR parameters (T_1, T_2, M_0) for mouse tissue using an NMR spectrometer in vitro [43]. Further, Kato et al. measured the temperature dependence of T_1 for copper sulfate solution with NMR-CT and determined the temperature with accuracy better than 0.5°C for a voxel of 1 cm [44]. In 1983, the temperature dependence of T_1 for blood was measured by Parker et al. with NMR-CT. They found that it was 1.4% per degree Celsius [45]. Based on this result, they could estimate the temperature of a blood sample with accuracy of 2°C in a 5-min scan. In 1983, Amemiya and Kamimura proved the possibility of real-time thermometry by measuring thermal equilibrium magnetization M_0 with field focusing (FF) NMR [46]. Recently, as a temperature-monitoring method using MRI, chemical shift phenomena have been introduced to measure temperature, but the accuracy has not yet reached a value good enough to warrant applying this method to hyperthermia treatment.

REFERENCES

[1] "Biological effects of electromagnetic fields and measurement," in *Japanese IEE Committee*, organized by Y. Kotsuka, Corona Publishing, 1995, pp. 215–216.

[2] T. Matsuda, *Hyperthermia*, Iryou kagaku Sha 1999, pp. 4–9.

[3] "Advanced technology for radiation therapy and measurement," in *Japanese IEE Committee*, organized by Y. Kotsuka, Corona Publishing, 2001, pp. 133–146.

[4] M. Hiraoka, Y. Tanaka, K. Sugimachi, Y. Kotsuka, et al., "Development of RF and microwave heating equipment and clinical application to cancer treatment in Japan," in A. Rosen, A. Vander Vorst, and Y. Kotsuka (Eds.), Special Issue on Medical Applications and Biological Effect of RF/Microwaves, *IEEE Trans. Microwave Theory Tech.*, Vol. 48, No. 48, pp. 1789–1799, Nov. 2000.

[5] M. Hiraoka, S. Jo, K. Akuta, et al., "Radiofrequency capacitive hyperthermia for deep-seated tumors," *I. Studies on Thermometry, Cancer*, Vol. 60, No.1, pp. 121–127, July 1987.

[6] Y. Kotsuka et al., "Development of double-electrode applicator for localized thermal therapy," in A. Rosen, A. Vander Vorst, and Y. Kotsuka (Eds.), Special Issue on Medical Applications and Biological Effect of RF/Microwaves, *IEEE Trans. Microwave Theory Tech.*, Vol. 48, No. 11, pp. 1906–1908, Nov. 2000.

[7] L. Hamada, H. Yoshimura, K. Ito, "A new heating technique for temperature distribution control in interstitial microwave hyperthermia," *IEICE Trans. Electron.*, Vol. E82-C, No. 7, pp. 1318–1325. July 1999.

[8] K. Saito, O. Nakayama, L. Hamada, K. Ito, "Analysis of temperature distributions generated by square array applicator composed of coaxial-slot antennas for hyperthermia," *Trans. IEICE*, Vol. 1, J82-B, No. 9, pp. 1730–1738, Sept. 1999.

[9] Y. Nikawa, M. Kikuchi, T. Terakawa, et al., "Heating system with a lens applicator for 430 MHz microwave," *Int. J. Hyperthermia*, Vol. 6, No. 3, pp. 671–684, May–June 1990.

[10] Y. Kotsuka et al., "Development of ferrite core applicator system for deep-induction hyperthermia," *IEEE Trans. Microwave Theory Tech.*, Vol. 44, No. 10, pp. 1803–1810, Oct. 1996.

[11] Y. Kotsuka, "On a method of inductive heating in hyperthermia," *8th Int. Cong. of Hyperthermic Oncology*, Apr. 2000, p. 121.

[12] J. R. Olson, "A review of magnetic Induction method for hyperthermia treatment of cancer," *IEEE Trans. Biomed. Eng.*, Vol. 31, No. 1, pp. 91–97, Jan. 1984.

[13] P. S. Ruggera et al., "Development of a family of RF helical coil applicator which produce transversely uniform axially distributed heating in cylindrical fat-muscle phantoms," *IEEE Trans. Biomed. Eng.*, Vol. 31, No. 1, pp. 98–105, Jan. 1984.

[14] J. G. Kern et al., "Experimental characterization of helical coil as hyperthermia applications," *IEEE Trans. Biomed. Eng.*, Vol. 35, No. 8, pp. 46–52, 1988.

[15] P. P. Antich et al., "Selective heating of cutaneous human tumors at 22.12 MHz," *IEEE Trans. Microwave Theory Tech.*, Vol. 26, No. 8, pp. 569–572, Aug. 1982.

[16] I. Kimura et al., "VLF induction heating of clinical hyperthermia," *IEEE Trans. Magn.*, Vol. 22, No. 6, pp. 1897–1900, 1986.

[17] T. Kobayashi, Y. Kida, "Interstitial hyperthermia of malignant brain tumors using implant heating system (IHS) and its future prospects," *Jpn. J. Hyperthermic Oncol.*, Vol. 11, pp. 68–75, 1995.

[18] Y. Kida, T. Kobayashi, "Hyperthermia of malignant brain tumor with implant heating system, histopathological study," *Jpn. J. Hyperthermic Oncol.*, Vol. 7, pp. 159–169, 1991.

[19] D. T. Tompkins et al., "Effect of implant variables on temperatures achieved during ferromagnetic hyperthermia," *Int. J. Hyperthermia*, Vol. 8, pp. 241–251, 1992.

[20] Z. P. Chen et al., "Errors in the two-dimensional simulation of ferromagnetic implant hyperthermia," *Int. J. Hyperthermia*, Vol. 7, pp. 735–739, 1991.

[21] I. Tohnai, Y. Goto, Y. Hayashi, M. Ueda, T. Kobayashi, M. Matsui, "Preoperative thermochemotherapy of oral cancer using magnetic induction hyperthermia," *Int. J. Hyperthermia*, Vol. 12, pp. 37–47, 1996.

[22] K. W. Chan et al., "Use of thermocouples in the intense fields of ferromagnetic implant hyperthermia," *Int. J. Hyperthermia*, Vol. 9, pp. 831–848, 1993.

[23] D. T. Tompkins et al., "Temperature-dependent versus, constant-rate blood perfusion modeling in ferromagnetic thermoseed hyperthermia: Results with a model of the human prostate," *Int. J. Hyperthermia*, Vol. 10, pp. 517–536, 1994.

[24] B. P. Partington et al., "Temperature distributions, microangiographic and histopathologic correlations in normal tissue heated by ferromagnetic needles," *Int. J. Hyperthermia*, Vol. 5, pp. 319–327, 1989.

[25] M. A. Burton et al., "In vitro and in vivo responses of doxorubicin ion exchange microspheres to hyperthermia," *Int. J. Hyperthermia*, Vol. 8, pp. 485–494, 1992.

[26] H. Matsuno, I. Tohnai, K. Mitsudo, et al., "Interstitial hyperthermia using magnetite cationic liposomes inhibit to tumor growth of VX-7 transplanted tumor in rabbit tongue," *Jpn. J. Hyperthermic Oncol.*, Vol. 17, pp. 141–150, 2001.

[27] S. Wada, K. Tazawa, I. Furuta, S. Takemori, T. Minamimura, H. Nagae, "Application of hyperthermia using dextran magnetite complex (DM) for head and neck cancer," *Jpn. J. Hyperthermic Oncol.*, Vol. 14, pp. 197–205, 1998.

[28] M. Shinkai, M. Suzuki, N. Yokoi, M. Yanase, W. Shimizu, H. Honda, T. Kobayashi, "Development of anticancer drugs—Encapsulated magnetoliposome and its combination effect of hyperthermia and chemotherapy," *Jpn. J. Hyperthermic Oncol.*, Vol. 14, pp. 15–22, 1998.

[29] M. Shinkai, M. Matsui, T. Kobayashi, "Heat properties of magnetoliposomes for local hyperthermia," *Jpn. J. Hyperthermic Oncol.*, Vol. 10, pp. 168–177, 1994.

[30] Y. Kotsuka et al., "New wireless thermometer for RF and microwave thermal therapy using an MMIC in an Si BJT VCO type," *IEEE Trans. Microwave Theory Tech.*, Vol. 47, No. 12, pp. 2630–2635, Dec. 1999.

[31] Y. Kotsuka, "Heating characteristics of small and high efficiency implant for thermal therapy," *Proc. URSI GA.*, Maastricht, The Netherlands, Aug. 2002.

[32] Y. Kotsuka, H. Okada, "Development of small high efficiency implant for deep local hyperthermia," *Jpn. J. Hyperthermic Oncol.*, Vol. 19, No. 1, pp. 11–22, Mar. 2003.

[33] W. Bush, "Über den Finfluss, wetchen heftigere Eryspelen zuweilen auf organlsierte Neubildungen dusuben," *Verh. Naturh. Preuss. Rhein. Westphal.*, Vol. 23, pp. 28–30, 1886.

[34] F. Westermark, "Über die Behandlung des ulcerirenden Cervix carcinomas mittels Konstanter Wärme," *Zentralbl. Gynkol.*, pp. 1335–1339, 1898.

[35] A. Westra, W. C. Dewey, "Variation in sensitivity to heat shock during the cell-cycle of Chinese hamster cell in vitro," *Int. J. Radiat. Biol.*, Vol. 19, pp. 467–477, 1971.

[36] J. Overgaard, P. Bichel, "The influence of hypoxia and acidity on the hyperthermia response of malignant cells in vitro," *Radiology*, Vol. 123, pp. 511–514, 1977.

[37] G. Crile, Jr., "Selective distribution of cancers after exposure to heat," *Ann. Surg.*, Vol. 156, pp. 404–407, 1962.

[38] E. Ben-Hur, B. Bronk, et al., "Thermally enhanced radio-response of cultured Chinese hamster ceels–Inhibition of repair of sublethal damage and enhancement of lethal damage," *Radiat. Res.*, Vol. 58, pp. 38–51, 1974.

[39] Special Issue on Microwave in Medicine with Accent on the Application of Electromagnetics to Cancer Treatment, *IEEE Trans. Microwave Theory Tech.*, Vol. 26, No. 8, 1978.

[40] W. C. Dewey, L. E. Hopwood "Cellular responses to combinations of hyperthermia and radiation," *Radiology*, Vol. 123, pp. 463–474, 1977.

[41] Y. Kamimura, Y. Amemiya, *Automedica*, Vol. 8, Gordon and Breach, 1987, pp. 295–313.

[42] C. J. Lewa, Z. Majewska, "Temperature relationships of proton spin-lattice relaxation time T_1 in biological tissues," *Bull Cancer*, Vol. 67, pp. 525–530, 1980.

[43] H. Kato, E. Kano, T. Sugahara, Y. Ujeno, T. Nishida, T. Ishida, "Possible application of noninvasive thermometry for hyperthermia using NMR," in E. Kano (Ed.), *Proc. Int. Conf. Cancer Therapy by Hyperthermia, Radiation and drugs*, Kyoto, Sept. 5, 1981, Mag. Bros. Inc., Tokyo, 1983, pp. 33–38.

[44] H. Kato, E. Kano, H. Tanaka, T. Ishida, "Non-invasive thermometry using NMR-CT, hyperthermic oncology," *Proc. Sixth Annual Meeting of Hyperthermia Group of Japan*, Nov. 4–5, 1983, pp. 162–163.

[45] D. L. Parker, V. Smith, P. Sheldon, L. E. Crooks, L. Fussell, "Temperature distribution measurement in two dimensional NMR imaging," *Med. Phys.*, Vol. 10, pp. 321–325, 1983.

[46] Y. Kamimura, Y. Amemiya, "An experimental study of temperature resolution of a thermometry using the field focusing NMR," *Trans. IECE Japan*, Vol. J68-C, pp. 562–569, 1985.

PROBLEMS

4.1. Explain the meanings of electronic polarization, atomic polarization, and orientation polarization.

4.2. Derive expression (4.23).

4.3. Derive Eq. (4.33) using expressions (4.30), (4.31), and (4.32).

4.4. Derive Eq. (4.38) from Maxwell's equations.

4.5. Calculate the first expression in (4.54) and derive the second expression in (4.54).

4.6. Explain the Seebeck effect.

4.7. Explain the principle of the optical fiber thermometer.

CHAPTER FIVE

EM Wave Absorbers Protecting Biological and Medical Environment

Recently, EM environments have become very complex because of the wide and rapid spread of many kinds of electric or electronic devices, as exemplified by recent cellular telephone progress. As a result, EM wave interference problems due to these devices have increased in frequency. Also, biological effects based on these kinds of EM wave radiation have been feared. As a countermeasure for protecting biological and medical environment as well as the measurement involved, knowledge of EM wave absorbers has become important. In this chapter, we study materials for EM wave absorbers, theory, and application.

5.1 FOUNDATION OF EM WAVE ABSORBERS

The start of the research on EM wave absorbers dates back to the mid-1930s. The first wave absorber was due to research in the 2-GHz band in Naamlooze Vennootschap Machinerieen in the Netherlands [1]. The wave absorber reported in 1945 was for military purpose [2]. It was used for protecting the periscope and snorkel of a submarine from search by radar. It consists of the structure that piled up "Wesch material," which is comprised of composite rubber and carbonyl iron, a resistance sheet, and plastic. This absorber has been called a "Jauman absorber." In the project organized by O. Halpen at the

RF/Microwave Interaction with Biological Tissues, By André Vander Vorst, Arye Rosen, and Youji Kotsuka

TABLE 5.1 Reflection Coefficients in EM Wave Absorber

Return Loss[a] (dB)	Electric Field Reflection	Electric Field Standing	Electric Power Reflection
−20	0.1	1.2	0.01
−30	0.03	1.06	0.001

[a] −20 dB means that 99% EM wave energy that was emitted to the absorber is absorbed. Similarly, −30 dB means that 99.9% EM wave energy is absorbed.

MIT Radiation Laboratory in the United States, an application-type wave absorber was developed [3]. In the same project, a kind of resonant-type "Salisbury screen absorber" was developed which was made by putting a resistive sheet with a surface resistance value of 377 Ω one-quarter wavelength away from a conductive plate. In 1953, a pyramidal wave absorber was developed by L. K. Neher. This absorber has been broadly used for anechoic chambers [4].

The EM wave absorber has been defined as an object that can absorb incident EM waves and convert these into a Joule heat or which can cancel the phase of the incident wave. The level of absorbing ability of the absorber is measured quantitatively using the return loss or reflection coefficient in decibels. Though a clear definition of what is a wave absorber has not been determined, the level of −20 dB in the reflection coefficient is considered to be a standard for a usual wave absorber, and it has been considered that the wave absorber should be better than this limit indicates. This value of −20 dB corresponds to a 0.1 value of the reflection coefficient of the electric field and a 0.01 value of the reflection coefficient of electric power, as shown in Table 5.1. It means that 99% of the total EM wave energy emitted to the wave absorber is absorbed.

5.2 CLASSIFICATION OF WAVE ABSORBERS

Electromagnetic wave absorbers are normally classified from the viewpoint of constituent material, structural shape, frequency characteristics, and application, respectively.

5.2.1 Classification by Constituent Material

The classification from a constituent material viewpoint is as follows:

1. *Resistive-Type Absorber.* This type of absorber uses resistive materials such as carbon black, graphite nichrome, and chromium as a basic material. These materials are used in a usual plate-type absorber or a film form. The high-frequency current that flows on the surface of the resis-

tive material is converted into Joule's heat, and the material can there-fore absorb the EM wave.

2. *Dielectric-Type Absorber.* This is a EM wave absorber which is composed mainly of some of the above-mentioned carbon materials, mixing them into forming urethane, forming styrene, rubber materials, and so on. The matching characteristic depends on the frequency dispersion character-istic of the complex relative permittivity of the absorber.

3. *Magnetic-Type Absorber.* This is an EM wave absorber essentially formed of ferrite, for which sintering ferrite, rubber ferrite, plastic ferrite, and so on, are generally used. It can absorb the EM wave because it depends on the frequency dispersion characteristics of the complex relative permeability.

5.2.2 Classification by Structural Shape

Electromagnetic wave absorbers are also classified on the basis of two struc-tural shapes: one based on number of layers constituting the wave absorber and the other based on appearance.

Classification by Number of Layers

Single-Layer Type. The absorber that is composed of only a single layer or single material is called a single-layer absorber. Generally, a conductive plate is attached to the back of this absorber. A typical example of this type is the ferrite wave absorber.

Two-Layer Wave Absorber. This absorber is composed of two different absorptive materials. This arrangement is chosen to obtain broadband absorp-tion characteristics and to improve them.

Multilayer-Type Wave Absorber. Generally, the absorber with more than three layers is called a multilayer wave absorber. In this case, the different layers are composed of absorbers with different material constants. When con-stituting a multilayer absorber, one can realize an absorber with broadband frequency characteristics.

Classification by Appearance
Wave absorbers are classified according to appearance.

Plane-Type Wave Absorber. The plane-type wave absorber has the shape of a plane surface in the incident plane of the EM wave, as shown in Figure 5.1*a*. Usually, this is a ferrite absorber.

Sawtooth-Type Wave Absorber. This wave absorber has the shape of a sawtooth in the incident plane of the EM wave, as shown in Figure 5.1*b*. This

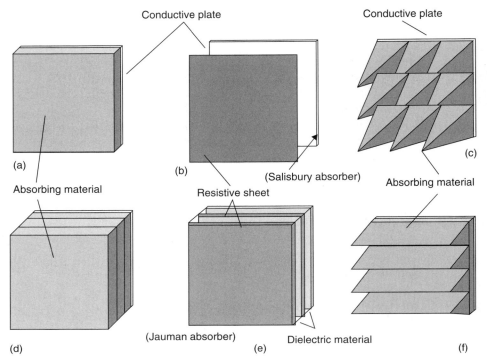

FIGURE 5.1   Classification based on appearance.

offers a means to realize a broadband absorber, as does the next type of absorber.

Pyramidal-Type Wave Absorber.   This absorber has a broadband frequency characteristic because the absorber has the structure of pyramidal taper in the incident plane of the EM wave, as shown in Figure 5.1c. A typical absorber of this kind is made of foam polyurethane rubber containing carbon powder.

5.3   FUNDAMENTAL PRINCIPLE

In this section, we describe the constitutive principle of a magnetic wave absorber such as a ferrite absorber from the viewpoint of transmission line theory and compare it to a resistive-type absorber.

It can easily be imagined that a pair of parallel conductive wires or a coaxial cable can transmit EM waves, referring to the example of a feeder that leads to the TV receiver from the antenna (see Sections 1.3.4 and 1.3.5). Heinrich Rudolph Hertz confirmed the existence of EM waves by his experiment in

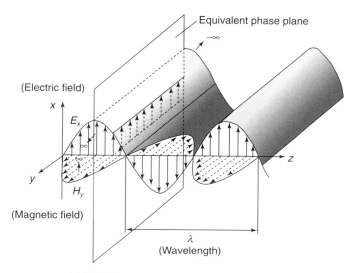

FIGURE 5.2   Illustration of plane wave.

1888. At the Germany Science Museum (Deutsches Museum), located in the shoal of the Isar River, Munich, Germany, a relic of his great achievement still exists. A pair of parallel conductive wires which tested the propagation characteristics of EM waves, such as the standing-wave distribution, is also exhibited there.

In many applications, the wave absorber has been designed by taking a plane wave into consideration. A plane wave is a wave for which the equiphase surface of the EM field perpendicular to the direction of propagation always constitutes a plane. In this case, the electric and magnetic fields are orthogonal to each other, as illustrated in Figure 5.2, and the wave does not have components of electric and magnetic fields in the direction of propagation, or the z axis. The wave with such a field distribution has a TEM mode (*transverse electromagnetic* mode, see Section 1.3.4). In fact, a TEM mode without electric and magnetic field components in the direction of propagation can be realized using a transmission line which consists of two conductors (at most, three conductors), as shown in Figure 5.3. This is obvious from the development of the coaxial line shown in the same figure. The field distributions in Figures 5.3*a,c* and *d* are used for the standing-wave measurement. The reason we investigate the phenomenon related to the EM plane wave by replacing it by the transmission line theory is based on the approach mentioned above.

Consider the propagation of a plane wave along the two parallel wires of the transmission line in Figure 5.4. Based on transmission line theory, the condition of perfectly absorbing the EM wave at the load impedance Z_R without any reflection, which results in thermal energy, is the case where Z_R is equal

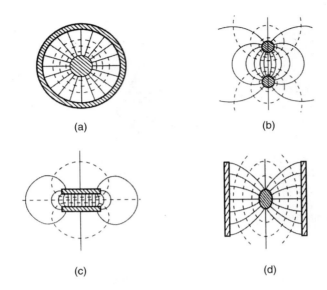

FIGURE 5.3   Examples of TEM mode.

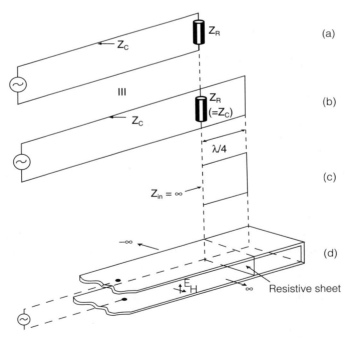

FIGURE 5.4   Principle of EM wave absorber based on transmission line theory.

to the characteristic impedance Z_c of the transmission line. This condition also holds even if the short circuit is connected a quarter wavelength apart from the load impedance Z_R, as shown in Figure 5.4b. This is based on the fact that the input impedance Z_{in}, when looking into the short circuit from the position a quarter wavelength apart, becomes infinite, as shown in Figure 5.4c. In short, although Z_{in} is connected in parallel with Z_R, actually Z_{in} is disconnected from Z_R because Z_{in} takes an infinite value. On the basis of the above considerations, let us consider a parallel-plate line equivalent to the present transmission line in which the TEM mode is propagated. When the medium between two parallel plates is assumed to be a vacuum and the interval between the two plates is narrow enough with respect to wavelength, the impedance corresponding to Z_c in the above-mentioned transmission line is now 120π ohms ($= 377\,\Omega$, which is called *free-space impedance*). Therefore, by mounting a thin resistive film with this 120π-ohm value of resistance at a position a quarter wavelength apart from a shorting terminal, the EM wave is absorbed: Matching is obtained. Since the EM field distribution when the EM wave propagates between this parallel-plate line is identical to the case in which a plane wave propagates in free space, the EM wave absorber can be constructed, by placing a resistive film and a shorting board (a conductive plate) in the space, as shown in Figure 5.5a. This is a basic configuration for the resistive-type absorber.

But now, from this basic structural principle for the resistive-type absorber, we can suppose a thin *magnetic resistive* film instead of the thin *resistive* film. The theoretical reason that develops the concept of magnetic resistive film is based on the duality principle. Actually, duality has been established between the above-mentioned resistive film and the magnetic resistive film. This duality

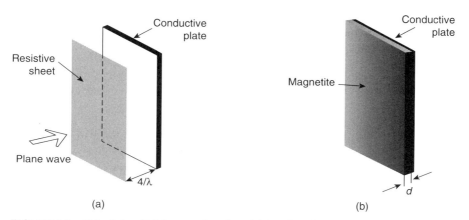

(a) (b)

FIGURE 5.5 Principle of EM wave absorber: (a) by electric resistive sheet; (b) by magnetic resistive sheet.

TABLE 5.2   Duality Principle

A	B
Series connection	Parallel connection
Impedance	Admittance
Voltage	Electric current
Open circuit	Short circuit etc.

is expressed in Table 5.2 if the transmission line circuit is taken as the present example. It means that, if one circuit is expressed as the characteristics of A in Table 5.2 and a new circuit obtained by changing A into the characteristics of B still represents the similar property to A, we call this nature *duality*. According to this relation, although the end impedance is infinite in the case of the resistive film absorber, a wave absorber by a magnetic resistive film is considered if the magnetic resistor is used so that the end impedance becomes zero. To obtain this, the thin magnetic absorber is attached to a conductive plate (in order to make the end impedance zero) and a magnetic wave absorber such as a ferrite absorber can be constructed: The EM wave is perfectly absorbed if the impedance looking into the magnetic resistive film is equal to 120π ohms. This is the basic structural principle of the magnetic wave absorber investigated from the theoretical viewpoint. Recently, however, various deformed types have also been proposed. They are based on the above fundamental structural principle and due to the large demand of EM wave absorbers.

5.4   FUNDAMENTAL THEORY OF EM WAVE ABSORBERS

5.4.1   Single-Layer-Type Wave Absorber

First, let us derive the expression of the reflection coefficient when a plane TE wave makes an oblique incidence with a single-layer absorber with thickness d. The coordinate system is defined as shown in Figure 5.6. In usual wave absorbers, medium I is considered a vacuum while medium II is composed of homogeneous absorptive material backed with a conductive plate. As shown in Figure 5.6, the incident wave, forming an incident angle θ_i with the z axis, is reflected by a conductive plate, while the outgoing or transmitted wave and the reflected wave coexist in medium II. The following expressions are derived from Maxwell's equations. The derivation is straightforward. It necessitates, however, some practice in expressing the laws of reflection and refraction. The reader may whish to look at this in more detail or refresh his or her knowledge. The derivation, essentially based on geometric arguments, is given in detail in a number of textbooks [5–8].

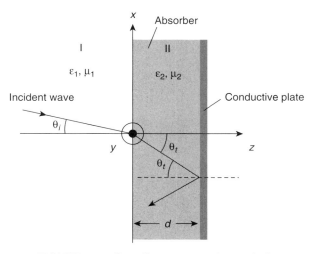

FIGURE 5.6   Coordinate system for analysis.

The derivation is done here for TE waves, while it can be done similarly for TM waves (*transverse electric* and *transverse magnetic* modes, respectively, see Section 1.3.4). In TE waves, the electric and magnetic fields in each medium that satisfy the wave equation derived from Maxwell's equations are expressed as follows.

Medium I:

$$E_{y1} = A \exp[-j\gamma_1(z \cos\theta_i - x \sin\theta_i)] + B \exp[j\gamma_1(z \cos\theta_r + x \sin\theta_r)] \qquad (5.1)$$

$$H_{x1} = -\frac{A \cos\theta_i}{Z_{c1}} \exp[-j\gamma_1(z \cos\theta_i - x \sin\theta_i)]$$

$$+\frac{B \cos\theta_r}{Z_{c1}} \exp[j\gamma_1(z \cos\theta_r + x \sin\theta_r)] \qquad (5.2)$$

Medium II:

$$E_{y2} = C \exp[-j\gamma_2(z \cos\theta_i' - x \sin\theta_i')] + D \exp[j\gamma_2(z \cos\theta_r' + x \sin\theta_r')] \qquad (5.3)$$

$$H_{x2} = -\frac{C \cos\theta_i'}{Z_{c2}} \exp[-j\gamma_2(z \cos\theta_i' - x \sin\theta_i')]$$

$$+\frac{D \cos\theta_r'}{Z_{c2}} \exp[j\gamma_2(z \cos\theta_r' + x \sin\theta_r')] \qquad (5.4)$$

where θ_i' and θ_r' designate the mean incident angle and reflected angle at the surface of a conductive plate in medium II, respectively, satisfying the relation $\theta_i' = \theta_r' = \theta_t$. We imposing the following boundary conditions to expressions (5.1)–(5.4):

In $z = 0$:

$$H_{x1}|_{z=0} = H_{x2}|_{z=0} \tag{5.5}$$

$$E_{y1}|_{z=0} = E_{y2}|_{z=0} \tag{5.6}$$

In $z = d$:

$$E_{y2}|_{z=d} = 0 \tag{5.7}$$

Then we obtain the expression of the reflection coefficient:

$$S = \frac{Z_{TE} - 1/\cos\,\theta_t}{Z_{TE} + 1/\cos\,\theta_i} \tag{5.8}$$

where

$$Z_{TE} = \frac{\mu_{r2}}{\sqrt{\varepsilon_{r2}\mu_{r2} - \sin\theta_i}}\tanh\left(j\frac{2\pi}{\lambda}\sqrt{\varepsilon_{r2}\mu_{r2} - \sin^2\theta_i}d\right) \tag{5.9}$$

and where ε_r and μ_r are the relative permittivity and relative permeability, respectively. The reflection coefficient $S = B/A$. In the course of this derivation, Snell's law has been used:

$$\frac{\sin\theta_t}{\sin\theta_i} = \frac{\gamma_1}{\gamma_2}$$

In the case of normal incidence of the TE wave, we can obtain the reflection coefficient by putting $\theta_i = 0$. Then,

$$S = \frac{\sqrt{\mu_{r2}/\varepsilon_{r2}}\,\tanh[j(2\pi/\lambda)\sqrt{\varepsilon_{r2}\mu_{r2}}d] - 1}{\sqrt{\mu_{r2}/\varepsilon_{r2}}\,\tanh[j(2\pi/\lambda)\sqrt{\varepsilon_{r2}\mu_{r2}}d] + 1} \tag{5.10}$$

Similarly, we can derive the reflection coefficient for a TM wave:

$$S = \frac{Z_{TM} - \cos\theta_i}{Z_{TM} + \cos\theta_i} \tag{5.11}$$

where

$$Z_{TM} = \frac{\sqrt{\varepsilon_{r2}\mu_{r2} - \sin^2\theta_i}}{\varepsilon_{r2}}\tanh\left(j\frac{2\pi d}{\lambda}\sqrt{\varepsilon_{r2}\mu_{r2} - \sin^2\theta_2}\right) \tag{5.12}$$

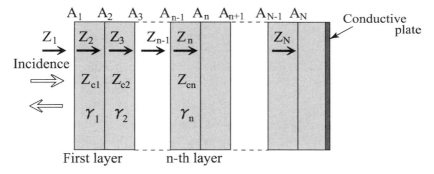

FIGURE 5.7   Configuration of multilayered absorber.

5.4.2   Multilayer-Type Wave Absorber

Normal Incident Case   Next, let us consider how to derive the reflection coefficient of an absorber composed of different layers of material with different constants and different thicknesses for each layer. Figure 5.7 illustrates the configuration of the multilayer wave absorber. Generally, this type of absorber can exhibit broadband absorbing characteristics or broadband matching characteristics. In Figure 5.7, the incident wave propagates from the left side to the right and goes into the absorber with normal incidence.

In the nth layer, the intrinsic impedance and propagation constant are denoted by Z_{cn}, and γ_n, respectively. These expressions are given as follows:

$$Z_{cn} = Z_0\sqrt{\frac{\mu_{rn}}{\varepsilon_{rn}}} = Z_0\sqrt{\frac{\mu'_{rn} - j\mu''_{rn}}{\varepsilon'_{rn} - j\varepsilon''_{rn}}} \qquad (5.13)$$

$$\gamma_n = j\frac{2\pi}{\lambda}\sqrt{\varepsilon_{rn}\mu_{rn}} = j\frac{2\pi}{\lambda}\sqrt{(\varepsilon'_{rn} - j\varepsilon''_{rn})(\mu'_{rn} - j\mu''_{rn})} \qquad (5.14)$$

where ε_{rn} and μ_{rn} are the complex relative permittivity and complex relative permeability, respectively. The reflection coefficient in this case can be calculated from the following expression when the value of the input impedance Z_1 or input impedance at the front plane is obtained:

$$S = \frac{Z_1 - Z_0}{Z_1 + Z_0} \qquad (5.15)$$

where $Z_0 = 120\pi$ ohms. Now, the value of Z_1 is calculated using the input impedance Z_{n-1} in the nth layer, where Z_{n-1} is expressed as the recurrence formula. To derive the recurrence formula Z_{n-1}, let us consider the expression of the impedance Z_N when looking into the Nth layer at the surface of the final layer A_N. The expression Z_N is denoted by the expression

$$Z_N = Z_{CN} \tanh \gamma_N d_N = Z_{CN} \tanh\left(j\frac{2\pi}{\lambda} \sqrt{\varepsilon_{rN}\mu_{rN}}\, d \right) \qquad (5.16)$$

Since the impedance looking into the nth layer at the surface of the nth layer is denoted by Z_n, Z_{n-1} is expressed by the following recurrence formula using the intrinsic impedance Z_{cn} in the nth layer:

$$Z_{n-1} = Z_{cn} \frac{Z_n + Z_{cn-1} \tanh \gamma_{n-1} d_{n-1}}{Z_{cn-1} + Z_n \tanh \gamma_{n-1} d_{n-1}} \qquad (5.17)$$

Note that this recurrence formula should be calculated from the side of a terminal conductive plate. Then, calculating Z_n successively from the impedance Z_N looking into the Nth layer, we can obtain the value of Z_1. Hence, the reflection coefficient S is obtained by substituting the free-space impedance Z_0 and Z_1 in expression (5.15).

Oblique Incident Case   The recurrence formula (5.17) can be used also when the plane wave with an incident angle θ is obliquely transmitted to the multiplayer-type wave absorber. However, the equations which calculate the input impedance Z_{cn}, propagation constant γ_n, and reflection coefficient S of the nth layer differ from the previous case, such as the following equations, depending on the TE wave or TM wave:

1. TE wave:

$$Z_{cn} = \frac{Z_0 \mu_{rn}}{\sqrt{\varepsilon_{rn}\mu_{rn} - \sin^2 \theta}} \qquad (5.18)$$

$$\gamma_n = j\frac{2\pi}{\lambda} \sqrt{\varepsilon_{rn}\mu_{rn} - \sin^2 \theta} \qquad (5.19)$$

$$S_{\text{TE}} = \frac{Z_1 - Z_0 \cos \theta}{Z_1 + Z_0 \cos \theta} \qquad (5.20)$$

2. TM wave:

$$Z_{cn} = \frac{Z_0 \sqrt{\varepsilon_{rn}\mu_{rn} - \sin^2 \theta}}{\varepsilon_{rn}} \qquad (5.21)$$

$$\gamma_n = j\frac{2\pi}{\lambda} \sqrt{\varepsilon_{rn}\mu_{rn} - \sin^2 \theta} \qquad (5.22)$$

$$S_{\text{TM}} = \frac{Z_1 - Z_0 \cos \theta}{Z_1 + Z_0 \cos \theta} \qquad (5.23)$$

The reflection coefficient can be calculated from Eqns. (5.20) and (5.23) if the recurrence formula (5.17) is calculated using the values of Z_{cn} and γ_n in each layer and if the input impedance Z_1 which looks into the absorber plane in the first layer is obtained.

5.4.3   Taper-Type Wave Absorber

Among the many wave absorbers used in anechoic chambers are the sawtooth and pyramidal wave absorbers. They are composed of a taper structure in the plane of incidence of the EM wave. Generally, in this case, a rigorous analysis of the reflection coefficient becomes very complex. Accordingly, in this text-book, we introduce an approximate method for calculating the reflection coefficient in the present case. This method is based on the idea that the dielectric constant at any taper portion is proportional to the area of the cross section at any taper portion, as shown in Figure 5.8. As a result, as shown in Figure 5.8, the equivalent dielectric constant at any taper portion is given by the expression:

$$\varepsilon_{re} = \left(\frac{\text{cross section of taper portion}}{\text{whole cross section of absorber}} \right) \times (\varepsilon_r - 1) + 1 \qquad (5.24)$$

According to this, the dielectric constant ε_r at any taper portion is approximated by the equivalent dielectric constant ε_{re} under the condition of dividing the taper portion as thin as possible. As a result, the analytical method of the multilayer absorber mentioned above is applied to the analysis of the reflec-

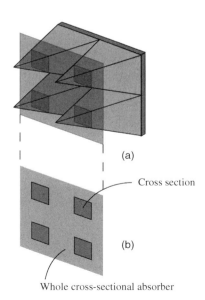

(a)

Cross section

(b)

Whole cross-sectional absorber

FIGURE 5.8   Approximation method for equivalent dielectric constant.

tion coefficient in the present case. Although this method is very simple, it exhibits a good approximation when the value of permeability μ_r is 1 and the complex permittivity ε_r takes a relatively small value, close to 1.

5.5 APPLICATION OF EM ABSORBER

The initial purpose of the wave absorber development was mainly the application to the anechoic chamber. A microwave oven, a household electric appliance, is an example of the early application of wave absorbers. For EM wave leakage prevention from the door of a microwave oven, means for installing the EM wave absorptive material have been developed. Later, the television ghost created by a wave reflected from a high-rise building and the false image obtained by a ship radar from a large bridge became problems. Electromagnetic wave absorbers have been developed as countermeasures to these problems. Recently, along with the rapid development of electronic equipment, the popularization of building indoor local-area network (LAN) systems, and mobile radio communications, EM wave absorbers are being be used for noise reduction. With this, the development of an absorber efficient in absorbing EM waves from microwaves to millimeter waves is demanded. In this section, we outline an example of current typical wave absorbers, the constitution method, characteristic example, features, and so on.

5.5.1 Quarter-Wavelength-Type Wave Absorber

A quarter-wavelength wave absorber is the most typical wave absorber based on the basic principle of EM wave absorption. Generally, it is made by attaching a conductive plane to the back of lossless dielectric material, which plays the role of a spacer, as shown in Figure 5.9, and the resistive material is coated on the surface of this spacer. This principle utilizes the property by which the input impedance Z_{in} looking into the absorber surface at the point from the conductive plate apart from a quarter wavelength ($\lambda/4$, where λ is the wavelength) becomes infinite, as indicated in Figure 5.9. This means that, when we put the resistive sheet with resistance R_S in the position which separates it from the conductive plate by $\lambda/4$, the input impedance Z_{in} equals R_s, as is clear from the equivalent transmission line shown in Figure 5.9c. The reflection coefficient S is denoted by the general expression

$$S = \frac{Z_{in} - Z_0}{Z_{in} + Z_0} \tag{5.25}$$

In this case, since one has $Z_{in} = R_s$ and $Z_0 = 120\pi$ ohms, the expression (5.25) becomes

$$S = \frac{R_s - Z_0}{R_s + Z_0} \tag{5.26}$$

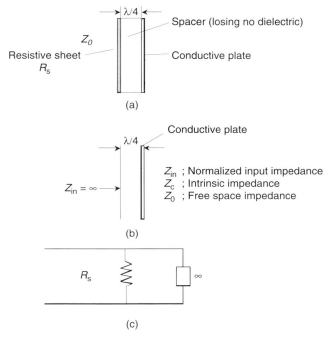

FIGURE 5.9 Explanation of a quarter-wavelength-type absorber.

By imposing the condition that the reflection coefficient S becomes zero, we obtain the relation $R_s = Z_0 = 120\pi$ ohms $(= 377\,\Omega)$. This is obviously the matching condition for the present absorber. This way, if the space between the conductive plate and the coated resistive plane is equal to a quarter wavelength and if, moreover, the input impedance R_s is equal to 120π ohms, a $\lambda/4$ absorber can be realized. Generally, as is clear from this constitution method, a $\lambda/4$ absorber is narrow band. Actually, since the spacer also has to some extent a dielectric constant, this dielectric constant should be taken into consideration when we determine the space between the conductive plate and the coated resistive plane. Because the constitution method of a $\lambda/4$ absorber is simple and it can be thinned at high frequency, it is now used abundantly as a millimeter-wave absorber.

As an example of $\lambda/4$-wave absorber, an EM wave absorber for a ship radar using resistance cloth woven with electroconductive fiber is shown in Figure 5.10. This absorber, composed of resistive cloth woven with a fiber at 3-mm intervals and with a resistance of $114\,\text{k}\Omega/\text{m}$, is used as EM wave absorbing material while foam polyethylene is used as the spacer. The weight is only $500\,\text{g/m}^2$. Normal incident characteristics for this ship radar absorber are shown on Figure 5.11. It is shown that each sample, A, B, and C, leads to a

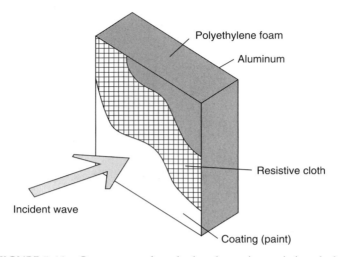

FIGURE 5.10   Quarter-wavelength absorber using resistive cloth.

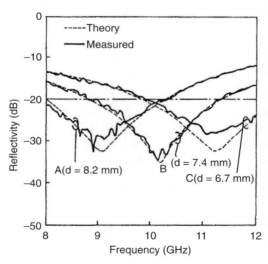

FIGURE 5.11   Frequency-matching characteristics when changing thickness of absorber.

matching frequency band over 2 GHz with a reflection coefficient below −20 dB. When the absorber is submitted to a TM wave with an oblique incidence, the reflection coefficient remains under −20 dB up to 30 degrees of obliqueness. This absorber was improved further to bear environmental factors using a resistance cloth laminated by chlorosulfone polyethylene rubber.

5.5.2   Single-Layer-Type Wave Absorber

Single-layer wave absorbers are composed of materials such as carbon and magnetic material such as ferrites. These materials are backed with a conductive plate as shown in Figure 5.1*a*. Accordingly, the design of a single-layer absorber for normal incidence is based on Eqn. (5.10). That is, if two constants among the dielectric constant ε_{r2}, relative permeability μ_{r2}, and thickness d of the absorber are determined, the remaining constant can be calculated by the following expression under the condition that the reflection coefficient goes to zero (matching):

$$S = \frac{\sqrt{\mu_{r2}/\varepsilon_{r2}}\,\tanh[j(2\pi/\lambda)\sqrt{\varepsilon_{r2}\mu_{r2}}\,d]-1}{\sqrt{\mu_{r2}/\varepsilon_{r2}}\,\tanh[j(2\pi/\lambda)\sqrt{\varepsilon_{r2}\mu_{r2}}\,d - 1]+1} \qquad \text{[Eq. (5.10)]}$$

When a dielectric material such as carbon is used and its permittivity value is known, the thickness d can be calculated assuming the relative permeability value to be 1. When ε_{r2} and d/λ_0 are taken together as parameters, a relationship between thickness d and dielectric constant ε_{r2} can be found. In magnetic materials such as ferrite, however, the parameters ε_{r2} and μ_{r2} always take values different from 1 and, moreover, they vary with frequency. Therefore, these values have to be precisely determined; then the thickness d can be calculated.

An example of a single-layer absorber uses sintering ferrite and rubber ferrite. Generally, since the sintering ferrite has a large relative permeability value in comparison with rubber ferrite, it is applied to the absorber in the frequency region from 30 MHz to 1.5 GHz. The thickness is usually 4–10 mm. Although this absorber is a single-layer absorber, the relative frequency bandwidth is comparatively large, ranging from 80 to 150%. For this application, rubber ferrite material is made by mixing powder of sintering ferrite and putting ferrite into the rubber. Therefore, this material has special features of flexibility and can easily be processed. However, the rubber ferrite does not have a large permeability compared to sintering ferrite because the ferrite powder has been mixed into the rubber. The frequency band of EM wave absorption is therefore above 1 GHz. The relative bandwidth is about 30% for usual rubber ferrite. Features of both materials are shown in Tables 5.3 and 5.4.

5.5.3   Multilayer Wave Absorber

We now introduce the EM absorber that is used for the countermeasure of TV ghosts as an example of a two-layered absorber. The EM wave absorption wall for the high-rise building is composed of three layers—granite, ferrite, and steel bars—and a concrete layer on the surface of the building, as shown in Figure 5.12. In this construction, the equivalent complex relative permeability is being adjusted by the change of the magnetic resistance of the ferrite layer by controlling the slit size between adjacent ferrite tiles. That is, the good matching characteristic in television frequency bands is realized by

TABLE 5.3 Examples of EM Wave Absorber Using Sintered Ferrite

Type of Ferrite	Thickness d (mm)	Frequency Range below $-20\,dB$ (MHz)	Central Matching Frequency f_0 (MHz)	$\Delta f/f_0$ (%)
Ni–Zn system	7.0	80–360	190	147
Mn–Mg–Zn system	7.5	250–800	410	134
Ni–Zn system	4.0	600–1270	860	78

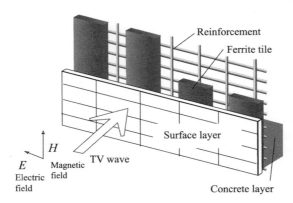

FIGURE 5.12 Example of multilayered absorber.

FIGURE 5.13 Tokyo metropolitan office.

optimizing the slit dimension and the thickness of the granite and ferrite. Figure 5.13 is a photo of a Tokyo agency building which conducted this countermeasure.

5.5.4 Pyramidal Wave Absorber

The pyramidal wave absorber is broadly used for anechoic chambers. This absorber is composed of material made by mixing carbon powder into forming urethane, forming styrene, rubber materials, and so on, as mentioned in Section 5.1. The matching characteristics depend on the frequency dispersion characteristic of the complex relative permittivity of the material. This kind of absorber usually possesses broadband matching characteristics at frequencies above 30 MHz. It is desirable for the pyramidal wave absorber for an anechoic

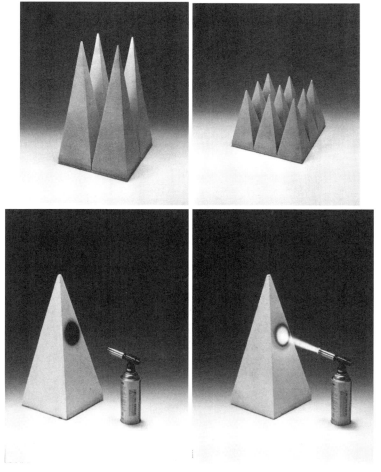

FIGURE 5.14   Examples of pyramidal absorber.

FIGURE 5.15   Examples of anechoic chamber.

TABLE 5.4   Characteristics of EM wave Absorber Using Rubber Ferrite (Ni–Zn System)

Volume Mixture Ratio of Ferrite (%)	Thickness d (mm)	Frequency Range Below −20 dB (MHz)	Central Matching Frequency f_0 (MHz)	$\Delta f/f_0$ (%)
45	5.8	2.7–3.75	3.2	33
36	5.8	3.4–4.5	4.0	28
33	5.0	4.6–6.0	5.3	26
31	4.8	5.5–7.4	6.35	30
35	4.2	6.4–8.6	7.75	28
45	3.4	9.0–10.2	9.7	12

chamber to have incombustible characteristics. Figures 5.14 and 5.15 show the absorber with incombustible characteristics and the inside of an anechoic chamber, respectively.

5.6   EM WAVE ABSORBERS BASED ON EQUIVALENT TRANSFORMATION METHOD OF MATERIAL CONSTANT

5.6.1   Microwave Absorber with Multiholes

To quickly respond to latest demands and to effectively use conventional ferrite materials, a simple method for controlling the matching frequency at microwaves is needed. In the case of ferrite absorbers, however, it has been a challenge to develop an EM wave absorber at the desired matching frequency. This is due to the fact that ferrite material is manufactured though a complex process involving such conditions as controlled sintered temperature, pressure, and a specific ratio of composite materials. Moreover, even if a ferrite absorber is produced by these methods, it has at most one or two matching frequency characteristics, because matching characteristics are mainly limited by the ferrite permeability characteristic and its thickness. To avoid these complex processes of producing absorbing materials, a new method for the effective use of ferrite for microwave absorbers has been proposed [9]. This absorber can be constructed by punching out small holes in the rubber ferrite. By doing so, the matching frequency characteristics of the absorber are broadly changed and improved using a conventional or single ferrite. We called these methods *equivalent transformation method of material constant*. This is because the matching frequency characteristic is modified easily, just by punching out small holes in the ferrite and by adjusting the hole size and the distance between adjacent holes, as if the ferrite material constant is changed. Figure 5.16 shows various kinds of slots for actual absorbers. In this section, an absorber with small holes as shown in Figure 5.17 is considered.

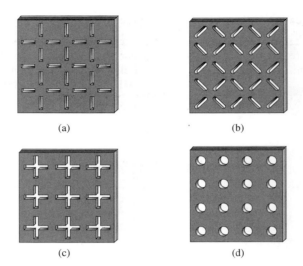

FIGURE 5.16 Various kinds of small-hole absorbers.

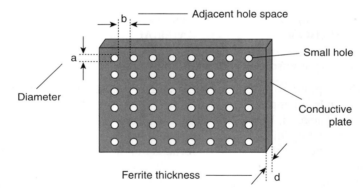

FIGURE 5.17 Fundamental construction of multihole microwave absorber.

Matching Characteristics

Effect of Hole Size. Figure 5.18 shows the theoretical matching characteristics based on a FDTD analysis when the size of the square hole is taken as a parameter, the other parameters being constant. In this case, the space between adjacent holes is 4 mm and the rubber ferrite thickness is 6.5 mm. Figure 5.19 shows the matching characteristics when the adjacent hole space is taken as a parameter. Throughout this investigation, a frequency dispersion equation of permeability has been used as the value of permeability, as

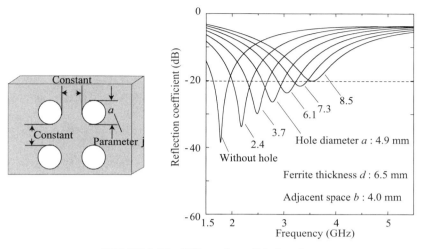

FIGURE 5.18   Effect of small holes sizes.

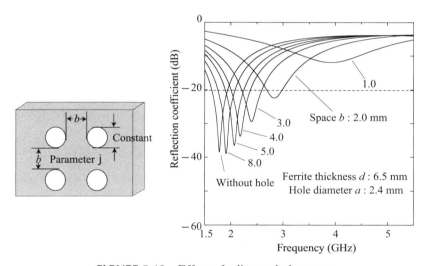

FIGURE 5.19   Effect of adjacent hole space.

explained in below under effect of permeability. To summarize the theoretical results, by simply punching holes in the rubber ferrite: (i) the matching frequency characteristic is shifted toward a higher frequency region as the hole size increases and (ii) the matching frequency characteristic is shifted toward a higher frequency region as the adjacent hole distance decreases. A general tendency of the present matching frequency characteristics is illustrated in

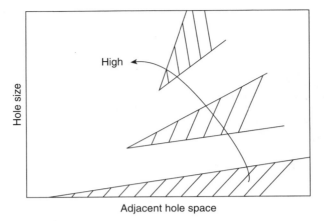

FIGURE 5.20   Relation of both hole sizes and adjacent hole spaces versus frequency.

Figure 5.20. Hatched-line regions represent the values of the reflection coefficient under −20 dB. This figure means that both the hole size and the adjacent hole distance need large sizes to obtain matching at a higher frequency.

Effect of Permeability.   To evaluate the matching characteristics, the following frequency dispersion equation of permeability is introduced [9]:

$$\mu_r' = 1 + \frac{Kf_1}{f_1 + jf} = \mu_r' - j\mu_r'' \tag{5.27}$$

where K is the static magnetic susceptibility when f is zero, f is the operating frequency, and the limit of Kf_1 is defined as the following expression derived from experimental investigation:

$$Kf_1 \leq 10 \qquad \text{GHz} \tag{5.28}$$

In this expression, if Kf_1 takes on a larger value, the imaginary part of the relative permeability tends to increase. This means that the effect of changing Kf_1 is mainly reflected in the value of the imaginary part of permeability. On the other hand, the change in value of K primarily affects the real part of the permeability: If K takes on a larger value, the real part of permeability is almost 1.0 in the actual frequency region. When K takes on a lower value, the real part of permeability becomes greater than 1.0. Figure 5.21 shows cases in which Kf_1 is taken as a parameter and K is assumed to take extreme values between 10 and 10^3. Figure 5.22 shows the matching characteristics with holes when Kf_1 is taken as parameter, the values of K being 10 and 10^3. It becomes clear that we obtain good matching characteristics when the value of K is low, even if Kf_1 is modified. Further, as the value of Kf_1 increases above 7 GHz

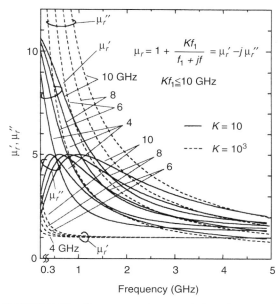

FIGURE 5.21   Frequency characteristics of permeability.

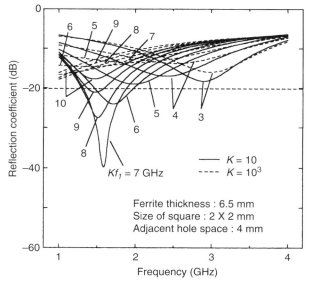

FIGURE 5.22   Matching characteristics with Kf_1 as parameter and K equals 10 and 10^3.

when the relative permittivity is 14, the matching frequency characteristics deteriorate in the present frequency region. Common rubber ferrite has a maximum value between $Kf_1 = 6\,\text{GHz}$ and $Kf_1 = 7\,\text{GHz}$ with a low K value. This is one reason why rubber ferrite is presently selected at microwaves.

Effect of Permittivity.   To investigate the effect of permittivity, general design charts have been established. Figure 5.23 illustrates the cases where the real part of relative permittivity takes the values 10 and 25 at the frequencies 2.45, 3.0, and 4.0 GHz. These permittivity values are the usual limits for rubber ferrite. If the rubber ferrite accounts for a large imaginary part of permittivity, we do not obtain a good matching characteristic. In the present study, rubber ferrite with a small imaginary part has been assumed from the outset. These data show an example, with values of $Kf_1 = 4.5\,\text{GHz}$ and $K = 3$ in Eqn. (5.27), while the ferrite thickness is 6 mm. From these design charts, comparing Figures 5.23a, d, we find that a larger permittivity value ($\varepsilon_r' = 25$) is effective when the matching frequency is lower. On the contrary, when the matching frequency is higher, Figures 5.23c, f show that a smaller permittivity value is effective for obtaining good matching characteristics.

Effect of Ferrite Thickness.   Next let us consider the ferrite thickness effect. Figures 5.24 and Fig. 5.25 illustrate the behavior of matching characteristics taking the ferrite thickness as a parameter. Figure 5.24 represents the changes of matching frequency characteristics when the ferrite thickness is decreased from the original thickness and hole sizes are increased, keeping hole adjacent spaces constant. It is found that the matching frequency characteristic shifts toward a higher frequency region as hole sizes are increased. One can find that there is an optimum hole size to obtain a good matching characteristic for each ferrite thickness. Similarly, Figure 5.25 shows the changes of matching frequency characteristics when the ferrite thickness is decreased and adjacent hole spaces are decreased, keeping hole size constant. Even in this case, it is found that an optimum adjacent hole space exists for each ferrite thickness.

Method of Perforation.   It is obvious that an accurate methodology for controlling the hole diameter for this rubber absorber is important. For this purpose, the authors investigated a laser perforation method. Figure 5.26 shows an example of holes with a diameter of 2 mm and adjacent hole spaces of 9 mm. The circular holes are formed using a CO_2 laser with a maximum power output of 200 W and a CW maximum pulse frequency of 10 kHz (Rofin Sinar Laser, SC × 20). The diameter accuracy is checked using a profile projector (Nikon, V-12). At present, holes with an error factor of less than ±0.1 mm can be formed. An error factor of this order does not raise problems in the present case. Further, smaller diameter perforation is also examined. We find that 0.1 mm hole diameter can be formed even when sintered ferrite is used. This accurate perforation method is important when the present absorber is used in the millimeter frequency region where a small diameter is

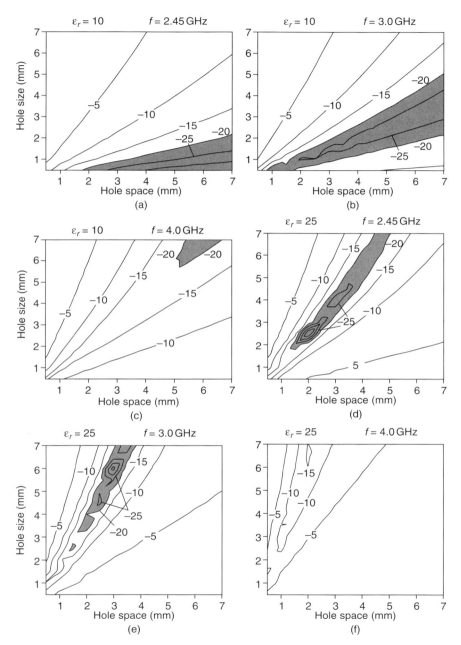

FIGURE 5.23   Matching characteristics with permittivity as parameter ($Kf_1 = 4.5$ GHz, $K = 3$, $d = 6.0$ mm).

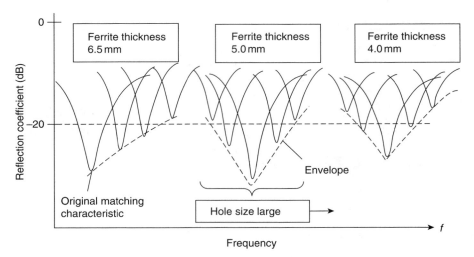

FIGURE 5.24 Matching characteristics with ferrite thickness as parameter ($f = 2.45\,\text{GHz}$, $Kf_1 = 4.5\,\text{GHz}$, $K = 3$, $\varepsilon_r = 14$).

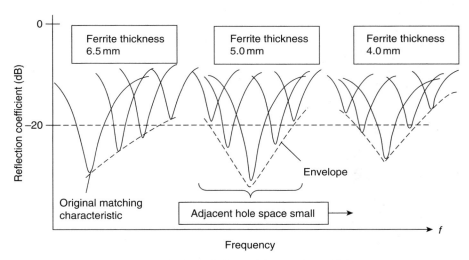

FIGURE 5.25 Matching characteristics with ferrite thickness as parameter ($f = 4.0\,\text{GHz}$, $Kf_1 = 4.5\,\text{GHz}$, $K = 3$, $\varepsilon_r = 14$).

required. Finally, the reason for the matching characteristics being changed and improved is considered based on the following principle: The equivalent values of permittivity and permeability decrease by punching small holes in the ferrite. As a result, the matching frequency rises to a higher frequency region. Figure 5.27 shows a simulation in which the value of permittivity is

FIGURE 5.26   Example of rubber ferrite absorber with multiholes using Rofin SC ×
20 laser ($d = 2.0$ mm).

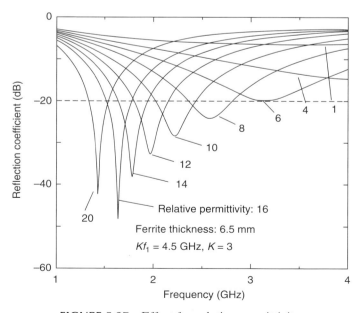

FIGURE 5.27   Effect for relative permittivity.

taken as a parameter; the values of the other parameters are given in the figure. We find that the matching frequency characteristics rise to the higher frequency regions even when only the permittivity value decreases.

5.6.2   Weakly Magnetized Ferrite Absorber

It is also possible to modify the material constant of the magnetic absorber by applying a static magnetic field H_{dc} to it [10]. When H_{dc} is applied to sintered ferrite, it is expected to reduce the matching thickness for the sintered ferrite absorber from 8 to 3 mm at the ultrahigh frequency (UHF) band under 750 g, for example. The matching frequency can also be modified significantly from 0.1 to 1.0 GHz by controlling the ferrite thickness and H_{dc} simultaneously.

As mentioned in Section 5.2, EM wave absorbers can be broadly divided into three types from the constituent material viewpoint: wave absorbers using a thin resistive sheet, dielectric material absorbers, and magnetic absorbers using such a ferrite material. Ferrite absorbers have a special feature that made it possible to achieve matching with a thinner matching thickness compared to the former one. But from the limit of *Snoek's principle*, a match for an ordinary sinter ferrite has not been obtained with a thickness below 4 mm without depending on matching frequency. Although a static magnetic field is applied to the present sintered ferrite, one should be aware that the magnetized ferrite enables this limitation to be circumvented.

Figure 5.28 is an example of matching characteristics when applying a static magnetic field H_{dc} which is due to a coil current. These characteristics were measured using a 20D coaxial waveguide with a solenoid coil surrounding the outer conductor and an iron cylinder shorting the inner and outer conductors, as shown in Figure 5.29. This shorting cylinder contributes in producing a uniform magnetic field and enforcing it like a magnetic pole. Measurement relations between coil current I and static magnetic field H_{dc} on the surface of the ferrite disk are shown in Figure 5.30. Because the ferrite plate is thin in the present investigation, the pure static magnetic field inside the ferrite seems to be very weak due to the self-demagnetizing force. In Figure 5.28, the dotted circles represent voltage standing-wave resistance (VSWR) characteristics in the absence of H_{dc} when the ferrite thickness is 6.2 mm and its central matching frequency f_0 is 0.15 GHz when H_{dc} is not applied. However, the matching frequency characteristics are able to change if H_{dc} is applied perpendicularly to the ferrite and the ferrite thickness is reduced simultaneously. As H_{dc} increases and the ferrite thickness is reduced simultaneously, the matching frequency shifts toward a higher frequency region.

Generally, with an EM wave absorber using a sintered ferrite, it has been pointed out that its matching thickness is limited, and this depends upon Snoek's limit when H_{dc} is not applied. Ordinarily, this matching thickness was 4 mm [10], but by applying a static magnetic field, empirical results suggest that it is possible to get a matching characteristic with a value smaller than the matching thickness of ferrite in the absence of H_{dc}. To put the magnetized

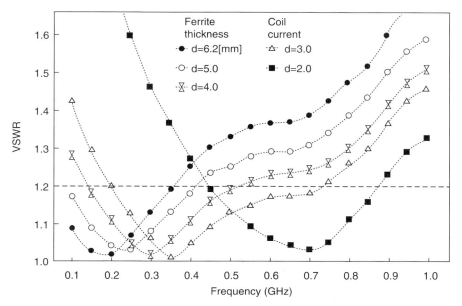

FIGURE 5.28   Matching characteristics by applying a static magnetic field H_{dc} (variable type of matching frequency).

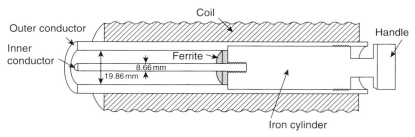

FIGURE 5.29   Coaxial waveguide used for measurement.

ferrite absorber into practical use, a method for controlling the appropriate intensity of H_{dc} is important. Concerning this problem, the following methods have been proposed. Figure 5.31 shows the method of controlling the intensity of H_{dc} applied to the ferrite. This absorber unit consists of three materials: a ferrite material, a conductive plate, and a magnet. In this construction, the value of H_{dc} that is applied to the ferrite is controlled by adjusting the thickness of the conductive material lying between the ferrite and the magnet. As the purpose of the conductive plate is to control the intensity of H_{dc}, non-magnetic material such as copper and brass can be used. This method has the

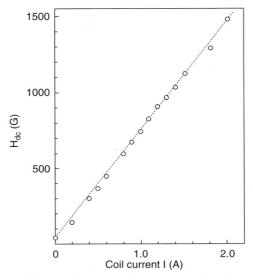

FIGURE 5.30   Relation between coil current and static magnetic field.

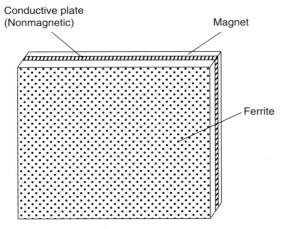

FIGURE 5.31   Method for controlling intensity of H_{dc}.

advantage that a static magnetic field which is applied by a magnet is not strictly regulated. This is because the H_{dc} field necessary for controlling the matching characteristic is obtained only by adjusting the thickness of the conductive material between the ferrite and the magnet.

5.6.3   Microwave Absorber with Surface-Printed Conductive Line Patterns

Another example of a microwave absorber based on the equivalent transformation of a material constant is a microwave absorber able to change a matching frequency characteristic and produce a thinner absorber. This has been done by printing small, periodical conductive line patterns on the surface of the absorbing materials using a single absorbing material [12]. The advantage of using conductive line patterns is that microwave absorbers can be made with characteristics capable of changing the matching frequency toward both a higher and a lower frequency region and yielding a matching characteristic of multipeaks. This idea is based on the principle that the conductive line patterns printed on the surface of absorbing material can be constructed so as to provide a behavior either of capacitance or inductance for the absorbing material, depending on wavelength.

These microwave absorbers are made by printed thin conductive lines on the surface of the rubber ferrite as shown in Figure 5.32. Figures 5.32*a*, *b* show patterns printed with a thin line lattice and cross patterns, respectively. Figures

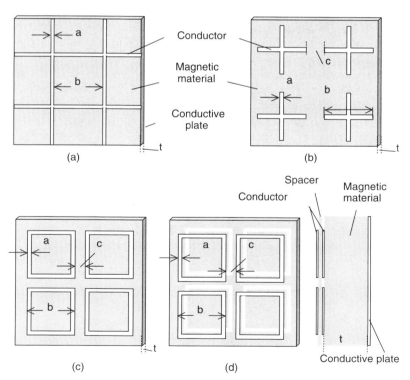

FIGURE 5.32   Examples of EM wave absorber conductive line patterns on surface: (*a*) lattice type; (*b*) cross type; (*c*) square-frame type; (*d*) double-layer type.

5.32c, d show the absorbers printed with periodical square frames and double-layer periodical square frames, respectively. The back of the rubber ferrite is attached to a conductive plate. These fundamental matching characteristics have been investigated using FDTD analysis. Matching characteristics corresponding to Figure 5.32 are shown in Figure 5.33. Figure 5.33a, b show the matching characteristics in the cases of a thin conductive line lattice and cross-line patterns, respectively. Figures 5.33c, d show the case of periodical squares and double-layer squares, respectively. The matching characteristics shown in Figure 5.33a corresponding to Figure 5.32a shift toward higher frequency regions as the conductive lattice size b is decreased. This is because these conductive patterns behave as if inductance is added to an equivalent transmission line circuit of the present absorber, which originally consists of resistance and inductance. However, the matching characteristics in the case of Figures 5.32b, c shift toward lower frequency regions as the space between the adjacent conductive patterns is decreased. This is because these conductive patterns behave as if capacitance is given at the end of the above-mentioned equivalent transmission line together with resistance and inductance. If a plate with a double-layer line pattern of squares is attached to the surface of the ferrite absorber, the matching characteristic starts to exhibit a twin-peak characteristic as shown in Figure 5.33d. Figure 5.34 shows a comparison of theoretical values with experimental values for a normal incidence in the case of periodical square patterns. This result shows the validity of the present analysis based on FDTD analysis.

To summarize the detailed matching characteristics for the present absorber, we consider the square-line patterns in Figure 5.32c as an example. When the size of conductive square b becomes large, considering the other parameters such as frame width, adjacent space, and absorber thickness as constants, the matching frequency characteristic tends to move toward a lower frequency. In the case where the frame width a is increased, keeping the other parameters constants, the matching frequency characteristic also shifts toward a lower frequency. Further, when the adjacent space c becomes narrow, keeping the other parameters as constants, the matching frequency characteristic also moves toward a lower frequency. As for the conductivity of the square frames, a good matching characteristic is obtained when the conductivity is larger than $10^4 \, \mathrm{S \, m^{-1}}$. Using these characteristics, a slim EM wave absorber can be designed. For example, when carbonyl iron is used as absorbing material in place of ferrite, a slim EM wave absorber with a thickness of 2 mm is designed at the frequency of 2.45 GHz, or the ISM band.

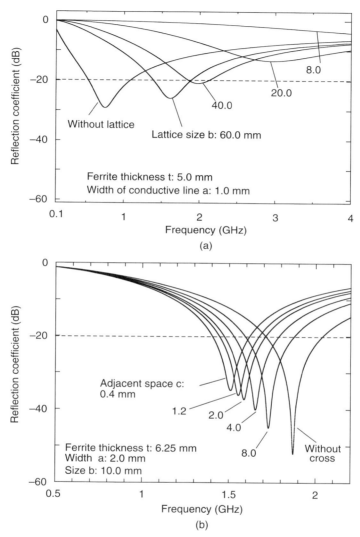

FIGURE 5.33 Matching characteristics of (*a*) lattice type; (*b*) cross type; (*c*) square-frame type; (*d*) double-layer type.

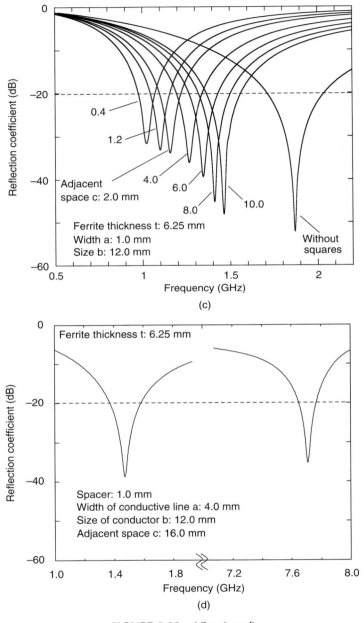

FIGURE 5.33   (*Continued*)

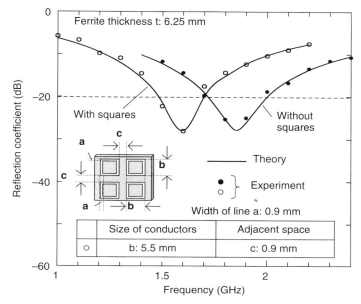

FIGURE 5.34   Comparison of theoretical values with measured values.

5.6.4   Integrated-Circuit-Type Absorber

In this section, we introduce a single-sheet broadband EM wave absorber based on a new concept of the equivalent material constant transformation method. This microwave absorber is composed of actual circuit elements of resistance, capacitance, and inductance. It covers the microwave and millimeter-wave regions. The three circuit elements are needed to finely control resonant conditions in the microwave circuit, that is, to obtain an optimum broadband matching condition. The structural concept of a microwave integrated-circuit absorber is based on the methods used for arranging three circuit constants, or resistance R, capacitance C, and inductance L, on a circuit board. The principle of constructing this absorber follows.

Figure 5.35a shows the equivalent circuit of a transmission line for the simplest EM wave absorber with space d and a conductive plate at the back of the absorber. To realize this circuit on a substrate backed with a conductive plate, the method of arranging a unit-circuit element, which consists of different conductive sections, is introduced as shown in Figure 5.35b: A unit-circuit element is composed of a high- and a low-conductive part to give resistance and inductance. The cross-shape pattern consists of two parts. One is a high-conductive part. The other is formed of low-conductive or low-resistive parts, as shown in the figure. When microwave currents flow on the surface of the high-conductive line part, this presents inductive characteristics. Since these inductance and resistance values are proportional to the length of a

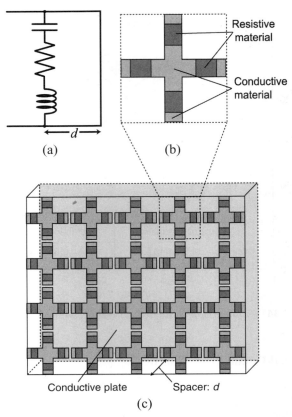

FIGURE 5.35   Configuration of EM wave absorber: (*a*) equivalent transmission line; (*b*) unit-circuit element; (*c*) total view of present absorber.

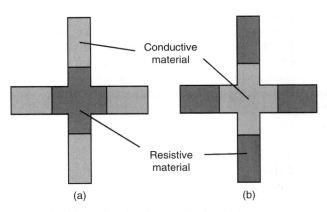

FIGURE 5.36   Various unit-circuit elements.

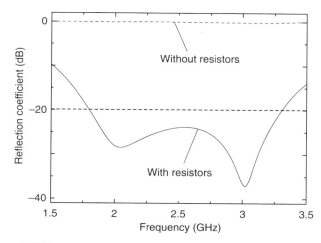

FIGURE 5.37   Experimental matching characteristics.

conductive line part, the resistance and the inductance can be modified by adjusting each length of a high- and a low-conductive part or their respective areas. Capacitance is obtained by changing the adjacent space between the tips of the unit-circuit elements in a cross shape. When a large capacitance is needed, however, the adjacent space between the tips of the unit-circuit elements becomes too narrow. To avoid this inconvenience, a unit-circuit element with a high conductivity at the tip of it is used as shown in Figure 5.36a. By introducing these structures, the variable factors for each circuit constant R, L, C can be provided.

As an example of an experimental reflection coefficient, a matching characteristic is shown in Figure 5.37. The solid line represents the matching characteristic of the present case; the resistance value is $89\,\Omega$. When the microchip resistors do not exist, that is, when the circuit is composed of only high-conductive strip lines, matching cannot be obtained, as is shown in Figure 5.37 by a dotted line. The tendency of this broadband matching characteristic agrees with the theoretical result obtained by FDTD analysis [13]. As an example at 60 GHz, the whole size b and the width a of the unit-circuit element are 2.85 and 0.6 mm, respectively (Fig. 5.38). Production of these fine unit-circuit elements becomes possible by applying integrated-circuit technology to the present absorber. This is why we call this an integrated-circuit EM wave absorber.

5.7   METHOD FOR IMPROVING RF FIELD DISTRIBUTION IN A SMALL ROOM

In the previous section, EM wave absorbers at high frequencies such as microwaves or millimeter waves were described. However, these absorbers

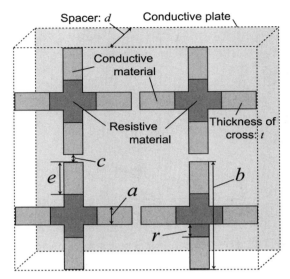

FIGURE 5.38   Construction of EM wave absorber model for analysis.

cannot be used in a relatively small room when compared to the wavelength of an RF radiator such as a medical treatment device using a high-power radiator, for instance a hyperthermia applicator at RF. Recently, not only hyperthermia but also many kinds of electronic devices have been introduced into the medical field. (See Chapter 6.) This trend has made it necessary to improve the operating environment from the viewpoint of electromagnetic compatibility (EMC). Also, the effects of radiated waves on a human body have attracted attention. It is, however, difficult to shield the environment between a patient and doctors or nurses who provide medical care simply by using conventional methods with wire gauze or other shielding materials. This is because site workers have to monitor the patient during the medical treatment. In addition, when the wavelength is larger than the room, an efficient shielding effect is not obtained by using these conventional methods. When the wavelength is larger than the room size, the field distribution is no longer that of plane-wave propagation. In this environment, the field distribution is in the form of a quasi-static field. Therefore, the idea of conventional EM wave absorbers is not available for the present case.

In this section, a method of "absorbing" or controlling the field distribution in a small room with respect to the wavelength will be given [14]. This idea is based upon the principle that an electric field concentrates in materials with a large permittivity. In particular, this method is effective in the case where radiated power slightly exceeds the value above guidelines [14]. Therefore, a countermeasure for artificially making strong and weak regions of electric fields in the room becomes necessary by adjusting the material constant and

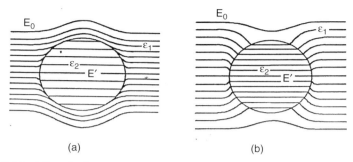

FIGURE 5.39 Electric flux density distribution on sphere with different dielectric material constants: (a) $\varepsilon_1 > \varepsilon_2$; (b) $\varepsilon_1 < \varepsilon_2$.

by changing the shape and arrangement of ceiling, wall, and so on; that is, this method is based on the principle of being able to concentrate the electrostatic field onto the surface of a material with a large permittivity value, as shown in Figure 5.39. Because the radiated wavelength is longer than the room size, the electric field in the present case is considered to be approximately an electrostatic field. Accordingly, if a region with partially weak or strong electric fields is realized, workers engaged in caring for patients may stay in the weak-field region or the safety zone. This method can effectively improve the field distribution in the room, particularly in a hyperthermia environment where the radiated fields slightly exceed the value in the guidelines.

Since it is difficult to measure and clarify the effects of controlling field distribution using an actual room, the field distribution in the room is analyzed using the *finite-element method* (FEM), taking an example of field distributions radiated from a RF capacitive hyperthermia applicator in a room that is relatively small compared to the wavelength of the RF radiator. The fundamental equation with a complex number for permittivity is expressed as

$$\frac{\partial}{\partial x}\left(\varepsilon\frac{\partial\phi}{\partial x}\right)+\frac{\partial}{\partial y}\left(\varepsilon\frac{\partial\phi}{\partial y}\right)=0 \tag{5.29}$$

where ϕ is the electric scalar potential, ω the angular frequency, σ the conductivity, and ε the complex permittivity. Figure 5.40 shows an example of the model used for the analysis. The width of the room surrounded by a concrete wall is 6 m and the height varies from 2.5 to 5 m. At the center of the room, a human model lies on a bed and is heated by a capacitive-type applicator with a pair of electrodes. The human model is assumed to consist of muscle in the present example. The bed is covered with cloth and its height is 1 m. The worker exposed to the electric field is able to take three positions at a, b, or c, respectively. A worker is also positioned at the left side between the wall and the bed. This is to compare the strength of the electric field with a worker also at the right side at the same position corresponding to the left. Mounted on the ceiling and left wall are semicircular cylinders made of ferrodielectric

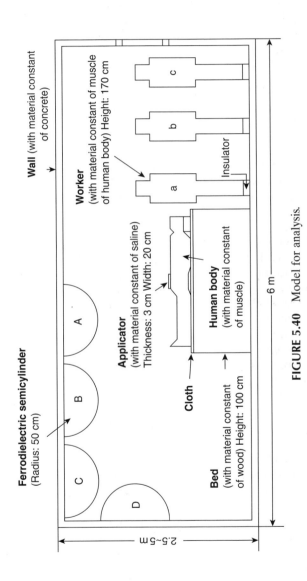

FIGURE 5.40 Model for analysis.

244

TABLE 5.5   Material Parameters

Material	Conductivity Sm^{-1}	Relative Permitivity
Human body (muscle)	0.62	81.0
Applicator (salt)	1.0×10^2	70.0
Wall (concrete)	0	5.3
Window (glass)	1.0×10^{-5}	8.5
Window frame (aluminum)	4.0×10^3	10.0
Bed (cloth)	1.0×10^{-5}	5.0
Ferroelectric material	1.0×10^{-5}	3.0×10^3–1.0×10^4

or conductive materials A, B, C, or D. The relative permittivity of the ferrodielectric material is assumed to take values from 3000 to 10,000. All these material parameters are tabulated in Table 5.5.

Figure 5.41a shows an example of visualized electric field distribution in the room when the ferrodielectric material is not mounted and a worker is standing at position b in Figure 5.40. The frequency is 13.56 MHz and the height of the ceiling is 3 m. Figure 5.41b shows the magnification of the field in the vicinity of the worker at position b. The solid line represents the vector electric field in the presence of a worker in the room. The dotted line shows the case when there is no worker in the room. This figure indicates that the direction of the vector is changed toward the worker and the electric field is concentrated on a worker when the worker is standing near the applicator.

Next, Figure 5.42a shows a visualized field distribution when ferrodielectric materials are not mounted. This field distribution is calculated from the ratio of the electric field for the cases when a worker is in the room and is not in the room. Dark color means a strong electric field. It is found that the electric field is concentrated on the worker. Figure 5.42b shows the case when the ferrodielectric materials are mounted on the ceiling and left-side wall and a worker is standing at the position b, as shown in Figure 5.40. The value of permittivity of the ferrodielectric material is 6000 and the height of the ceiling is 3 m. It is found that the electric field has a tendency concentrated on the ferrodielectric material. This field distribution with the ferrodielectric materials is compared to one without ferrodielectric materials. On the base of these investigations, it is suggested that the electric field in the room can be controlled by mounting ferrodielectric materials. This method can effectively improve the field distributions in the room, particularly in the case where the radiated field slightly exceeds the value in the guidelines.

Regarding the effect of the height of the ceiling, as the result of theoretical calculations taking the ceiling height as a parameter, the field intensity in a room has a tendency to be proportionally decreased as the ceiling height increases. Therefore, it is found that the suppressing effect of the electric field on a worker decreases as the ceiling height increases. On the basis of these investigations, it is suggested that the electric field in the room can be controlled by mounting ferrodielectric materials or conductors.

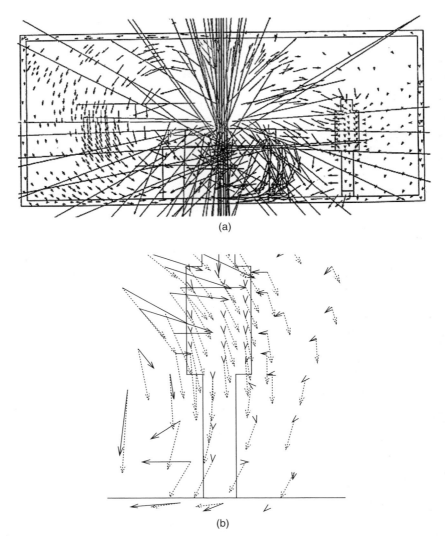

(a)

(b)

FIGURE 5.41 Examples of electric field distributions. (*a*) Field distribution without ferrodielectric materials. (*b*) Magnification of field in vicinity of worker in (*a*). (Solid and dashed lines show the cases of presence or not of worker, respectively).

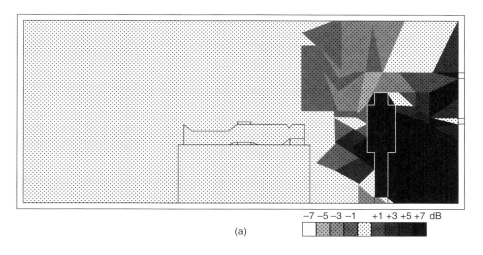

(a)

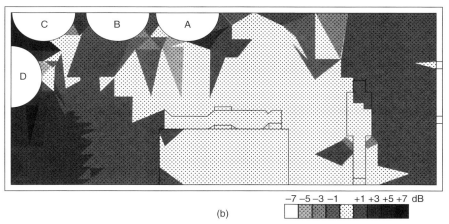

(b)

FIGURE 5.42 Visualized electric field distribution: (*a*) field distribution without ferrodielectric materials; (*b*) field distribution with ferrodielectric materials.

REFERENCES

[1] Naamlooze Vennootschap Machinerieen, French Patent 802, 728, Feb. 19, 1936.

[2] H. A. Schade, "Schornsteinfeger," U.S. Tech. Mission to Europe, Tech. Rep. 90-45 AD-47746, May 1945.

[3] O. Halpren, M. J. Johnson, Jr., "Radar, summary report of HARP project," *OSRD Div.14*, Vol.1, Pt. π, Chs. 9–12.

[4] W. H. Emerson, "Electromagnetic wave absorber and anechoic chambers through the years," *IEEE Trans. Antennas Propag.*, Vol. AP-21, No. 4, pp. 484–490, July 1973.

[5] P. Lorrain, D. Corson, *Electromagnetic Fields and Waves*, San Francisco: Freeman, 1970.

[6] S. Ramo, J. R. Whinnery, T. Van Duzer, *Fields and Waves in Communication Electronics*, New York: Wiley, 1965.

[7] A. Vander Vorst, *Transmission, Propagation et Rayonnement*, Brussels: De Boeck, 1995.

[8] E. Jordan, *Electromagnetic Waves and Radiating Systems*, Englewoods Cliff, NJ: Prentice-Hall, 1950.

[9] M. Amano, Y. Kotsuka, "A method of effective use of ferrite for microwave absorber," *IEEE Trans. Microwave Theory Tech.*, Vol. 51, No. 1, pp. 238–245, Jan. 2003.

[10] Y. Kotsuka, H. Yamazaki, "Fundamental investigation on a weakly magnetized absorber," *IEEE Trans. EMC*, Vol. 42, No. 2, pp. 116–124, May 2000.

[11] M. Amano, Y. Kotsuka, "Fundamental investigation on matching characteristics and thinned WM-wave absorber with periodical thin conductive patterns," *J. Magnetics Soc. Jpn.*, Vol. 27, pp. 583–589, 2003.

[12] M. Amano, Y. Kotsuka, "Matching characteristics of EM-wave absorber with metallic patterns on the front and back surfaces," *Trans. IEICE,* B, Vol. J86, No. 7, July 2003.

[13] Y. Kotsuka, M. Amano, "Broadband EM-wave absorber based on integrated circuit concept," *IEEE MTT-S Int. Microwave Symp. Dig.*, Vol. 2, pp. 1263–1266, June 2003.

[14] Y. Kotsuka, T. Tanaka, "Method of improving EM field distribution in a small room with an RF radiator," *IEEE Trans. EMC*, Vol. 41, No. 1, Feb. 1999.

PROBLEMS

5.1. Calculate the reflection coefficient of a single-layer magnetic absorber when the absorber thickness is 6.0 mm and the relative permittivity is 12.0 and the relative complex permeability $\mu_r = 2.08 - j1.44$ in the case of a normal incident wave at a frequency of 2.0 GHz (Fig. P5.1).

5.2. Calculate the reflection coefficient at a frequency of 3.0 GHz of a single-layer magnetic absorber when the absorber thickness is 2.0 mm and the relative permittivity is 20.0 and the relative complex permeability $\mu_r = 4.0 - j3.0$.

5.3. Calculate the reflection coefficient of a two-layer absorber at 2.0 GHz which is composed of a dielectric material with a thickness of 1.0 mm on the surface of a magnetic material with a thickness of 6 mm in the case of a normal incident wave and when the relative permittivity of the dielectric material is 6.0, the relative permeability of the magnetic material $\mu_r = 2.08 - j1.44$, and its relative permittivity is 12.0 (Fig. P5.3).

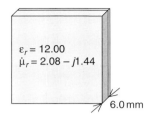

FIGURE P5.1

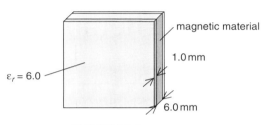

FIGURE P5.3

5.4. Calculate the reflection coefficient of a two-layer absorber at 3.0 GHz, when dielectric material is of thickness 0.5 mm and is on the surface of a magnetic absorber of thickness 2.0 mm when the relative permittivity of the dielectric material constant is 6.0 and the relative permeability of the magnetic absorber $\mu_r = 4.0 - j3.0$ with a relative permittivity of 20.

5.5. Calculate the reflection coefficient at 3.0 GHz of a two-layer absorber that is composed of two magnetic materials with a thickness of 1.0 mm on the surface of a magnetic material with a thickness of 2 mm in the case of a normal incident wave. The magnetic material with the thickness of 1.0 mm has a relative permeability $\mu_r = 1.6 - j1.20$ and a relative permittivity of 12.0, while the magnetic material with the thickness of 2.0 mm has a relative permeability $\mu_r = 4.0 - j3.0$ and a relative permittivity of 20.0.

CHAPTER SIX

RF/Microwave Delivery Systems for Therapeutic Applications

6.1 INTRODUCTION

We define RF as a frequency range between a few kilohertz and a few tenths of megahertz and microwaves as a frequency range between hundreds of megahertz and about 30 GHz.

Many papers have been published in the last 15 years dealing with the RF/microwave delivery systems in use for various medical applications. Some of the applications are as follows [1]:

1. RF/microwave in cardiology
2. RF/microwave to treat benign prostatic hyperplasia (BPH)
3. RF/microwave alone and in combination with radiation or chemotherapy in cancer treatment
4. microwave for endometrial ablation
5. microwave-assisted liposuction
6. microwave balloon technology
7. RF for the treatment of gastroesophageal reflux disease (GERD)
8. RF for pain management.

In addition, we outline in this book three ideas about possible future research and continued development with the purpose of pointing the reader

RF/Microwave Interaction with Biological Tissues, By André Vander Vorst, Arye Rosen, and Youji Kotsuka
Copyright © 2006 by John Wiley & Sons, Inc.

in the direction of utilizing RF/microwaves in medicine. These three new areas are (1) endoscopic light source and microwaves for *photodynamic therapy* (PDT) [2, 3], (2) microwave-assisted *anastomosis* [4], and (3) *thermally molded stent* for cardiology, urology, and other medical and veterinary applications [1].

Although most of the above applications are currently approved for use on humans across the world, some of the early design issues still persist, namely the design of a microwave antenna in the near field with changing *coupling impedance*; that is, the antenna inserted into tissue for ablation purposes will see a change from normal tissue to scar tissue, for example. In addition, the importance of choice of the correct coaxial cable for a specific microwave application should not be overlooked. Cable heating along the transmission line could cause thermal injury to healthy tissue outside the intended treatment area. Simplified rules for the design of coaxial cables followed by measured results of a very small low-loss coaxial cable that can be inserted into arteries and veins will be discussed.

6.2 TRANSMISSION LINES AND WAVEGUIDES FOR MEDICAL APPLICATIONS

6.2.1 Coaxial Cable

The history and the theory of coaxial cable development are for the most part trivial to the electrical engineering community. When the cable is introduced into a biological vessel, however, its complexity changes. Because of characteristics such as very low loss, flexibility, heat capacitance, and material compatibility with tissue, the coaxial cable is a key component in the microwave delivery system and requires special attention.

Cable Specifications   At high frequencies, a two-wire transmission line might couple part of the energy being delivered to the surrounding tissues within its path. The EM energy, however, is completely confined within the coaxial cable. Thus, a coaxial cable is the high-frequency equivalent of a two-wire transmission line necessary to deliver electrical energy from a generator to a load while confining the microwave field within a specifically delineated structure.

In microwave balloon angioplasty (MBA) [1], the role of the coaxial cable is to deliver the microwave power from a generator to an interstitial antenna inserted into an angioplasty balloon catheter that has been positioned in an artery, following whatever tortuous path as necessary to reach the treatment site. The small size inherently leads to high loss of the microwave power being transmitted down the length of the cable. The lost microwave power is converted to heat, which raises the temperature of the cable, the catheter, and, to an undetermined extent, the artery and surrounding tissue. Less power is available for radiation from the antenna into the plaque, arterial walls, or vass at the treatment site. Another important electrical parameter in a coaxial cable is the characteristic impedance, which is determined by the dimensions and the dielectric material. Imperfections in the cable or connectors, or antenna mismatch, produce power reflections that reduces the radiated power as well as introduce hot spots in the cable itself.

Design Consideration   The first consideration in the design of a coaxial transmission line is the maximum allowable outer conductor diameter, which, for MBA application, is 0.022 in. or less. Next, the dielectric material must be chosen with consideration of its dielectric loss and physical properties. The optimum choice (lowest loss) dielectric is air; however, foam dielectric materials are available that approach the dielectric constant of air. The center conductor's diameter is then adjusted to achieve the desired impedance characteristics. An optimum geometry exists for achieving minimum attenuation in a coaxial transmission line. The *characteristic impedance* of the cable is directly related to the geometry of the conductor's diameter.

Power Loss   Power loss in a coaxial cable is related to the attenuation constant α. This constant has two components: α_c, due to conductor loss, and α_d, due to dielectric losses, where $\alpha = \alpha_c + \alpha_d$. For those low-loss dielectrics used in microwave cables, the dielectric loss is negligible (<2%) compared to the conductor losses. For copper conductors surrounded by an insulator having a dielectric constant ε, the losses in decibels per 100 ft or in decibels per meter can be determined from

$$\alpha_c = \frac{0.214\sqrt{F}}{Z_0} \frac{1}{D}\left(1 + \frac{D}{d}\right) \qquad \text{dB/100 ft}$$

or   (6.1)

$$\alpha_c = \frac{0.023\sqrt{F}}{Z_0} \frac{1}{D}\left(1 + \frac{D}{d}\right) \qquad \text{dB/m}$$

where F is the frequency in megahertz, D is in inches or meters, and

$$\alpha_d = 2.77 F\sqrt{\varepsilon}\tan\delta \qquad \text{dB/100 ft}$$
$$\alpha_d = 0.0909 F\sqrt{\varepsilon}\tan\delta \qquad \text{dB/m} \qquad (6.2)$$

where D is the inner diameter (i.d.) of the outer conductor, d is the outer diameter (o.d.) of the inner conductor in inches, and $\tan\delta$ is the dielectric loss factor.

If the outer diameter is held constant because of mechanical size constraints, as is the case considered here, it can be determined from the above expressions that [5].

1. The loss has a broad minimum at a *D/d* ratio of 3.59 for any dielectric at any frequency.

TABLE 6.1   Cable Losses

ε	α_c		α_d		Z_0	D/d	α_{total}		Dielectric
	dB/ft	dB/m	dB/ft	dB/m			dB/ft	dB/m	
1.0	0.253	0.8300525	0.068	0.2230971	77	3.59	0.321	1.0531496	Air
1.36	0.29	0.9514436	0.079	0.2591864	66	3.59	0.373	1.2237533	Air-filled Teflon
2.1	0.367	1.2040682	0.0985	0.3231627	53	3.59	0.46	1.5091864	Teflon
1.0	0.279	0.9153543	0.068	0.2230971	50	2.3	0.347	1.1384514	Air
1.36	0.308	1.0104987	0.079	0.2591864	50	2.64	0.387	1.269685	Air-filled Teflon
2.1	0.368	1.2073491	0.0985	0.3231627	50	3.35	0.468	1.5354331	Teflon

2. For a constant impedance Z_0, the loss decreases as ε approaches unity while the diameter of the inner conductor increases to maintain the constant impedance.

These effects are shown in Table 6.1 for various cables having an outer conductor of 0.025 in. with various dielectric constant materials (typical loss factor of 0.0001) operating at 2450 MHz.

The following is an example of a cable–antenna assembly design and testing. The assembly consists of lengths of miniature coaxial cable terminated in a gap, whip, or helical antenna (Fig. 6.1) in order to radiate microwave power to treated tissue during balloon angioplasty experiments. The gap is approximately 0.025 in. wide and is located approximately 0.31 in. from the short-circuited end of the cable. Two cable sizes have been used for most of these assemblies. The basic characteristics of these two commercially available semi-rigid coaxial cables are given in Table 6.2.

A typical test procedure for the cable–antenna assembly included the following steps. After assembly, the cable and antenna were first checked for continuity and leakage. Recording the total DC resistance with the gap shorted (short circuited) is also recommended as a simple check on the cable–antenna potential microwave loss. The cable–antenna assembly was tested under high-power conditions at 2450 MHz to verify its ability to handle the power without breakdown or excessive heating. The cables were surrounded by a test catheter and a simulated arterial environment during this test. For example, the cables were wrapped in wet sponges over their full lengths during the tests. The gap antennas were inserted into a plastic pipette containing deionized water. The pipette was wrapped in a saline-soaked sponge for RF loading purposes. The temperature at the outer pipette surface was monitored as evidence of the RF heating from radiated power. The wetness and contact of the sponges varied considerably, causing the temperature rise to be only qualitative. All units were subjected to an input of 30 W for 30 s, monitored for excessive reflected power, and inspected for damage after removal. The results are presented in

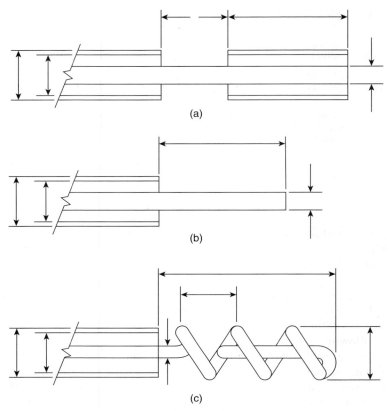

FIGURE 6.1   Antenna configuration: (*a*) gap antenna; (*b*) whip antenna; (*c*) helical antenna.

TABLE 6.2   Basic Characteristics of Semirigid Coaxial Cables By Two Manufacturers

Specifications	Micro-Coax	Precision Tube
Type no.	UT 20-M	CA50020
Outer conductor diameter	0.023 in. (0.584 mm)	0.020 in. (0.508 mm)
Dielectric diameter	0.015 in. (0.381 mm)	0.015 in. (0.381 mm)
Center conductor diameter	0.0045 in. (0.1143 mm)	0.0044 in. (0.112 mm)
Attenuation at 2.4 GHz	1.13 dB/ft (3.71 dB/m)	1.33 dB/ft (4.36 dB/m)
Rated power at 2.4 GHz	12.5 W at 20°C	1.8 W at 40°C
Comments	Ultra malleable, stiff; dielectric core is a loose fit-can slip	Weak, easily kinked; push with extreme caution

TABLE 6.3   High-Power Test at 2450 MHz

Cable Number	Outer Diameter in.	Outer Diameter mm	Total Length in.	Total Length m	Power (W) Incident	Power (W) Reflected	Time (s)	Temperature Rise (°C) at End of 30-s Period
A1	0.020	0.508	45	1.143	30	0.2	30	11.5
A2	0.020	0.508	44	1.118	30	0.2	30	17.1
A3	0.020	0.508	48	1.219	30	0.3	30	6.5
B1	0.023	0.584	48	1.219	30	0.9	30	13.3
B2	0.023	0.584	48	1.219	30	0.5	30	6.4

Table 6.3, which also lists the temperature rises measured during one set of postassembly tests. The discrepancy in temperature rise measured in cables B1 and B2 is probably due to the pressure exerted on the phantom.

Low-Loss Fully Flexible Coaxial Cable   Material other than copper can be utilized in medical applications if the skin effect phenomenon in microwave frequencies is understood. A coaxial cable having a silver-coated stainless steel center conductor was utilized in our flexible cable–antenna delivery system.

Skin Effect   In the design of a coaxial cable at microwave frequencies, the guided-wave portion requires special attention. In a coaxial cable, the inner surface of its outer conductor and the outer surface of its inner conductor provide the boundaries for the propagation wave. There could be a sharp transition in the properties of the medium at the surface of the boundaries, specifically, a nonconductive medium, followed by a highly conductive surface, such as the center conductor of a coaxial wire that could be hollowed or made from other material such as silver-coated stainless steel. The metals used in practical guides have finite conductivity; therefore, dynamic fields are propagated to a thickness δ into the guide walls (skin depth). A simple case is that of a TEM wave propagating in a highly conducting medium such as copper and silver, where

$$\delta = \frac{1}{\sqrt{\pi F \mu \sigma}} \quad \text{m} \tag{6.3}$$

where F is the frequency and μ and σ are the permeability and the conductivity of the medium, respectively. At depth δ, the electric field intensity has been attenuated to 37% of the initial value (i.e., its value at the surface of the metal, and the power transported has been attenuated to 13.6% of its initial value). (See Section 1.5.) For all practical purposes, waves can be said to be completely attenuated at depths equal to 5δ. From Eqn. (6.3), it is clear that a minimum layer of 5δ of a highly conductive material is needed to cover the outside of the inner conductor and the inside of the outer conductor for a

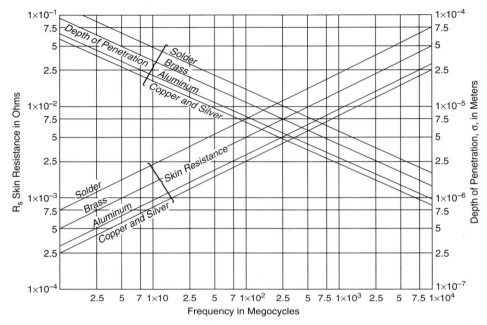

FIGURE 6.2   Skin resistance and depth of penetration for several metals [6].

coaxial cable. Depth of penetration δ (in meters) and skin resistance (in ohms) are plotted in Figure 6.2.

Coaxial Cable for Microwave Balloon Angioplasty   Based on the information above, a microwave cable having the following clinical device specifications was developed:

1. *Device requirements*
   A. Electrical
      i. Attenuation at 2.45 GHz, 6 dB maximum (1.0 dB/ft, or 3.28 dB/m)
      ii. Rated power at 2.45 GHz, 80 W
      iii. Distal tip (antenna)
   B. Mechanical: must be flexible enough to track through catheter around the aortic arch and across the lesion
   C. Functional: must be possible to locate proximal end of antenna under fluoroscopy to line up with thermocouple
2. *Device description*: coaxial cable with whip or gap antenna
   A. Center conductor
      i. Material: Ag-coated stainless steel (SS)

      ii. Dimensions

         a. SS diameter 0.005 ± 0.0003 in.

         b. Coated wire diameter 0.006 in. maximum

         c. Coating thickness 0.0003 ± 0.0001 in.

         d. Length 70.0 in. maximum

  B. Dielectric

      i. Material: Goretex expanded PTFE (polytetraflouroethylene)

      ii. Dimensions: diameter 0.017 in. maximum wrapped (0.015 in. under compression of outer conductor)

  C. Outer conductor

      i. Material

         a. Option 1: Ag-coated copper-rolled flat wire

         b. Option 2: Ag-coated SS-rolled flat wire

      ii. Dimensions

         a. Width 0.007 ± 0.002 in.

         b. Thickness 0.0015 ± 0.0001 in.

         c. Ag thickness 0.0003 ± 0.0001 in.

         d. Wrapped diameter 0.022 in. maximum

  D. Connector

      i. Material: SS/copper

      ii. Type: female SMA (subminiature version A) per MIL-C-39012

  E. Coating

      i. Must withstand 100°C

      ii. Thickness 0.002 in. maximum

3. *Device design*

  A. Dielectric extruded over center conductor

  B. Two layers of Ag-coated SS ribbon either overlap wound over dielectric to form outer conductor or wound clockwise and counterclockwise

  C. SS hypotube slid over proximal end of outer conductor coils for support and soldered to coils

  D. Connector attached to proximal end

  E. Whip antenna formed at distal end by removing outer conductor coils

  F. Distal end of cable to be coated with an insulator

  G. Proximal end of antenna to be radio-opaque

4. *Biocompatibility and sterility*

  A. Material testing

      i. All material to pass USP (United States Pharmacopeia) biological tests for class VI plastics

         a. Sensitization assay (does not promote allergic reaction)

    b. Cytotoxicity test

    c. Thrombogenicity test

    d. Hemolysis (breaking red cells) test

    e. Genotoxicity test

  ii. Material verification

  iii. Heavy metals

  iv. Leachables

B. Sterility

  i. EtO residuals outgas time

  ii. Sterile cycle

  iii. Pyrogen-pyrogen inhibition: detects bacterial endotoxins

  iv. Bioburden and spore recovery: natural microbial population of product (quarterly monitoring program)

C. Shelf life: 6 months from date of manufacture

6.2.2 Circular Waveguide [7–9]

Fundamentals A catheter, that is, a cylindrical tube made of plastic (Fig. 6.3a), can become a circular waveguide when the inside of the wall is metallized. When filled with a material with ε close to that of deionized water having dielectric constant of 80, this metallized tube becomes a loaded, thus reduced waveguide. Although not utilized in our experiments (but used by others), for the sake of completeness it is worthwhile analyzing the capabilities of the *metallized catheter* that becomes a transmission medium for EM energy.

The dominant mode in a circular waveguide is the TE_{11} mode. This mode is shown in Figure 6.3b, along with a number of other possible modes of transmission. The field equations are expressed by using the cylindrical coordinate system for the TE_{11} wave (dominant mode).

In the following equations, B is an arbitrary constant determining amplitude, $J_1(\cdot)$ is the Bessel function of the first order, ω is the circular oscillation frequency of an EM field, c is the speed of light, and a is the radius of a circular tube:

$$H_z = BJ_1\left(u'_{1,1}\frac{r}{a}\right)\cos\theta\cos(\omega t - \beta z)$$

$$H_r = B\frac{\beta a}{u'_{1,1}}J_1\left(u'_{1,1}\frac{r}{a}\right)\cos\theta\sin(\omega t - \beta z)$$

$$H_\theta = B\frac{\beta a^2}{\left(u'_{1,1}\right)^2 r}J_1\left(u'_{1,1}\frac{r}{a}\right)\sin\theta\sin(\omega t - \beta z) \tag{6.4}$$

$$E_z = 0 \qquad E_r = \mu\frac{\omega}{\beta}H_\theta \qquad E_\theta = -\mu\frac{\omega}{\beta}H_r \tag{6.5}$$

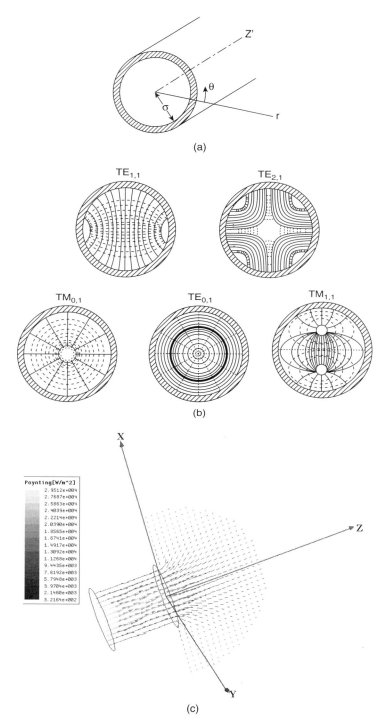

FIGURE 6.3 (*a*) Circular waveguide [7]. (*b*) Field configurations of various modes in a circular waveguide. Solid lines indicate electric field, broken lines magnetic field [7]. (*c*) Simulation of power density distribution (pointing vector) of a truncated waveguide.

$$u'_{1,1} = 1.841 \qquad u_1 = (u'_{1,1})^2 = 3.39 \qquad \beta^2 = \left(\frac{\omega}{c}\right)^2 \varepsilon \left(\frac{u'_{1,1}}{a}\right)^2 \tag{6.6}$$

For water, $\varepsilon = 80$; for air, $\varepsilon = 1$. In the previous equations, the subscript z denotes field components in axial direction, and the transverse field components are decomposed along the two orthogonal directions, r and t_s, which represent the radial and tangential directions, respectively.

The notation applied to a circular waveguide is shown in Figure 6.3a. For any mode of transmission in a circular waveguide, there may be axial field components (z), and there are the transverse field components, which may be resolved into two directions, tangential (θ) and radial (r). All of these components vary periodically along a circular path concentric with the wall and vary in a manner related to a Bessel function of order m along a radius. Any particular mode is identified by the notation $TE_{m,n}$ or $TM_{m,n}$, where m is the order of the Bessel function representing the field variation and n represents the nth zero of $J_n(\cdot)$ (for TM) and $J'_n(\cdot)$ (for TE). The cutoff wavelength in a circular waveguide for all modes depends upon the ratio of the diameter to the wavelength. For the $TE_{m,n}$ wave, the cutoff wavelength is given by the formula

$$\lambda_c = \frac{2\pi a}{u'_{m,n}} \tag{6.7}$$

where a is the radius of the guide. The constant $u'_{m,n}$ is the nth root of the equation $J'_m(u) = 0$. Some of the lower values of $u'_{m,n}$ are

$$u'_{0,1} = 3.832 \qquad u'_{0,2} = 7.016$$
$$u'_{1,1} = 1.841 \qquad u'_{1,2} = 5.332$$
$$u'_{2,1} = 3.054 \qquad u'_{2,2} = 6.706$$
$$u'_{3,1} = 4.201 \qquad u'_{3,2} = 8.016$$

For the mode TE_{11}

$$\lambda_c = \frac{2\pi a}{1.841} = 3.41a \qquad \lambda_g = \frac{\lambda}{\sqrt{1-(\lambda/\lambda_c)^2}} \tag{6.8}$$

where λ_g is the wavelength of an air-filled hollow pipe and λ is the wavelength of a plane wave in a dielectric medium with $\varepsilon \approx 80$.

Similarly, the cutoff wavelength of a $TM_{m,n}$ wave is given by

$$\lambda_c = \frac{2\pi a}{u_{m,n}} \tag{6.9}$$

Some of the lower values of $u_{m,n}$ are given by

$$u_{0,1} = 2.405 \qquad u_{0,2} = 5.520$$
$$u_{1,1} = 3.832 \qquad u_{1,2} = 7.016$$
$$u_{1,2} = 5.135 \qquad u_{3,2} = 8.417$$

For 3–5 GHz, the tube radius (a) is generally in the range of 3–5 mm. A 3-mm-size tube filled with high dielectric constant material can be utilized in many medical applications, including MBA, and when truncated will radiate energy into the tissue at 3–5 GHz.

Simulation of the power density distribution (pointing vector) of a truncated circular waveguide is shown in Figure 6.3c.

It is worthwhile mentioning that in a flexible circular waveguide (flexible catheter) the waveguide can introduce mode conversion/reconversion as a consequence of bending or shape transformation (circular waveguide to elliptical waveguide), which in turn increases the waveguide's loss. In our particular application, this is not critical.

Power Capacity of a Circular Waveguide   For the $TE_{1,1}$ mode, the relationship between the power capacity (P) and the maximum allowable field strength E_{max} is

$$\frac{P}{E_{max}^2} = 1.99 \times 10^{-3} a^2 \left(\frac{\lambda}{\lambda_g} \right) \tag{6.10}$$

where $E_{max} \approx 30,000$ V/cm in an air-filled guide under standard sea-level conditions of temperature, pressure, and humidity and between 3–5 times higher in a deionized-water ($\varepsilon_2 \approx 80$) filled guide. In standard applications, the maximum power that the guide will safely carry is a fraction of the theoretical maximum.

6.3   ANTENNAS

6.3.1   Fundamentals

To effectively deliver microwave energy to heat myocardial tissue and create histological changes, it is essential to control the size and location of the radiated field, thus controlling the affected tissue volume. In our studies, three types of microwave antennas at the tip of a coaxial cable were investigated: (1) the whip antenna (Fig. 6.1a), (2) the gap antenna (Fig. 6.1b), and (3) the helical antenna (Fig. 6.1c). Temperature distributions along the tissue cylinder in the direction of the antenna length for whip and gap antennas were measured to have a Gaussian shape with peaks adjacent to the junction in the whip antenna, and a fall of temperature rise, and therefore tissue damage, was observed near the tip of the antennas. These two types of antennas were ideal for the treatment of most cases of atherosclerotic arteries and were utilized through all of the MBA studies. The helical antenna—that has some end-fire characteristics, if designed correctly, thus facilitating heating next to the antenna tip—was not used for MBA since the very small coil size (0.5 mm diameter) needed for coronary angioplasty failed to show any disadvantages in preliminary testing in vitro and is so much more difficult to produce.

6.3.2   Antenna Configurations

This section is intended to provide a short summary of the general types of antennas used in our MBA experiments. Two major types of transmission/antennas systems are considered:

1. Insulated dipole and monopole (coaxial type of antenna).
2. Truncated circular waveguide (that could be loaded with a high-dielectric-constant material). This option was not utilized in our experiments, but the theory of circular waveguides was summarized in Section 6.2.2.

Electric Dipole   The antenna equivalent to an electric dipole (Fig. 6.4) consists of thin wires terminated in small spheres [7]. If energized by a harmonic source across the gap, the charge on the spheres is given by

$$q = q_0 e^{j\omega t} \tag{6.11}$$

The magnitude of the dipole moment of the antenna is

$$P = q_0 l e^{j\omega t} = p_0 e^{j\omega t} \tag{6.12}$$

where l is the length of the dipole and p_0 is the static dipole moment. Since $l \ll \lambda$, the current at any instant may be taken to be the same at all points along the antenna. The EM field set up by a dipole in spherical coordinates is given by

$$E_r = \frac{1}{2\pi\varepsilon}\left(\frac{1}{r^3} + \frac{jk}{r^2}\right)\cos\theta p_0 e^{j(\omega t - kr)}$$

$$E_\theta = \frac{1}{4\pi\varepsilon}\left(\frac{1}{r^3} + \frac{jk}{r^2} - \frac{k^2}{r}\right)\sin\theta p_0 e^{j(\omega t - kr)}$$

$$H_\phi = \frac{j\omega}{4\pi}\left(\frac{1}{r^2} + \frac{jk}{r}\right)\sin\theta p_0 e^{j(\omega t - kr)} \tag{6.13}$$

FIGURE 6.4   Dipole antenna [7].

where k is the *wave number*, with $k = 2\pi/\lambda$. Also,

$$E_\phi = 0 \quad \text{and} \quad H_r = H_\theta = 0 \tag{6.14}$$

The static field varies inversely with r^3, the induction field varies inversely with r^2, and the radiation *field varies* inversely with r. At a distance $r > 1/k = \lambda/2\pi$ the radiation field becomes the leading term, and at a larger distance, the static and the induction fields are negligible relative to the radiation field. However, it is only at distances much greater than $r = \lambda/2\pi$ that one can neglect the static and induction fields.

The *dipole impedance* consists of a resistive component and a reactive component in which the resistive component corresponds to the power delivered by the dipole. The power delivered to the dipole in turn divides into the ohmic losses in the conductors and the radiated power. In an ideal case (no ohmic loss), the resistive component of the impedance is its radiation resistance. The reactive component corresponds to energy stored in the antenna near-field region.

If P is the average power radiated per unit time and R_r is the *radiation resistance*,

$$\overline{P} = \tfrac{1}{2}|I_0|^2 R_r \tag{6.15}$$

The radiation resistance is given as

$$R_r = \frac{k^2 l^2}{6\pi}\left(\frac{\mu}{\varepsilon}\right)^{1/2} \tag{6.16}$$

If we assume the dipole length to be a small fraction of λ, R_r in turn will be small, and this resistance will produce a low *efficiency radiation*. This is why an effective antenna length should be an appreciable fraction of a wavelength.

In summary, Figure 6.4 shows an elementary length l of wire carrying a current I that is assumed to be of the same amplitude and phase through the element. The most commonly used assumption is to consider the source of radiation to be an isotropic point source, modified in directivity by a gain factor, from which emanates an ever-increasing spherical wavefront that can be considered to be an equiphase plane wave in the far-field region. In the case of an interstitial antenna, intended to heat only the most proximal tissue, these assumptions are not valid. The much more complicated near-field and intermediate-field regions must be considered.

The *near-field region*, where a different set of simplifying assumptions can be used, is considered to be within radial distances of less than 0.01λ; the *intermediate field*, where simplifying assumptions are not used, extends from the near-field limit to approximately 5λ. A further complication is that the wavelength λ is a function of the square root of the dielectric constant of the medium in which the antenna is immersed. The effective dielectric constant of water and muscle tissue modified by the plastic of the catheter will be on the

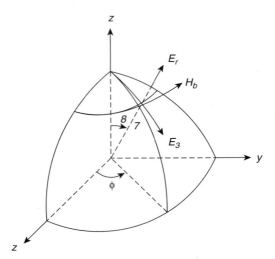

FIGURE 6.5 Isotropic spherical waves, spherical coordinates [7].

order of 64, resulting in a wavelength of approximately 15 mm at 2450 MHz. The complicated intermediate region will, therefore, extend from 0.15 to 75 mm, encompassing the region of interest for this analysis that extends from the surface of the angioplasty balloon radially outward into the muscle tissue for a distance of several millimeters—typically from 1.5 to 6.0 mm (0.1λ to 0.4λ). The reason the intermediate region is difficult to analyze is the need for considering both radial and tangential electric fields, shown by the electric dipole field in Figure 6.5. These two fields are comparable in magnitude in the range of 0.1λ to 0.4λ, and the contributions of both vector fields must be spatially combined. The electric field density as a function of kr is shown in Figure 6.6. Figure 6.7 depicts the intermediate field as a function of ωt.

6.4 RF AND MICROWAVE ABLATION [1, 10–79]

6.4.1 Fundamentals

The benefits of using heat in therapeutic medicine have been well recognized for 150 years. Busch [10] and by Coley [10, 11] described tumor regression after a patient developed a high fever. Since the beginning of the twentieth century, physicians have treated various types of tumors using heat alone (heat was found to kill tumor cells more aggressively than normal cells) and later utilizing a combination of X-rays and systemic heat. Also used for the treatment of cancer was the combination of heat and chemotherapy. The effects of heat and the duration of heat delivery were both found to be critical. Cellular toxicity was increased by either increasing the temperature or increasing

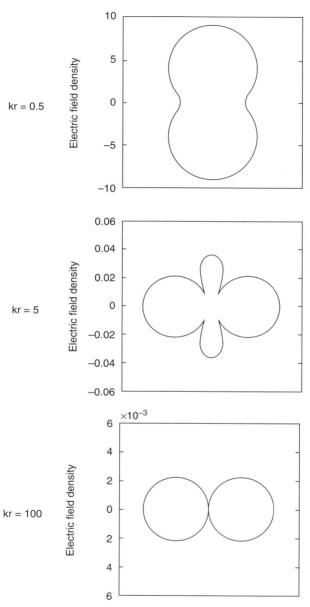

FIGURE 6.6   Electric field density as a function of *kr*.

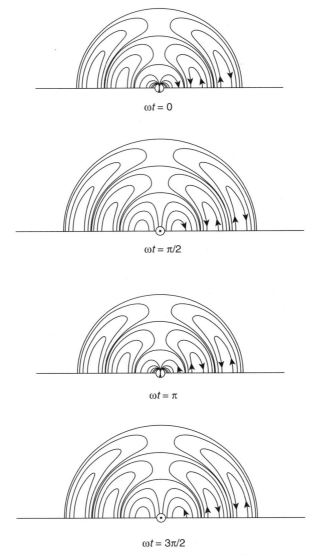

$\omega t = 0$

$\omega t = \pi/2$

$\omega t = \pi$

$\omega t = 3\pi/2$

FIGURE 6.7   Intermediate far field [7].

the duration of heat delivery. In the last 15 years, the use of RF/microwaves in medical diagnostics and therapy has increased sharply in parallel to the dramatic increase in the use of RF/microwaves in communications.

In the field of cardiology, RF/microwaves are used to treat various types of arrhythmias. By selective tissue ablation, either the source or the pathways responsible for the abnormal cardiac rhythm can be eliminated. In urology, similar systems are used to treat BPH, a condition resulting from overgrowth of the prostate leading to obstruction of urine flow. In the last few years, several otolaryngological centers around the world have been utilizing RF to treat upper airway obstruction and alleviate sleep apnea. Although the treatment of cancer by means of RF and microwaves has met with various degrees of success and is well documented, many oncological centers are not yet using these modalities. However, renewed interest in high-temperature localized heating for the treatment of liver, kidney, and breast cancers has led to results showing a prolongation and improvement in the quality of life.

In the case of renal cancer, an RF *needle ablation* device is being utilized to treat patients with kidney tumors 0.5–3.0 cm in size. Such RF needles are inserted directly into the tumor and are heated to approximately 100°C for 10 min. In the case of RF treatment for hepatic neoplasm (liver cancer), a multitude of needles having thermistors at their tips for temperature control are being utilized. Research is currently being conducted on the utilization of microwave energy for the treatment of bladder cancer by means of transurethral microwave needle ablation (TUMNA), a procedure that utilizes a simple device under local, or lower lumbar, anesthesia. The patient is examined by cystoscopy to confirm the position and tumor size. The needle antenna is advanced into the bladder, then inserted into the tumor. Power as high as 100 W, depending on the tumor size, is delivered for about 60–90 s.

Radio frequencies at around 500 kHz are also used in the treatment of other medical conditions such as GERD and for the tightening of damaged ligaments. It is important to mention that new therapeutic applications of RF/microwaves are continuously being developed. Microwave-assisted liposuction, for example, is a technology that has undergone successful in vivo animal testing and is awaiting human trials (Fig. 6.8*a*).

6.4.2 RF Development

The utilization of RF in medicine dates back to the 1920s, when research was conducted by W. T. Bovie, who pioneered the use of RF in surgery for both cutting and coagulating [73]. In the 1950s, S. Arino and B. J. Covos were the first to utilize and commercialize an RF system for use in neurosurgery [74, 75]. The RF generator is a source of RF voltage that is applied between two or more electrodes connected to a tissue to be ablated. The heating distribution is a function of the RF current density, where the greatest heat is generated in the region of the highest current density. The lesion size is a function of the size of the electrodes and the resulting electrode–tissue interface.

Since the publication of *New Frontiers in Medical Device Technology* [1] and of Chapter 2.14 in *RF and Microwave Handbook* [78], other RF applications of about 500 KHz have been developed. However, instrumentation such as the RF generator has remained essentially unchanged in most applications (Fig. 6.8*b*), although some of the improvements are important. Some of these

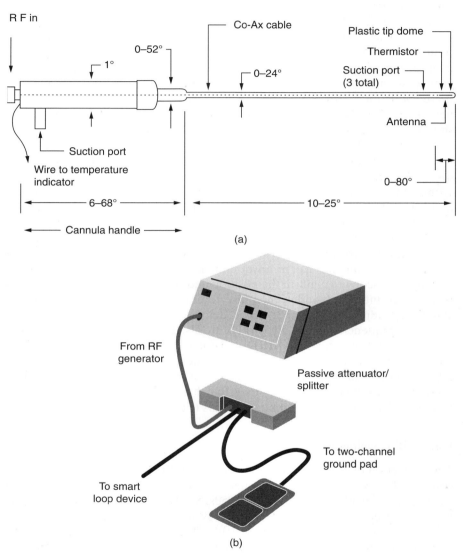

FIGURE 6.8   (*a*) Microwave-aided liposuction cannula. (*b*) RF ablation system.

applications are for liver cancer (utilizing a number of RF probes inserted into the solid tumor), breast cancer, fecal incontinence, pain relief, and trigeminal neuralgia (a condition that manifests itself as severe pain as a consequence of injury to the trigeminal nerve in the face and head). The RF procedures now

Coaxial cable terminated with gap type antenna

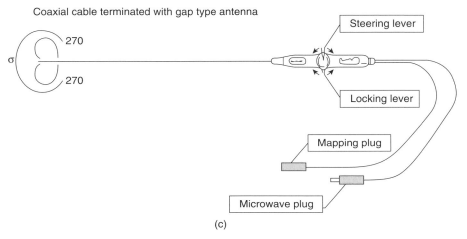

(c)

FIGURE 6.8 (c) Microwave ablation catheter.

in use as therapy for trigeminal neuralga provide immediate relief from pain and enjoy a success rate of 82% about 2 years following the procedure. Neuralgia has reoccurred in about 42% of patients, requiring a repeat procedure.

Radio-frequency ablation for pain management in cancer patients has also increased in popularity. A study has shown that 95% of patients with metastases to the bone have experienced a reduction in pain after such therapy [79, 80]. On a scale of 1–10, the average level of pain dropped from 5.8 before treatment to 1.8 at 12 weeks.

Another successful RF technique is the procedure to eliminate saphenous varicose vein reflux, often the underlying anatomical cause of varicose veins. The RF catheter-based endovascular occlusion of a refluxing saphenous vein provides a minimally invasive technique for eliminating saphenous varicose vein reflux. Vein occlusion is accomplished by means of RF heating of the vein wall to 85°C. The RF power delivered to the target tissue causes collagen contraction and denaturation of the endothelium. The RF closure procedure is performed under ultrasound guidance and replaces a surgical solution to this problem. Conventional surgery for varicose veins is associated with significant surgical morbidity and significant patient dissatisfaction [81].

Microwave endovascular occlusion utilizes a narrow balloon catheter similar to the one used for MBA [72–89].

6.4.3 Cardiac Ablation

In the discussion of RF/microwave catheter therapeutic modalitites, we have chosen to concentrate on cardiac ablation. To date, RF cardiac ablation is the "gold standard" in the treatment of a large number of supraventricular tachycardias. We have chosen to detail its success and include the potential for success in utilizing microwave cardiac ablation in new ways.

Radio-frequency ablation and to a lesser degree the *microwave ablation technique* (Fig. 6.8c) have become the treatments of choice in many types of cardiac arrhythmias. The actual instrumentation used as well as the catheter based techniques are similar to those of all other ablation procedures, as mentioned in this book and in the literature. From the detailed cardiac ablation techniques described in this book, the reader will gain a good understanding of the ablation techniques used in the treatment of other organs.

The heart is composed of three types of cardiac tissue; atrial muscle, ventricular muscle, and specialized excitatory and conduction tissues. The atrial and ventricular muscles of the heart are normally excited in synchrony. Each cardiac cycle begins with the generation of action potentials by the sinoatrial (SA) or sinoauricular node located in the posterior wall of the right atrium. These action potentials spread through the atrial muscle by means of specialized conduction tissues. The action potentials do not normally spread directly from the atrial to the ventricular chambers. Instead, the action potentials conducted in the atrial musculature reach the atrioventricular (AV) node and its associated fibers, which receive and delay the impulses. Potentials from the AV node are conducted to the His-Purkinje bundle. This structure carries the impulses to the ventricular musculature to cause the synchronous contraction of the ventricular muscles.

The term *paroxysmal tachycardia* refers to abnormal episodes involving a sudden increase in heart rate. Such tachycardia can result from an irritable focus in the atrium, the AV node, the bundle of His, or the ventricles. These episodes of tachycardia either may be initiated and sustained by a reentrant mechanism, termed a "circus" movement, or may be caused by repetitive firing of an isolated ectopic focus. While these episodes of tachycardia are usually amenable to treatment by medication, under certain circumstances surgical ablation of the abnormal focus or abnormally conducting tissue may be necessary.

The utilization of cardiac-based close chest treatment of cardiac arrhythmia started in the early 1980s. Direct current is delivered to the tip of an electrode with the goal of creating a localized injury to remove the cause of the specific arrhythmia. The delivery of 300–400 J from a standard cardiac defibrillator through a catheter resulted in 3000 V in the vicinity of the site to be ablated. The believed mechanisms of myocardial damage due to very high voltage are as follows: Electrolysis of water into hydrogen and oxygen gas increases resistance, resulting in arcing. Once arcing develops, there is a tremendous rise in temperature, which results in the creation of a shock wave. Several investigators have modified the catheter/probe and the DC energy source to minimize the uncontrolled size of the damaged tissue. Improvements in the control of energy delivery increase the safety and efficiency of the techniques. To date, the RF ablation technique is the standard in treating many of the supraventricular arrhythmias. It is estimated that, internationally, tens of millions of patients experience supraventricular tachycardia (SVT). Many patients do not have symptoms or have very mild symptoms which are well tolerated. On the other hand, it is worthwhile mentioning that SVT can lead to syncope and even sudden death. The three main mechanisms responsible for SVT are (1) atrial fibrillation (AF), (2) atrioventricular nodule reentrant tachycardia (AVNRT), and (3) accessory-pathway-mediated tachycardia (APMT). In AF, the electrical activity of the upper chambers of the heart becomes disorganized, causing a rapid electrical signal to impinge upon the AV node, which in turn creates a rapid ventricular response (Fig. 6.9a). Untreated patients with SVT may suffer a minor heart attack and/or congestive heart failure, even a stroke.

The most common form of SVT is AVNRT. In AVNRT there are two longitudinal disassociated pathways within the AV node. AVNRT will occur when the electrical impulse propagates down one pathway and then travels in retrograde up the other. Figure 6.9b depicts an accessory pathway between the atrium and ventricle. This form of SVT is know as Wolf–Parkinson–White (WPW) syndrome. Evidence of this accessory connection can be recognized on a surface electrocardiogram when preexcitation occurs and is defined by a short PR interval and a "delta wave" in the up slope of the QRS interval. As shown in Figure 6.10a, an electrical signal propagated down the AV node and up through the abnormal accessory pathway results in a rapid heart rate. Two kinds of therapies are being utilized to resolve SVT: (1) acute therapy and (2) chronic therapy. Acute therapy consists of pharmacological treatment and DC cardioversion, while chronic therapy includes pharmacology, surgical ablation, and catheter microablation (Fig. 6.10b).

The goal of any of the cardiac ablation techniques is to modify the electrical conduction system of the heart of patients having arrhythmia by converting electrically active cardiac tissue to electrically inactive scar tissue, thereby blocking the pathway or extra electrical circuit and eliminating the possibility of arrhythmia. In the case of RF ablation resistive heating occurs, leading to tissue coagulation and thus permanent tissue damage (Figs. 6.11a–d).

It is important to mention that the heat generation in the RF system occurs within the tissues themselves, not within the catheter tip. Lesion size is a function of the power level, time of delivery, electrode tissue interface, and size of the electrode. The last two parameters determine the tissue impedance presented to the RF generator. The concept of percutaneous cardiac tissue ablation using electrode catheters was discovered in 1979 when an attempted

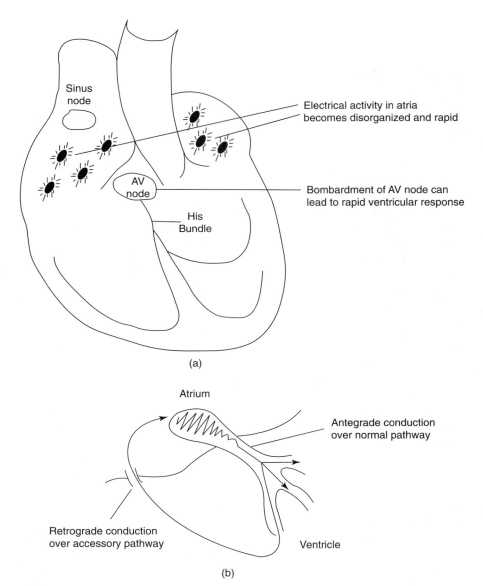

FIGURE 6.9 (*a*) Atrial fibrillation. (*b*) Arrhythmic circuit associated with the Wolff–Parkinson–White syndrome. In this syndrome, there is a connection between the atrium and ventricle outside the normal AV nodal pathway (accessory pathway). A tachycardia circuit can develop if an impulse conducts antegrade (forward) via the normal AV node pathway and is able to conduct retrograde (reverse) from the ventricle to the atrium via the accessory pathway. Catheter ablation successfully treats these arrhythmias because it interrupts accessory pathway conduction without interfering with normal AV nodal conduction.

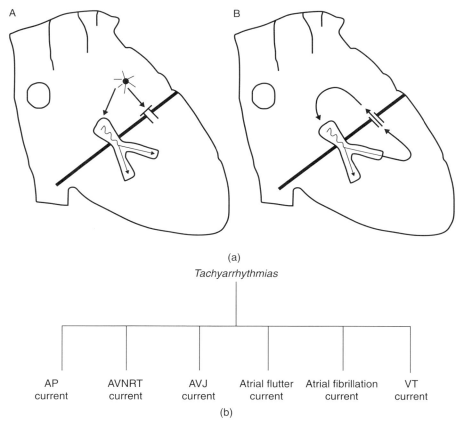

(a)

FIGURE 6.10   (*a*) Accessory pathways: A, premature impulse blocks antegradely in bypass tract; B, reentrant impulse established. (*b*) RF catheter ablation is the therapy of choice.

electrical cardioversion resulted in complete AV block; this occurred when the defibrillator was accidentally discharged through the bundle of His. The first human catheter ablations were performed in 1981 using direct current [13–16]. The problems occuring during DC ablation propelled the idea of RF ablation in 1988. The first RF catheter ablation was performed in men by Lumberg [40, 41]. Currently RF ablation is routinely used to treat atrioventricular junction (AVJ), accessory pathway (AP), AVNRT, and some types of ventricular tachycardia.

General anasthesia is not required during RF ablation, and there is no risk of barotrauma (Fig. 6.12). The success of RF ablation is based on the fact that it is a relatively easy procedure (when compared with surgery), during which a steerable catheter is advanced toward the target tissue and controlled, discrete, well-demarcated lesions are produced. The procedure is minimally invasive and requires a recovery time of just one day. The patient can resume normal activities after about two days. Figure 6.13 depicts the various vascular approaches used. Since this procedure cures the patient, drug therapy is no longer required. The estimated success rate for the three most common SVTs (AVJ, AP, and AVNRT) is better the 95%. The RF ablation system is similar in appearance to DC systems used in the past and to the other RF ablation systems used to treat other organs. The system (Fig. 6.8*b*) includes an RF catheter connected to an RF generator and a switching box designed to enable the physician to switch from the recording mode to the diagnostic mode. In the diagnostic mode the catheter is normally positioned in the right atrium, His bundle, coronary sinus, and right-ventricle apex positions to monitor activation times and initiate tachycardia utilizing various protocols. Precise mapping is required in order to determine the cause of the arrhythmia prior to use of the ablation catheter. In the ablation mode, RF current (high density) is delivered via the RF catheter's distal electrode to a back plate (low current density) placed on the patient's back. Currently, other systems are available in which two electrodes are inserted into the heart, one placed at the ablation

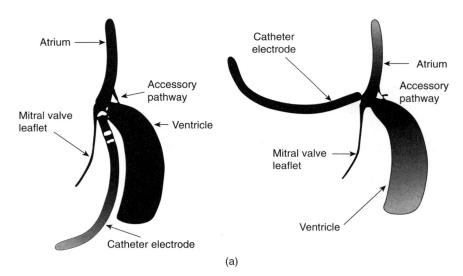

(a)

FIGURE 6.11 Diagrams of electrode positions used in RF catheter ablation of accessory pathways. (*a*) For ventricular approach a catheter is passed retrograde across the aortic valve and positioned under the mitral leaflet. (*b, c*) For atrial approach a catheter is passed across the interatrial septum (transseptal catheterization) and positioned on top of the mitral valve leaflet. Electrical mapping confirms the site of the accessory pathway prior to the delivery of RF energy [90].

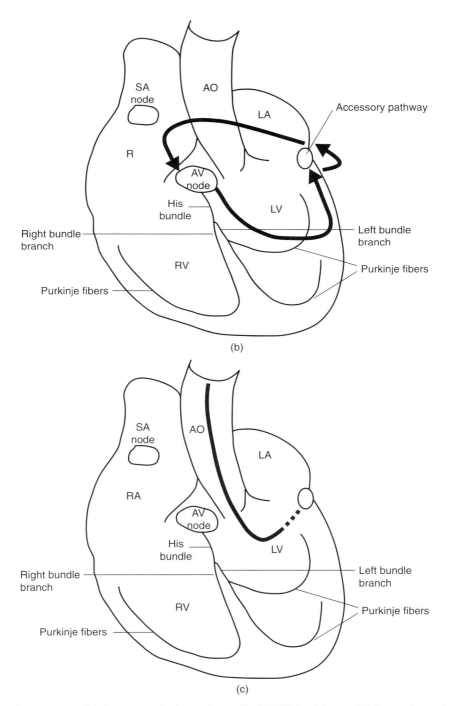

FIGURE 6.11 (*b*) Supraventricular tachycardia (Wolff–Parkinson–White syndrome). (*c*) Ablation catheter approach.

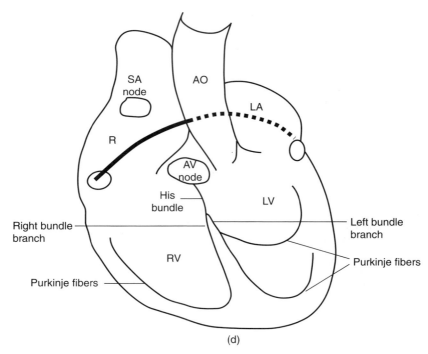

(d)

FIGURE 6.11   (*d*) Transeptal ablation catheter approach.

point and the other used for the current return path. Three major parameters are constantly being monitored: temperature, power output, and impedance. These parameters can be controlled to ensure optimal lesion generation and patient safety. It is only in the last five years that microwave catheter ablation has been explored as an option for the treatment of chronic atrial fibrillation and ventricular tachycardia. The application of microwave energy at 915 or 2450 MHz was researched in animals in the early 1980s with the purpose of creating a large myocardial lesion. This early research was later followed by further studies exploring temperature profiles that were measured around the antenna and the surrounding volume of tissue. The volume of heating for the microwave catheter system was reported to be 10 times greater than that of the RF catheter system at the same surface temperature.

The medical procedure uses a microwave catheter that includes a flexible coaxial transmission line and a terminal antenna. The position of the antenna in the chamber of the heart is adjusted with the aid of the displayed action potential. When the antenna is adjacent to the desired location, microwave energy is applied through the connection to the coaxial line (Fig. 6.14). Eval-

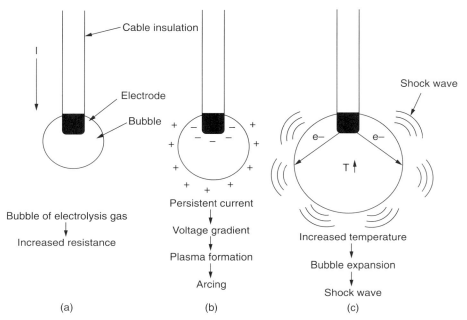

FIGURE 6.12   Mechanism of DC ablation. When direct current from a standard defibrillator is delivered to the tip of an electrode catheter, a series of events occurs, resulting in the generation of a shock wave. (*a*) The charge delivered to the electrode tip first results in the electrolysis of water into hydrogen and oxygen gas. An insulating bubble develops resulting in an increased impedance. (*b*) Current continues to flow to the tip of the catheter despite the rise in impedance, which results in a voltage gradient across the bubble. Arcing occurs. (*c*) Once arcing occurs, there is a tremendous rise in temperature, causing the bubble to expand, generating a shock wave (e^- = flow of electrons). (Adapted from *Clinical Cardiology*, 1990 [7].)

uation of a helical and whip antenna in a tissue-equivalent phantom was performed at 915 and 2450 MHz.

As of this writing, RF cardiac ablation is the most useful technique for a select type of arrhythmias, as discussed above. However, arrhythmias such as atrial fibrillation and ventricular tachycardia remain difficult to treat. Microwave ablation techniques, having deeper reach into the tissue, are currently being researched as a potential solution. As the microwave power is coupled into the tissue volume, the electrical dipole in the tissue will oscillate and create heat by a process known as dielectric heating.

To simulate the endocardial environment present in catheter ablation, a flow phantom model was tested. The model consists of phantom material that simulates cardiac muscle suspended in a saline perfusion chamber. Flow across the surface of the phantom is controlled by a perfusion pump, thus simulating

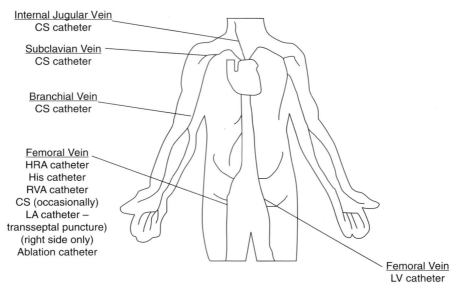

FIGURE 6.13 Vascular approaches for electrophysiology (EP) study.

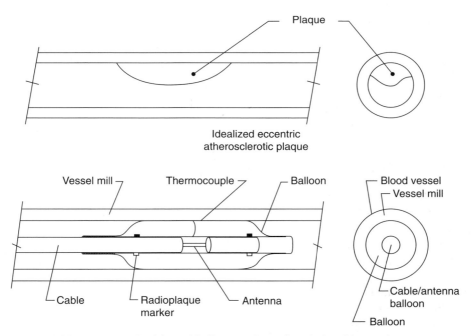

FIGURE 6.14 Position of balloon catheter in relationship to plaque.

blood flow. Each catheter design was analyzed by placing the catheter in the perfusion chamber, on the surface of the phantom material. The goal of the study was to characterize the temperature profile and size of the lesions produced by various RF and microwave catheters. These data were then correlated with data obtained from animals to test the validity of the results obtained from the phantom model.

6.5   PERFUSION CHAMBER [90]

6.5.1   General Description

A perfusion chamber was constructed from 3/8-in. clear acrylic slabs. It consisted of two halves which are each 10 cm in length, 4 cm in width, and 1.5 cm in depth. The lower half was filled with phantom material through which fine glass capillary tubes were placed every 5 mm along the length and every 2.5 mm along the depth. The phantom was separated from the top chamber by a thin layer of cellophane to prevent it from being dissolved by the chamber flow. The catheter of interest was placed at the center of the chamber and adequate contact of the entire catheter tip with the phantom was verified by visual inspection. The complete perfusion chamber was immersed in a saline bath maintained at a contact temperature of 37°C. The top portion of the chamber was perfused with 37°C saline at 4 L min^{-1} via the two side ports (Fig. 6.15a).

Muscle-equivalent phantoms for RF or microwave were made from mixtures of TX150, polyethylene powder, NaCl, and water and poured into the lower half of the perfusion chamber [91, 92]. The phantom material was allowed to set for 30 min prior to evaluation. The RF ablation catheters evaluated were a 4-mm-tip Steerocath and a large-tip 8-mm catheter.

A catheter for microwave ablation was constructed with a 12-mm helical antenna [16]. The microwave ablation catheter consists of a steerable 10F catheter with a central flexible coaxial cable with an overall diameter of 2.413 mm (0.095 in.). At the tip of the catheter the coaxial cable terminates in a helical antenna. The coaxial cable has an attenuation of approximately 29 dB/100 ft. Before the catheter was tested in the phantom material, it was evaluated on a network analyzer to confirm that the catheter radiated energy at 915 MHz.

Temperature measurements were obtained using a 12-channel Luxtron fiber-optic thermometry system. The fiber-optic probes were inserted into the fine glass capillary tubes at various positions to record the temperature. These measurements were taken at depths of 0, 2.5, 5, and 7.5 mm; at lengths of −5, 0, 5, 10, 15, 20, 25, and 30 mm from the tip along the length of the catheter; and at distances of 0, 2.5, 5, 7.5, 10, 12.5, and 15 mm from the catheter at each side. Data were recorded for a total of 90 s during each run: 30 s at baseline, 30 s during power delivery, and 30 s during cooling.

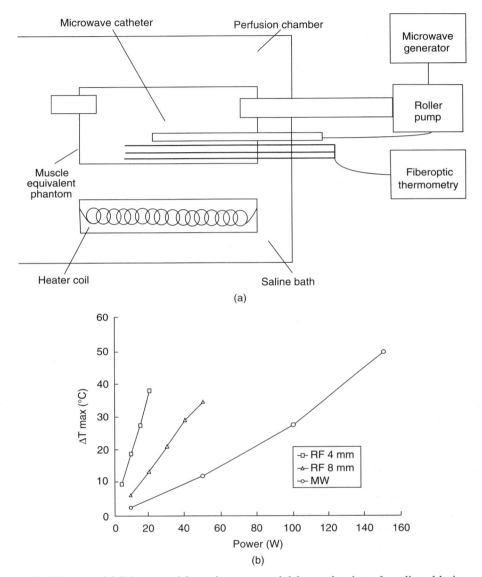

FIGURE 6.15   (*a*) Diagram of flow phantom model for evaluation of cardiac ablation catheters. The catheter to be evaluated is placed on the surface of the phantom, which is immersed in a perfusion chanber. Cardiac output is simulated by perfusing saline across the surface of the phantom by means of a roller pump. (*b*) Dosimetry of power versus temperature for RF and microwave catheters. The maximal change in temperature is at the surface, near the tip for RF catheters and at the base for the microwave catheter [90].

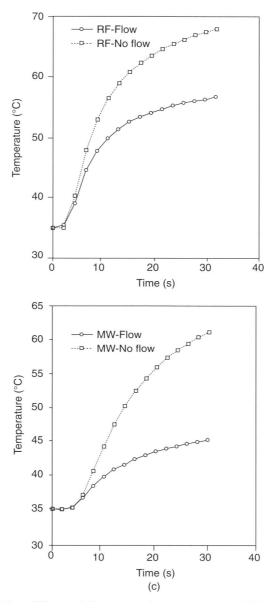

FIGURE 6.15 (*c*) Top: Effects of flow on surface temperature: RF catheter (4 mm). The changes in surface temperature as a function of time are plotted following delivery of 30 W of RF. Measurements were made in a static model (RF—no flow) and during flow of 4 L/min (RF—flow). Flow caused a decrease in both peak surface temperature and the rate of rise of surface temperature following power delivery. Bottom: Effects of flow on surface temperature: microwave. There was a decrease in peak surface temperature during flow (MW—flow) as compared to the static measurements (MW–no flow). During flow, the peak surface temperature was lower for MW than RF (46°C versus 57°C) [90].

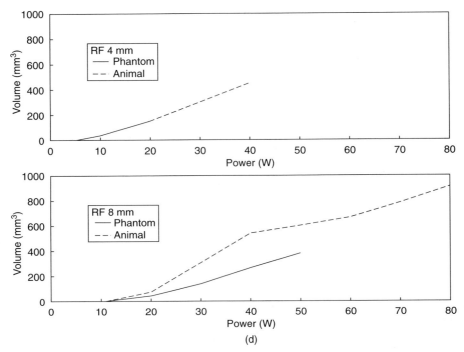

FIGURE 6.15 (*d*) Comparison of lesion volumes obtained with phantom and with canine left ventricular models for RF catheters [90].

6.5.2 Dose–Response Curve

The relationship between delivered power and tip temperature was evaluated for a 4-mm and an 8-mm RF catheter as well as for a microwave catheter (with a 12-mm helical antenna). There was a linear relationship between power and change in surface temperature (Fig. 6.15*b*). The correlation coefficient (*r*) was 0.998 for the 4-mm electrode, 0.997 for the 8-mm electrode, and 0.997 for the microwave catheter. More power was required to achieve the same change in temperature with an 8-mm electrode than with a 4-mm catheter. The power density required to achieve a similar surface temperature was 2.59 times higher for an 8-mm RF catheter. The microwave catheter required still higher power in order to achieve the same changes in temperature.

6.5.3 Depth and Rate of Heating

The time course of heating at various depths was analyzed by plotting the change in temperature as a function of time. The change in temperature was expressed as the percent of the maximal surface temperature achieved during

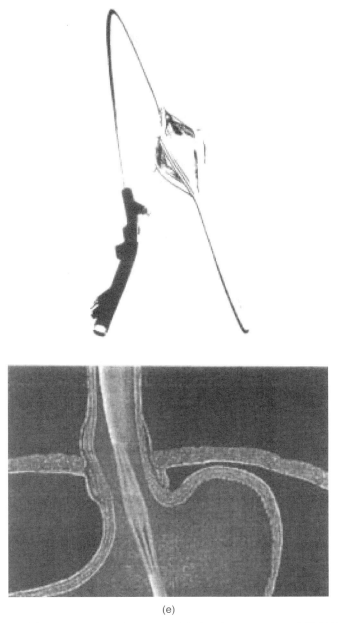

(e)

FIGURE 6.15   (*e*) Catheter used for nerve ablation in treatment of GERD [78].

power delivery. This analysis permitted a comparison of the time course of heating among the three catheters tested. Following power delivery there was a 3–4-s delay in temperature rise for all catheters tested. Peak temperature was achieved several seconds after power was turned off. The effect was delayed at depths of 2.5 and 5 mm. A greater depth of heating was observed with both the 8-mm RF and the microwave catheter. Following delivery of power, the surface cooled more rapidly than the deeper levels studied due to the flow of 37°C saline across the phantom surface.

In RF ablation, assuming that the current flow into the tissue is omnidirectional, the average decay rate of the power deposited is $Pr \sim 1/r^4$, where r is the distance from the electrode. In the case of microwaves, the power density decays as $Pr \sim \exp(-2\alpha r)/r^2$, where α is the tissue attenuation constant [93].

6.5.4 Effects of Flow on Surface Temperature

The effects of cardiac blood flow on surface temperature during cardiac ablation were evaluated by comparing a static and a flow phantom model. Static measurements were performed using the previously described phantom model but without perfusion of saline across the surface of the phantom. These values were compared to those obtained with saline perfusion of 4 L min^{-1} (Fig. 6.15c). Perfusion of saline caused a marked reduction in surface temperature for both RF and microwave ablation catheters. In the static model, the peak surface temperature measured 30 s after delivery of 30 W of power was 68°C for the 4-mm RF catheter and 61°C for the microwave catheter. Flow caused a reduction in peak surface temperature to 57°C with the RF catheter and 46°C with the microwave catheter. In addition, the rate of the rise of surface temperature was attenuated by flow. These data suggest that static phantom models will overestimate lesion size, since lesion size is proportional to surface temperature during ablation.

6.5.5 Lesion Volume

Lesion volume was plotted as a function of power for both the RF and microwave catheters. Lesion volumes were obtained from in vivo ablation of canine left ventricular myocardium as well as from measurements taken during phantom experiments. There was a direct relationship between lesion volume and delivered power for each electrode tested. Power delivery, and hence lesion size, was limited in the RF catheters by the development of a rise in impedance seen at higher power. A higher power was required to optimize lesion size with the 8-mm versus the 4-mm RF electrode (80 W versus 40 W). The maximal lesion volume with an 8-mm RF electrode was approximately twice that achieved with a 4-mm electrode (914 ± 362 mm versus 446 ± 150 mm^3, $p < 0.01$).

Lesion size was also calculated following the delivery of microwave energy. Microwave energy in vivo was then compared to estimated lesion size in the phantom model (Fig. 6.15*d*). There was a good correlation between the phantom and animal data obtained from the 4-mm RF catheter. For the 8-mm catheter, the phantom model underestimated lesion size when power was increased beyond 20 W. However, the phantom model accurately predicted that it would take approximately twice the delivered power for an 8-mm catheter to create a lesion similar in size to a 4-mm catheter.

The flow phantom model provides a simple in vitro method for analyzing cardiac ablation catheters. Heating of cardiac tissue to a level of cell death is the principal mechanism of successful cardiac ablation. Therefore, understanding the temperature profile of a particular catheter design is important in order to assess its effects on lesion size. The advantages of this phantom model are (1) the ability to measure the temperature at any point along the catheter shaft and (2) the control of chamber flow and thus its effects on surface cooling. This flow phantom model is superior to previously described static phantom models which do not take into account the surface-cooling effects of cardiac blood flow.

6.5.6 Limitations

The lesion size derived from the phantom model was only an estimate of the lesion size in vivo. The measurement in the phantom assumed that the myocardium was flat. In addition, data points were only obtained at 2.5 and 5 mm below the surface of the phantom. More closely spaced thermometry probes would likely have yielded more accurate information. However, closer spacing of the thermometry probes is not feasible in this model. Furthermore, it was assumed that a temperature of 47°C resulted in irreversible cell death. This assumption also influenced the results. Nonetheless, this system accurately reflects trends in temperature profiles that result from changes in catheter design or type.

Recently, in the abstract entitled "A successful option for chronic atrial fibrillation" [94, 95], the authors reported that their preliminary clinical study showed that microwave ablation represents an effective and safe option to treat chronic AF on patients with mitral valve disease and coronary artery disease. The authors predicted that microwave ablation would create larger and deeper lesions and thus would be effective in patients with deeper arrhythmogenic foci, such as those present in ventricular tachycardia. In addition, microwave energy can be delivered through tissue coagulation, which is not the case when the RF modality is used. Microwave heating is developed by the oscillation of molecular dipoles in tissue by EM fields at microwave frequency. Microwave ablation does not depend on good antenna tissue contact, whereas in the RF ablation technique, the RF electrode–tissue interface is of prime importance.

Although the technique and general instrumentation of RF/microwave ablation were discussed in detail in Section 6.4.3, the authors have elected to discuss two additional areas of importance: GERD [96–103] and endometrial ablation [104]. These two areas are consistently gaining approval and patients' satisfaction worldwide. In addition, in the case of endometrial ablation, a circular waveguide at 10 GHz is being utilized, a unique approach at a frequency higher than that allocated in the United States for medical applications. In the case of GERD, new areas of research are currently being developed to alleviate *Barrett's esophagus* (a condition that might be a precursor to esophageal cancer) using both microwave and light sources (see Section 6.9).

6.6   RF GASTROESOPHAGEAL REFLUX DISEASE

Gastroesophageal reflux disease [96] results from the chronic backward flow of stomach contents into the esophagus and is considered to be the most common disease of the gastrointestinal track. The reflux of stomach contents, which includes acid as well as bile and digestive enzymes, causes irritation of the esophagus that can result in heartburn, chest pain, swallowing difficulty, as well as other clinical problems. It also can cause Barrett's esophagus.

Barrett's esophagus (BE) is a lesion of the lining of the esophagus characterized by the replacement of the normal stratified squamous epithelium by a metaplastic columnar epithelium. It is an acquired condition that develops predominantly from chronic GERD. It is characterized endoscopically by proximal displacement of the squamocolumnar junction. Barrett's mucosa appears as red islands or fingerlike projections of columnar epithelium. Patients with BE are at risk for the development of dysplasia. On a background of dysplastic epithelium, esophageal adenocarcinoma may then develop. Barrett's esophagus with dysplasia therefore has been identified as a premalignant condition. The estimated risk for developing adenocarcinoma has been reported to be anywhere from 1 case in 16 to 1 case in 441 patient-years of follow-up, or 30–169 times greater than that in the general population.

It is estimated that 14 million adults in the United States require medical intervention to control the symptoms of GERD. The majority of Patients with GERD have normal lower esophageal sphincter (LES), the muscular valve at the junction of the esophagus and stomach that prevents reflux from occurring. Yet, in patients with GERD, the sphincter muscles relax frequently to cause reflux. The most common cause of GERD, involving over 80% of patients, is an abnormal neurological reflex termed transient lower esophageal relaxation (TLESR). An TLSER is prompted when there is stretching of the stomach wall, which is the case at the end of a meal. The RF procedure to alleviate the problem was developed by Conrad Stewart Medical, Inc. The RF system consists of an RF generator similar to one used in RF ablation, for example, and a flexible balloon catheter that helps to position needle electrodes against the wall of the LES/CARDIA (LES/gastric cardia). Thermocouples located at the tip of each needle facilitate the temperature measurement while performing the procedure. The RF generator has four channels allowing for independent control of the four needle electrodes. A preset temperature for every electrode assures the safety of the operation. The generator automatically stops delivering power to the electrodes exceeding the preset temperature. The procedure can be performed on an outpatient basis in less than 60 min. The physician advances the balloon catheter as shown in Figure 6.15e, where the needle electrodes are positioned against the wall of the LES/CARDIA. Needles are then deployed into the tissue, while the RF is being delivered to each needle. An irrigation system is then deployed to maintain the integrity of the mucosa. The delivery of RF energy is repeated in order to create well-defined coagulative lesions along the length of the lower esophageal sphincter. Over the next few weeks the coagulative tissue shrinks, which in turn creates a tighter LES and a less compliant CARDIA, and the tighter valve provides increased resistance to reflux of stomach contents into the esophagus. Since the development of the Stratar procedure of GERD, U.S. clinical studies have demonstrated a positive outcome in patients who have undergone the procedure, with an estimated success rate of 80%.

6.7 ENDOMETRIAL ABLATION

6.7.1 Microwave Endometrial Ablation

The success of using RF microwave in treating BPH and cardiac arrhythmias is probably one of the major factors behind the development of microwave endometrial ablation [104–107], an alternative procedure to hysterectomy in cases of abnormality and uterine bleeding. Microwave endometrial ablation is currently being utilized for uterine bleeding (menorrhagia). Menorrhagia is defined as blood loss of more then 80 mL per menstrual cycle. Microsulis, Inc., of the United Kingdom, was the first to introduce and research the microwave system at 9.2 GHz to treat menorrhagia. The 9.2 GHz was chosen in order to achieve a controlled tissue penetration not exceedings 6 mm. The treatment is relatively simple. The patient receives general anesthesia followed by cervical dilation to 9 mm, allowing the microwave applicator to reach the uterine base where the ablation procedure starts. The average temperature of 95°C is maintained throughout the treatment. The treatment time is between 1 and 4 min, depending on uterine parameters such as size and endometrial thickness.

6.7.2 RF Endometrial Ablation

NovaSure claims success with its RF endometrial ablation technique [108]. Prior to the RF activation, NovaSure's three-dimensional, fan-shaped, expandable bipolar device is inserted into the uterine cavity where suction is applied to draw the uterine lining close to the RF applicator. Power as high as 100 W for between 40 and 100 s is applied during the ablation procedure. The desired treatment temperature in the RF case is around 75°C. A success rate of better than 97% is reported utilizing the NovaSure procedure. It is worthwhile mentioning that some limited success was observed with complete ablation in women with large and distorted uterine cavities.

6.8 MICROWAVE MEASUREMENT TECHNIQUES: EXAMPLES

6.8.1 Introduction

This section addresses useful microwave measurement techniques that are consistently gaining approval in the medical arena [1, 109–120].

Research in the area of *microwave radiometry* continues, with the hope of accomplishing a better design for receiving systems (including antennas) operating over a range of frequencies in order to increase temperature-measuring accuracy.

Radiometry is defined as the measurement of natural EM radiation or emission from the body at microwave frequencies. It may be applied to the detection and diagnosis of pathological conditions in which there are disease-related temperature differentials, such as in the early detection and diagnosis of breast cancer. Other applications, such as in the detection of tumors in other sites and in monitoring temperature during hyperthermia, are currently being investigated.

The use of microwaves to monitor vital life signs such as respiration and heartbeat was researched in the early part of the 1980s. Papers in this area have been appearing recently in the literature, reporting success where modern processing techniques are utilized.

The basis of *noninvasive monitoring* devices is the motion detection capability of simple microwave radarlike circuits in which phase displacements between transmitted and reflected signals can be related to physical movement of the reflecting surfaces. To monitor the pulsating motion of the heart or that of any artery or organ of the human body with a radio signal, it is necessary to either sense the phase-varying reflections from the body surface caused by the underlying organ action or directly monitor this action by causing most of the radio signal to penetrate the skin, fat, or muscle tissue which overlies the artery or organ of interest. Microwave frequency signals are most appropriate for use since the physical displacements of the arteries and organs, although quite small, is significant in relation to the short wavelengths of the microwave signals, which are generally on the order of centimeters in air and further reduced in tissue by the dielectric properties of the tissue.

Researchers have reported on the use of microwaves to measure chest wall motion in response to left-ventricle activity. Others have demonstrated the use of microwave Doppler radar to monitor the movement of arterial walls, from which the patient's pulse rate could be clearly determined.

Measurement of microwave attenuation in the human torso, EM wave effects on biological materials and systems, mapping of internal organs, and diagnosis of pulmonary edema have also been reported.

We have elected not to discuss accomplishments in the area of microwave balloon angioplasty since that subject has been covered in detail in other books [1] and articles. However, we will discuss possible applications of microwaves in cardiology, for example, that were developed previously and in parallel with the development of microwave balloon angioplasty.

The following possible applications of microwave technology in cardiology will be reviewed:

Measurement of blood perfusion in heart muscle by means of microwave energy.

Measurement of coronary arterial lumen by means of microwave apparatus.

Effect of microwave therapy upon the functional state of the cardiovascular system in patients with hypertension.

6.8.2 Method of Measuring Blood Perfusion (Flow) in Heart Muscle by Use of Microwave Energy

A method is described by which perfusion of tissue is measured by irradiating the tissue with microwave energy and measuring the rate of temperature decay in that tissue as an index of fluid perfusion [121].

It is known that the rate of blood flow within a tissue can be measured by a process of heating the tissues with a device, such as a probe, that is in contact with or in close proximity to the tissue being examined. Then the decrease in temperature (microwave power "off") is recorded by a thermocouple positioned in or near the probe. The recorded temperature changes are representative of the blood flow in the tissue. The heated device and thermocouple effectively act as a flowmeter for determining the blood flow as a function of the rate at which heat is carried away from the tissue.

Heated probes and thermocouples used for the determination of blood flow were first Introduced by F. A. Gibbs in 1933 [122] for the purpose of measuring flow in blood vessels. Heated probes and thermocouples were later used as flowmeters by C. F. Schmid and J. C. Pierson for measuring blood flow in solid organs [123]. Further investigation by J. Grayson and his colleagues [124, 125] demonstrated that a heated probe with a thermocouple could be used in accordance with a certain relation, known as *Carslaw's equation*, to measure the thermal conductivity k of any solid, semisolid, or liquid in which the heated probe and thermocouple were inserted.

"Heated" thermocouples or thermistors used in flowmeters function to provide heat by conduction into the tissue in immediate contact with the heating device and then measure the temperature of that tissue. Determination of fluid (blood) perfusion heretofore was limited by the heating of tissue essentially in contact with a heated device.

In the microwave technique, fluid perfusion of tissue is determined by measuring the thermal conductivity of the tissue by irradiating the tissue with a

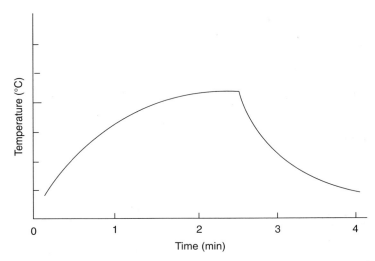

FIGURE 6.16   Blood perfusion of muscle as function of temperature and time.

microwave signal having a predetermined repetition rate, amplitude, and frequency to elevate the temperature of a volume of the tissue to a predetermined temperature and measuring the rate of decay of the temperature of tissue (Figs. 6.16 and 6.17). This decay is indicative of the thermal conductivity of the volume of tissue and thus the fluid perfusion of that volume. Figure 6.17 depicts the system used, where the temperature change was measured with the help of a thermocouple.

The temperature elevation in the volume of tissue should not be so great as to damage or injure the tissue. Moreover, the tissue should not be heated so much as to dilate the blood vessels within it, since such dilation may artificially affect the blood perfusion and thereby preclude an accurate determination of the state of the tissue. Since the purpose of the process is diagnostic, it is critically important that the blood perfusion be determined accurately with a minimum of disturbance to the tissue. Thus, the increase in temperature within tissues should be no more than 1°C, and preferably less than 0.5°C.

6.8.3   Lumen Measurement of Arteries Utilizing Microwave Apparatus

This section describes a microwave-aided balloon angioplasty catheter in which the balloon bears a metallization pattern that allows measurements to be made of the expanded lumen [126, 127]. Figure 6.18 illustrates the general steps involved in the procedure. The procedure begins with the insertion of the catheter (cable/antenna) and placement of the distal end adjacent to an obstruction to be treated, then measured. As illustrated in Figure 6.18a, the

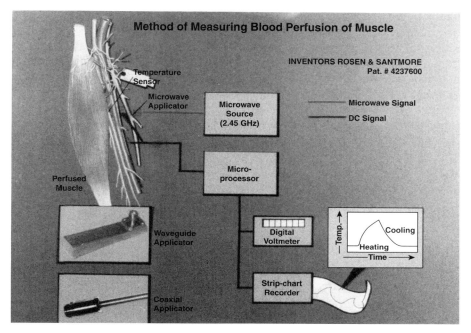

FIGURE 6.17   Method of measuring blood perfusion of muscle.

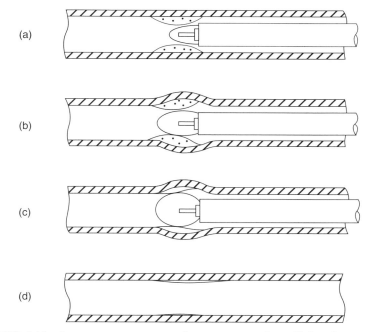

FIGURE 6.18   Lumen measurement of coronary arteries utilizing S parameters.

balloon membrane is then partially inflated against the resistance of the obstruction. Figure 6.18b illustrates the measurement of the radius (configuration) of the balloon by measurement of the microwave signal reflected from the antenna (S_{11}). As depicted in Figure 6.18b, the expansion of the balloon might be insufficient, indicating that further treatment is required. Further inflation of the balloon and additional microwave radiation at the same power level or higher are depicted in Figure 6.18c. Figure 6.18d illustrates a condition in which the power was turned off as a consequence of the desired opening having been achieved using S_{11} as a guide. Figures 6.19a,b depict S_{11}

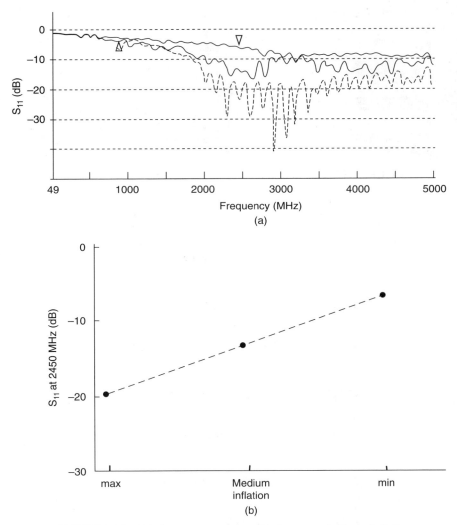

FIGURE 6.19   (a) S_{11} measurements. (b) S_{11} versus balloon inflation.

variation measured as a function of balloon extension in an artery at around 2.45 GHz.

For the sake of completeness, it is our intention to mention (rather than detail) the following additional research conducted throughout the world utilizing microwave in cardiology:

1. The effect of microwave therapy upon the functional state of the cardiovascular system in patients with hypertension [128].
2. The pacemaker protective undershirt [129].
3. Control of arrhythmia in the isolated heart by means of microwaves [130].
4. The application of microwave to acupoints for the treatment of coronary heart disease [131].

1. In his paper entitled "The effect of microwave therapy upon the functional state of the cardiovascular system in patients with hypertension," A. D. Fastykovsky of the Department of Physiotherapy and Resort Treatment, Kiev, Ukrainian SSR, describes the effect of microwave treatment at 2.375 GHz on the order of 20–50 W [128] in 95 patients with hypertension and other cardiovascular disease. The results on hypertension are only cited. Microwave energy was applied to the portal zone by means of a rectangular emitter having the dimensions of 30 × 9 cm from a distance of 7–10 cm. The power was delivered for 10 min daily for a course of treatment of 12–15 days. The results cited in the paper are summarized as follows: Prior to the treatment, the mean systolic arterial pressure measured at the brachial arteries was 167.7 + 2.39 mm Hg and diastolic pressure of 95.8 + 1.0 mm Hg, whereas following the treatment it decreased to 139.0 + 2.28 mm Hg and 87.6 + 0.89 mm Hg, respectively ($P < 0.001$).

2. The subject of a pacemaker protective cloth is a relatively mature one. Research in this general area was done in the United States as well as in Europe and is summarized in the paper entitled "Pacemaker protective undershirt" [129]. In that paper, the authors describe a protective undershirt made of metallized textile fabric which retains most textile properties such as lightness and comfort and at the same time has the characteristics of a reliable shielding material to EM radiation. In experiments conducted at the Pacemaker Center UCL, Belgrade, Yugoslavia, the authors report having demonstrated that the function of a pacemaker could be affected by a microwave field but that, by introducing the protective undershirt, EM interference with the operation of the pacemaker ceased.

3. Evidence of low-level microwave effects on both living and isolated hearts has been reported in the last 25 years. Investigation in the early 1980s of noninvasive irradiation to control heartbeat resulted in failure. More recently, C. C. Tamburello et al. of the Department of Electrical Engineering, Palermo, Italy, have investigated the effect of microwave irradiation on iso-

lated chick embryo hearts. In their paper entitled "Analysis of microwave effects on isolated hearts" [130] and in other publications [132, 133], the authors report that their experiments have shown low-intensity pulse-modulated microwave energy at 2.45 GHz to have affected the heartbeat of isolated chick embryo heart: The heartbeat is said to have changed with the increase in modulation frequency above the natural unperturbed heart rate. In the presence of arrhythmia, the authors report having demonstrated that the heartbeat could be regulated with the proper pulse repetition frequency. During the experiments, the hearts were irradiated with a 2.45-GHz pulsed source having a peak power of 10 mW and a duty cycle of 10%. The estimated incident peak power density was 3 mW cm^{-2}.

4. Authors of a number of papers on the application of microwaves and millimeter-waves to acupoints for the treatment of various ailments have suggested the utilization of millimeter waves (53 GHz) to reduce the frequency of angina attacks. The irradiation zones selected are the acupoints based on the acupuncture theory of traditional Chinese therapy [131, 134]. At the beginning of the treatment, the patients continued with their prescribed medical therapy (drugs), with reduction of the medical dose as a function of the length of the millimeter-wave therapy. The procedures took place once or twice a week for 25 min at a power level of 0.1 mW.

6.9 FUTURE RESEARCH

The section on future research is relatively short, since its purpose is not to provide an exhaustive description of the possible utilization of microwaves in medicine. Rather its purpose is to describe to the reader the result of one uncompleted research program in which better solutions for "tissue welding" were sought (National Institutes of Health funded grant R41-HL55758, "Microwave System for Tissue Anastomoses," William P. Santamore, Principal Investigator) and another program which is awaiting investigation on the subject of "photodynamic therapy." We believe that such examples will trigger new ideas in the minds of students and other researchers.

6.9.1 Microwave Tissue Welding [135]

In Vitro and In Vivo Experimental Studies   Initial studies utilizing endoscopic techniques are revolutionizing many surgical procedures, and further expansion of the endoscopic approach is inevitable. Although minimal access surgery is advantageous to patients, the technical problems imposed by the limited access are promoting existing tissue closure technologies. Laser tissue welding offers several potential advantages over suture anastomosis or closure. These include faster healing, less inflammatory response, a better flow surface, and the ability of the weld site to grow.

While experimental tissue welding has proved feasible, clinical application of the technique has been limited by the low initial weld strength and by the need for precise apposition of tissue edges before welding. A fibrinogen-based "solder" for laser welding has resolved some of these problems. For tissue welding, the addition of a photosensitizing agent to the solder provides greater spatial control over laser energy deposition, thereby minimizing collateral tissue damage.

The initial step in the formation of the weld is tissue heating as a result of absorption of laser light by tissue pigments or by water. This heating produces structural changes in extracellular matrix proteins. These changes may involve uncoiling of the secondary or tertiary molecular structures. Following these structural changes, bonding may then occur as the final step in the formation of the weld. Because of its prominent role as a structural protein, collagen is likely to be involved in the welding process.

Since tissue welding is heat activated, microwaves could be used. Accordingly, a prototype microwave system to create tissue anatomosis was designed, developed, and tested. The microwave system consisted of a generator similar to the one used in cardiology and a catheter terminated by an antenna. The prototype system was built and tested on the bench top.

Series of studies consisted of both in vitro and in vivo experiments. The in vitro studies determined the strength of the biological solder and found several approaches to increase this strength. The in vitro studies also examined doping of the biological solder. By doping this solder with polar molecules, we concentrated the microwave energy primarily where the heat was needed, at the anastomoses site. The in vitro studies compared microwave tissue welded anastomoses to standard sutured anastomoses. We observed tear strengths of over 1 kg for the microwave tissue welded anastomoses, comparable to those of sutured anastomoses.

In several animals, we created microwave tissue anastomoses that gave high initial flow rates. However, in some studies, the anastomoses became obstructed by an intraluminal thrombus. The cause of the thrombus was an inadequate aperture between the two anastomosed vessels. Problems with excessive fluid surrounding the site of tissue welding and with creating an opening between the vessels still need to be resolved.

In addition to having all the advantages of laser anastomoses, microwave anastomoses can offer several advantages over laser anastomoses. Since all the solder will be heated at the same time, the anastomosis should be accomplished faster. Since the microwave energy will only concentrate in the doped solder, there is no need to aim the microwave energy. A microwave system should be cheaper to build and safer to use. In any case, a microwave system for tissue welding will provide the clinician with an additional alternative.

All surgeries involve some type of tissue reapproximation or closure. Even if a microwave system for tissue closure is only suitable for a few specific applications, the technique could still have a large market. With the increasing use of endoscopic approaches, new anastomotic techniques are needed. Tissue welding is an exciting new approach. Since tissue welding is heat activated, microwaves could be used.

Previous Approaches to Anastomosis   As stated before, laser welding has been used to create vascular anastomoses. Successful laser tissue welding relied on precise apposition of the tissue edges and avoidance of thermal injury due to excess energy delivery and nonspecific absorption. Early disruption due to the diminished strength of welds immediately after their creation has prevented more widespread use of this technology. A laser-fibrinogen-reinforcing suture anastomosis does not require as precise an apposition as does an anastomosis primarily made by laser. (The biological solder fills in the spaces between the tissues to be welded.) The addition of a fibrinogen solder has been shown to increase the strength of such welds.

It is believed that the main problem with these tissue welds is the small surface area available for bonding (Fig. 6.20). The initial approach was to increase the surface area for the anastomoses and to control heating with microwave energy (Fig. 6.21). The solder was placed not only between the two arteries but also from a sleeve around the vessel. This approach provided the greater surface area. The biological solder would be activated by microwave energy applied via a catheter within the arteries. This approach required that the biological solder have some intrinsic strength and the microwave energy would be selectively absorbed by the biological solder.

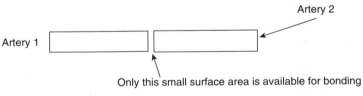

FIGURE 6.20   Small surface area available for bonding.

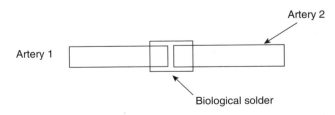

FIGURE 6.21   Sleeve around vessels to provide greater surface area.

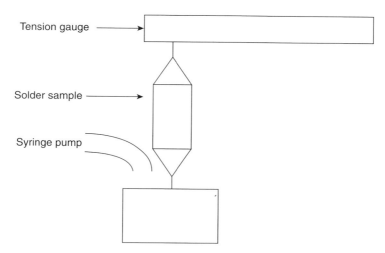

FIGURE 6.22   Apparatus for tension strength measurements.

Bench-Top Tests   The objective of these tests was to determine the strength of the biological solder. The biological solder was primarily an albumin solution. The concentration of the albumin tested was 30, 40, and 50%. The materials added to the albumin solution were silk thread, silk thread with knots, gauze, or sponge. A Tissue-Tek II disk ($25 \times 20 \times 5\,mm^3$) was used to contain 1 mL of albumin solution.

After microwave heating, the sample was cut in a 1-cm width and the tension strength was measured by the apparatus depicted in Figure 6.22. Via a syringe pump, water was infused into a bucket suspended from the biological solder sample. While the tension was recorded, water was infused until the sample ripped apart (tear strength).

From in vitro experiments we found that 40% chicken albumin solution had the highest tear strength. Then a 40% albumin solution with different materials [silk threads (al+silk), knotted silk threads (al+sk+n), sponge (al+sp), and a gauze (al+gauz)] was used. Among the materials added, we found that a 40% albumin solution combined with sponge had the highest tear strength (Fig. 6.23).

Doping Biological Solder   To create tissue anastomoses with microwaves, the tissue temperature must be kept below the threshold for damage, while the biological solder is heated above 60°C. This process is depicted in Figure 6.24 for an arterial anastomosis. The microwave antenna is positioned inside the artery, and solder is placed on the outside of the vessel and in any small gap between the arteries. The graphs on the left side show the desired temperature profiles. If microwaves heated by convection, these temperature

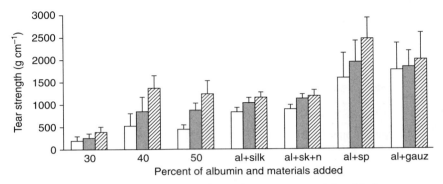

FIGURE 6.23   Tear strength results versus materials added.

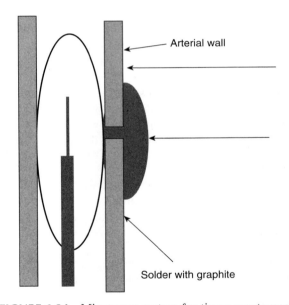

FIGURE 6.24   Microwave system for tissue anastomosis.

profiles would be impossible. However, microwaves heat by radiation, not by convection. Thus, the energy absorbed by a material will determine its temperature. By doping the solder with molecules that readily absorb microwave energy, the temperature in the solder will rise faster and higher than the arterial temperature.

How much microwave energy a material absorbs depends on the complex interaction of microwave with the material. There are literally tens of thousands of molecules with higher microwave absorption than biological tissue and the biological solder. However, selection of the best agents is not easy. In the proposed studies, graphite was utilized, because of its limitations as a biocompatible or biodegradable material, and because it is cheap and easy to use.

For these initial experiments different concentrations of albumin were first tested. The results showed that as the concentration of albumin increased, the microwave time required to cause coagulation decreased. The coagulation, similar to that of egg whites could be easily observed. One disadvantage of a high albumin concentration is that the biological solder becomes a thick paste. This thick paste is hard to apply to an artery. Thus, for practical reasons, we used a 40% albumin solution for the remainder of our experiments.

We then examined two different doping materials: graphite and ferric oxide. We tested the time required to coagulate the biological solder with different concentrations of graphite and ferric oxide added to the solution. Adding either graphite or ferric oxide greatly decreased the time to coagulation; however, this effect was not linear. The biggest changes occurred with the addition of small concentrations of graphite or ferric oxide. Higher concentrations had less effect on shortening the coagulation time further. Thus, we selected a 1–2% concentration of graphite or ferric oxide for the rest of the experiments.

We also measured the temperature in the biological solder. The solder was placed on the outside of the artery, and the microwave antenna/catheter was placed within the artery. A thermocouple was placed in the biological solder. The thermocouple was positioned perpendicular to the microwave antenna to prevent interference. A number of power settings and times were examined. The temperature of the solder rose quickly when the microwave energy was applied. In several initial studies, we noticed that the artery was getting warm. Thus, we started to infuse deionized, distilled water through the artery while applying the microwave energy. We also redesigned our antenna/catheter system so that the microwave energy only radiated in one direction. These simple procedures kept the artery cool.

In Vitro Vessel Anastomosis   Canine carotid arteries were anastomosed end to end with microwaves using 40% albumin and 1% ferric oxide as solder. The tear strength of microwave anastomosis compared was to the hand-sewing anastomosis. The tear strength of the microwave anastomosis was $954 \pm 132\,g$ versus $795 \pm 152\,g$ for the hand-sewn anastomosis. There were problems, however: While we were able to create strong anastomoses in vitro with this approach, the volume of material that was needed to surround and secure the artery anastomoses would likely generate significant inflammatory responses in vivo. Thus, we abandoned this approach.

We then hypothesized that by placing the vessels side by side, we could in effect create an end-to-end anastomosis. This approach would provide the needed large surface area for anastomosis strength without use of the extra material. This anastomosis approach is depicted in Figures 6.25 and 6.26.

The biological solder is placed on the coronary artery. The internal thoracic artery is then placed on the coronary artery. The microwave catheter is placed through the cut distal end of the internal thoracic artery. The microwave power is applied, thereby welding the vessels together. Note that this welding occurs without interrupting flow through the coronary artery.

The microwave catheter is removed and a hole punched through the internal thoracic and coronary arteries. The distal end of the internal thoracic artery is occluded with a clip.

Using the approach depicted in Figures 6.25 and 6.26, canine carotid arteries were anastomosed end to end with microwaves using 40% albumin and 1% graphite as solder. The tear strength of microwave anastomosis was measured by the techniques outlined above. The tear strength was 898 ± 121 g. Although this strength was more than adequate, we noticed that if the stress was applied perpendicular to the long axis of the vessels, the vessels could be peeled apart at low forces. Thus, in subsequent anastomoses, we placed two sutures at each end of the anastomosis. These sutures help to increase the tear strength (1008 ± 147 g), but more importantly, the sutures prevented this peeling.

In Vivo Experiment   We performed three acute canine experiments. In each animal, the carotid artery was anastomosed to the jugular vein in the side-to-

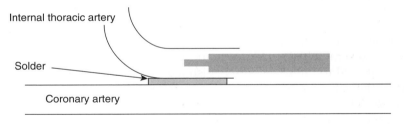

FIGURE 6.25   Placing vessels side by side for anastomosis.

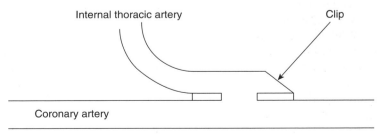

FIGURE 6.26   End-to-end anastomosis.

side fashion, depicted in Figures 6.25 and 6.26. Note that we did not have a hole punch to cut the carotid artery and jugular vein after creating the ansto-mosis. So, we had to cut each vessel before we created the anastomosis and then try to align the holes.

We noticed that the vessel surface needed to be kept dry for a better anas-tomosis. We measured flow through the carotid artery before and after the anastomosis. After the anastomosis, the flow rate was close to baseline. However, the flow tended to decrease over a 1-h period. This flow decrease was associated with the development of a thrombus in the vessel. At the end of the study, we noticed that the hole in the jugular vein was not in line with the hole in the carotid artery.

Technical problems do remain:

1. Microwave tissue welding does not work very well in a wet environment. Possible solutions include using a small blower and turning up the power on the microwave unit.

2. At present, we have no device for making the hole between the two arteries.

6.9.2   Endoscopic Light Source and Microwaves for Photodynamic Therapy

Photodynamic therapy (PDT) is a treatment modality using a photosensitiz-ing drug which is activated by specific wavelengths of light. The light-excited drug then interacts with molecular oxygen to produce a toxic oxygen species, known as singlet oxygen, which mediates cellular death. The appeal of PDT in oncology is that the photosensitizers are absorbed to a greater extent in tumor tissues as compared to normal tissues. In addition, PDT can be used before, after, or concurrently with traditional cancer therapy modalities such as chemotherapy, radiation therapy, or surgery. Photodynamic therapy has been approved by the Food and Drug Administration for the treatment of both obstructing and preinvasive lesions of the lung and esophagus. The efficacy of these treatments, however, is limited by the supply of oxygen to the treated tissue. The consumption of oxygen during PDT is a rapid process that can quickly deplete available oxygen in the illuminated tissue [136, 137].

The presence of oxygen is a critical factor in determining the effectiveness of the photodynamic effect. The PDT of anoxic tissues produces no effect [138] and the presence of hypoxia can diminish the PDT effect [139, 140].

Several approaches have been used to increase the availability of oxygen to the treated tissue during PDT. Minimizing oxygen depletion is one such approach that can be achieved either by reducing light fluence rate or by frac-tionating light delivery. At high fluence rates, mathematical modeling indicates that oxygen consumption can outpace the rate of oxygen diffusion from cap-illaries, thus resulting in a decrease in the oxygen level in surrounding tissue. Lowering the fluence rate has been shown in preclinical studies to improve the efficiency of tumor response to PDT. Unfortunately, decreasing fluence rate to levels leading to improvement in tumor response can result in treatment times requiring hours of illumination, thus making this approach clinically impractical. Another approach to minimizing oxygen depletion is by fractionating light delivery. Alternating intervals in the application of light ("on" and "off") allow for reoxygenation of hypoxic tissues. The mechanism for reoxygenation is not well understood and the optimal fractionation schedule has not been determined. Also, the addition of "light-off" intervals would lengthen treatment sessions, thus affecting the practicality of light delivery in the clinic setting. Another approach that may increase the availability of oxygen during PDT involves increasing blood flow and therefore oxygen supply to the illuminated tissues. Increased blood flow and tissue oxygenation can be achieved by local tissue heating. In the past, technical difficulties in delivering heat to tissues have limited the use of hyperthermia in the treatment of cancer, but the development of imprintable microwave antennas and devices to place these antennas next to or within affected tissues will allow the further investigation of hyperthermia and its potential synergistic effect with PDT.

In the past it was proposed that a self-contained esophageal LED laser-diffusing device be built and tested that would, through the use of an imprintable microwave antenna, provide local tissue heating and thus increase blood flow and tissue oxygenation to improve the PDT effect.

6.9.3 Microwave Balloon Catheter

The techniques of utilizing microwave balloon catheters were specifically researched in the areas of cardiology, urology, and breast cancer, to mention only a few [1, 141]. Microwave balloon angioplasty was the first microwave application in cardiology. It was developed with the ultimate goal of decreasing both acute and long-term restenosis. Like percutaneous transluminal balloon angioplasty (PTCA), MBA employs a balloon catheter that is advanced to the site of arterial stenosis. While PTCA uses only the pressure generated by balloon inflation to dilate the affected artery, MBA takes advantage of the volume-heating properties of microwave emitters. In MBA a microwave cable–antenna assembly is threaded through the catheter, with the antenna centered in the balloon portion of the catheter. By heating the tissue as the balloon is inflated, it was hoped that a patent vessel would be created that would be resistant to both acute and chronic reocclusion. Early in vivo studies at 2.45 GHz were conducted to assess the effects of various energy levels upon normal and atherosclerotic rabbit iliac arteries. Research on the therapeutic potential was subsequently conducted on atherosclerotic rabbit iliac arteries using microwave energy to raise the balloon surface temperature to 70–85°C. When compared to simultaneously performed conventional angioplasty, MBA at 85°C produced significantly wider luminal diameters, both immediately after angioplasty and 4 weeks after the procedure. Microwave balloon catheters in the treatment of BPH are currently utilized in a number of medical centers in the United States.

With balloon catheters it is possible both to produce high therapeutic temperatures throughout the prostate gland without causing tissue burning and to create biological stents in the urethra in a single treatment session. Compared to conventional microwave catheters, the distances microwaves have to travel through the prostate to reach the outer surface of the gland are reduced by the use of balloon catheters, as is the radial spreading of the microwave energy. Furthermore, compression of the gland tissues reduces blood flow and its cooling effect within the gland. Also, since catheter balloons make excellent contact with the urethra, much better than do conventional catheters, the urethra is well cooled by the cooling liquid within the balloon and is therefore well protected from thermal damage.

Other microwave balloon systems have been used, such as a balloon microwave catheter in the treatment of cancer. The above accomplishments will help to ensure our success in developing a microwave antenna printed directly onto the balloon, as discussed previously.

A microwave antenna that is imprinted onto the surface of a balloon catheter should be built for use in future research. The antenna will allow uniform heating of tissues along the entire length of the balloon. It will provide heat around the entire circumference of the balloon. It will also be designed to tolerate multiple balloon inflations and deflations without disruption of current flow through the conducting material.

6.9.4 Thermally Molded Stent for Cardiology, Urology, and Other Medical and Veterinary Applications

Every year, several hundred thousand people die suddenly in the United States from acute myocardial infarction, and many more suffer from chronic heart problems. A major contributing factor in both acute and chronic heart problems is a reduction in nutrient blood flow to the muscles of the heart, resulting from a reduction of blood flow through the coronary blood vessels. The reduction in flow may be caused by deposits of atherosclerotic plaque on the walls of the blood vessel, which causes a narrowing of the lumen or channel of the blood vessel. When the lumen is sufficiently narrowed, the rate of flow of blood may be so diminished that a spontaneous formation of thrombus or clot occurs by a variety of physiological mechanisms. As is known, once a blood clot has started to develop, it extends within minutes, in part because the proteolytic action of thrombin acts on prothrombin normally present, tending to split it into additional thrombin, which causes additional clotting. Thus, the presence of atherosclerotic plaque not only reduces the blood flow to the heart muscle, which it nourishes, but also is a major predisposing factor in coronary thrombosis.

Among the treatments available for the conditions resulting from plaque formation are pharmacological means such as the use of drugs, for example nitroglycerin, for dilating the coronary blood vessels to improve flow. In those cases too far advanced to be manageable by drugs, surgical treatment may be indicated. One of the surgical techniques commonly used is the coronary bypass, in which substitute blood vessels shunt or bypass the blockage. The bypass operation, though effective, is expensive and subject to substantial risks.

Percutaneous transluminal balloon catheter angioplasty is an alternative form of treatment. This method involves the insertion of a deflated balloon into the lumen of an artery partially obstructed by plaque and the inflation of the balloon in order to enlarge the lumen. The lumen remains expanded after removal of the catheter. The major problem with this technique is restenosis of the narrowed vessel by recurrence of the arterial plaque. The causes of restenosis are not fully understood but appear to be related to one or more of the following: (a) the natural elasticity of the walls of the vas, which tends to result in closure of the vas following expansion by the balloon; (b) deposition of additional plaque or other matter at the site of the original stenosis, due to those conditions which led to the original depositions; and (c) deposition of additional material at the site of the balloon treatment due to damage or changes to the arterial walls resulting from expansion by the balloon.

Stents are often used in medicine for supporting an existing natural or surgical channel or cavity (vas) in the body to aid in holding the vas open. Stents are commonly used during angioplasty, such as the coronary balloon angioplasty or the treatment of BPH. Due to its structural rigidity the stent, when expanded, tends to hold the vas open.

A biodegradable thermoplastic hellow stent containing an anti-restenosis agent, surrounding the balloon of a microwave-aided angioplasty catheter can be a topic for future research. The catheter is placed in position in a vas, and thermal energy such as electrical, microwave laser, or RF energy is produced at or radiated from the distal end of the catheter to heat and thereby expand the stent against the walls of the vas. The heating is terminated following the expansion to allow the stent to harden or set in the expanded position. The balloon is collapsed and removed, leaving the expanded stent.

REFERENCES

[1] A. Rosen, H. Rosen (Eds.), *New Frontiers in Medical Device Technologies*, New York: Wiley, 1995.

[2] A. Rosen, H. D. Rosen, R. A. Hsi, D. Rosen, "Topics in RF/microwaves and optics in therapeutic medicine, Part I, Opportunities for microwaves and photonics in medical diagnostics and therapy," *MWP 2003 Proceedings*, International Topical Meeting on Microwave Photonics, Hungarian Academy of Sciences, Budapest, Hungary, Sept. 10–12, 2003.

[3] A. Rosen, H. D. Rosen, R. A. Hsi, D. Rosen, G. Happawana, G. Evans, A. Fathy, L. Stern, "Advances in RF/microwaves and light sources for applications in ther-

apeutic medicine, new technologies and applications," *Mikon 2004 Proceedings*, XV International Conference on Microwaves, Radar, and Wireless Communications, Poland, May 17–21, 2004.

[4] A. Rosen, W. Santamore, "Microwave system for tissue anastomoses," NIH grant R41-HL55758.

[5] E. Franke, "Minimum attenuation geometry for coaxial transmission line," *RF Design*, pp. 58–62, May 1989.

[6] H. J. Reich, P. F. Ordung, H. L. Krauss, J. G. Skalnik (Eds.), *The Van Nostrand Series in Electronica and Communications*, New York: Van Nostrand, 1953.

[7] S. Silver (Ed.), *Microwave Antenna Theory and Design*, New York: Dover, 1965.

[8] J. A. Kong, *Electromagnetic Wave Theory*, 2nd ed., New York: Wiley, 1990.

[9] T. Moreno, *Microwave Transmission Design Data*, New York: Dover, 1958.

[10] W. Busch, "Ueber den Eingluss, welchen heftigere Erysepeln zuweilen auf organisirte Neubeldungers ausuben," *Verhanal d Naturh Verd Pruess* (*Rheinl u Westphal, Bonn*), Vol. 23, pp. 28–30, 1866.

[11] W. B. Coley, "The treatment of malignant tumors by repeated inoculations of erysipelas; with a report of 10 original cases," *Am. J. Med. Sci.*, Vol. 105, pp. 487–511, 1893.

[12] P. Walinsky, A. Rosen, A. J. Greenspon, "Method and apparatus for high frequency catheter ablation," U.S. Patent 4,641,649, Feb. 10, 1987.

[13] J. J. Gallagher, R. H. Svenson, J. H. Kassell, et al., "Catheter technique for closed chest ablation of the atrioventricular conduction system," *New Engl. J. Med.*, Vol. 306, p. 194, 1982.

[14] M. M. Scheinman, F. Morady, D. S. Hess, et al., "Catheter induced ablation of the atrioventricular junction to control refractory supraventricular arrhythmias," *J. Am. Med. Assoc.*, Vol. 248, p. 851, 1982.

[15] G. Fontaine, J. L. Tonet, R. Frank, et al., "Clinical experience with fulguration and antiarrhythmic therapy for the treatment of ventricular tachycardia. Long term followup of 43 patients," *Chest*, Vol. 95, p. 785, 1989.

[16] D. E. Haines, "Current and future modalities of catheter ablation for the treatment of cardiac arrhythmias," *J. Invas. Cardiol.*, Vol. 4, p. 291, 1992.

[17] E. G. C. A. Boyd, P. M. Holt, "An investigation into the electrical ablation technique and a method of electrode assessment," *PACE*, Vol. 8, p. 815, 1985.

[18] G. H. Bardy, F. Coltorti, T. D. Ivey, et al., "Some factors affecting bubble formation with catheter-mediated defibrillator pulses," *Circulation*, Vol. 73, p. 525, 1986.

[19] G. H. Bardy, P. L. Sawyer, "Biophysical and anatomical considerations for safe and efficacious catheter ablation for arrhythmias," *Clin. Cardiol.*, Vol. 13, p. 425, 1990.

[20] D. E. Ward, M. Davies, "Transvenous high-energy shock for ablating atrioventricular conduction in man: Observations on the histological effect," *Br. Heart J.*, Vol. 51, p. 175, 1984.

[21] F. C. Kempf, R. A. Falcone, R. V. Iozzo, et al., "Anatomic and hemodynamic effects of catheter-delivered ablation energies in the ventricle," *Am. J. Cardiol.*, Vol. 56, p. 373, 1985.

[22] J. L. Jones, C. C. Proskauer, W. K. Paull, et al., "Ultrastructural injury to duck myocardial cells in vitro following 'electric' countershock," *Circ. Res.*, Vol. 46, p. 387, 1980.

[23] J. L. Jones, E. Lepeschkin, R. E. Jones, et al., "Response of cultured myocardial cells to countershock-type electrical field stimulation," *Am. J. Physiol.*, Vol. 235, p. H214, 1978.

[24] D. C. Westveer, T. Nelson, J. R. Stewart, et al., "Sequelae of left ventricular electrical endocardial ablation," *J. Am. Coll. Cardiol.*, Vol. 5, p. 956, 1985.

[25] B. B. Lerman, J. L. Weiss, B. H. Bulkley, et al., "Myocardial injury and induction of arrhythmia by direct current shocks delivered via endocardial catheters in dogs," *Circulation*, Vol. 69, p. 1006, 1984.

[26] G. H. Bardy, P. L. Sawyer, G. Johnson, et al., "The effects of voltage and charge of electrical ablation on canine myocardium," *Am. J. Physiol.*, Vol. 257, p. 1534, 1989.

[27] D. Cunningham, E. Rowland, A. F. Rickards, "A new low energy power source for catheter ablation," *PACE*, Vol. 9, p. 1384, 1986.

[28] E. Rowland, D. Cunningham, A. Ahsan, et al., "Transvenous ablation of atrioventricular conduction with a low energy power source." *Br. Heart J.* Vol. 62, p. 361, 1989.

[29] G. T. Evans, Jr., M. M. Scheinman, the Executive Committee of the Registry. "The Percutaneous Cardiac Mapping and Ablation Registry: Final summary of results," *PACE* Vol. 11, p. 1621, 1988.

[30] G. T. Evans, Jr., M. M. Scheinman, G. H. Bardy, et al., "Predictors of in-hospital mortality afer DC catheter ablation of atrioventricular conduction," *Circulation* Vol. 84, p. 1924, 1991.

[31] S. J. Kalbfleish, J. J. Langberg, "Catheter ablation with radiofrequency energy: Biophysical aspects and clinical applications," *J. Cardiovasc Electrophysiol* Vol. 3, p. 173, 1992.

[32] S. K. S. Huang, "Use of radiofrequency energy for catheter ablation of the endomyocardium: A prospective energy source," *J. Electrophysiol* Vol. 1, p. 78 1987.

[33] S. K. S. Huang, "Advances in applications of radiofrequency current to catheter ablation therapy," *PACE* Vol. 14, p. 28, 1991.

[34] W. Haverkamp, G. Hindricks, H. Gulker, et al., "Coagulation of ventricular myocardium using radiofrequency alternating current: Biophysical aspects and experimental findings," *PACE* Vol. 12, p. 187, 1989.

[35] D. E. Haines, D. D. Watson, "Tissue heating during radiofrequency catheter ablation: A thermodynamic model and observations in isolated perfused and superfused canine right ventricular free wall," *PACE* Vol. 12, p. 962, 1989.

[36] D. E. Haines, D. D. Watson, A. F. Verow, "Electrode radius predicts lesion radius during radiofrequency heating. Validation of a proposed thermodynamic model," *Circ. Res.* Vol. 67, p. 124, 1990.

[37] D. E. Haines, A. F. Verow, "Observations on electrode-tissue interface temperature and effect on electrical impedance during radiofrequency ablation of ventricular myocardium," *Circulation* Vol. 82, p. 1034, 1990.

[38] K. H. Hoyt, S. K. Huang, F. I. Marcus, et al., "Factors influencing trans-catheter radiofrequency ablation of the myocardium," *J. Appl. Cardiol.* Vol. 1, p. 469, 1986.

[39] W. M. Jackman, X. Wang, K. J. Friday, et al., "Catheter ablation of atrioventricular junction using radiofrequency current in 17 patients. Comparison of standard and large-tip electrode catheters," *Circulation,* Vol. 83, p. 1562, 1991.

[40] J. J. Langberg, M. Chin, D. J. Schamp, et al., "Ablation of the atrioventricular junction with radiofrequency energy using a new electrode catheter," *Am. J. Cardiol.* Vol. 67, p. 142, 1991.

[41] J. J. Langberg, M. Chin, M. Rosenqvist, et al., "Catheter ablation of the atrioventricular junction with radiofrequency energy," *Circulation,* Vol. 80, p. 1527, 1989.

[42] J. E. Olgin, M. M. Scheinman, "Comparison of high-energy direct current and radiofrequency catheter ablation of the atrioventricular junction," *J. Am. Coll. Cardiol.* Vol. 21, p. 557, 1993.

[43] M. E. Josephson, J. A. Kastor, "Supraventricular tachycardia: Mechanisms and management," *Ann. Intern. Med.* Vol. 87, p. 346, 1977.

[44] W. M. Jackman, X. Wang, K. J. Friday, et al., "Catheter ablation of accessory atrioventricular pathways (Wolff-Parkinson-White syndrome) by radiofrequency current," *New Engl J. Med.* Vol. 324, p. 1605, 1991.

[45] A. J. Greenspon, P. Walinsky, A. Rosen, "Catheter ablation for the treatment of cardiac arrhythmias," in A. Rosen and H. Rosen (Eds.), *New Frontiers in Medical Device Technology*, New York: Wiley, 1995, pp. 61–77.

[46] H. Calkins, J. Sousa, R. El-Atassi, et al., "Diagnosis and cure of the Wolff-Parkinson-White syndrome or paroxysmal supraventricular tachycardia during a single electropysiologic test," *New Engl. J. Med.* Vol. 324, p. 1612, 1991.

[47] M. Schluter, M. Geiger, J. Siebels, et al., "Catheter ablation using radiofrequency current to cure symptomatic patients with tachyarrhythmias related to an accessory atrioventricular pathway," *Circulation,* Vol. 84, p. 1644, 1991.

[48] J. F. Swartz, C. M. Tracy, R. Fletcher, "Radiofrequency endocardial catheter ablation of accessory atrioventricular pathway atrial insertion sites," *Circulation,* Vol. 87, p. 487, 1993.

[49] M. E. Josephson, "Supraventricular tachycardias," in *Clinical Cardiac Electrophysiology Techniques and Interpretations*, 2nd ed., Philadelphia: Lea & Febiger, 1993, p. 181.

[50] M. A. Lee, F. Morady, A. Kadish, et al., "Catheter modification of the atrioventricular junction using radiofrequency energy for control of atrioventricular nodal reentry tachycardia," *Circulation* Vol. 83, p. 827, 1991.

[51] W. M. Jackman, K. J. Beckman, J. H. McCelland, et al., "Treatment of supraventricular tachycardia due to atrioventricular nodal reentry, by radiofrequency catheter ablation of slow-pathway conduction," *New Engl. J. Med.* Vol. 327, p. 313, 1992.

[52] M. Haissaguerre, F. Gaita, B. Fischer, et al., "Elimination of atrioventricular nodal reentrant tachycardia using discrete slow potentiais to guide radiofrequency energy," *Circulaton,* Vol. 85, p. 2162, 1992.

[53] G. N. Kay, A. E. Epstein, S. M. Dailey, et al., "Selective radiofrequency ablation of the slow pathway for the treatment of atrioventricular nodal re-entrant tachycardia," *Circulation,* Vol. 85, p. 1675, 1992.

[54] M. R. Jazayeri, J. J. Hemple, J. S. Sra, et al., "Selective trans-catheter ablation of the fast and slow pathways using radiofrequency energy in patients with atrioventricular nodal reentrant tachycardia," *Circulation,* Vol. 85, p. 1318, 1992.

[55] M. Wathen, A. Natale, K. Wolfe, et al., "An anatomically guided approach to atri-oventricular node slow pathway ablation," *Am. J. Cardiol.* Vol. 70, p. 886, 1992.

[56] F. Morady, M. M. Scheinman, L. A. DiCarlo Jr., et al., "Catheter ablation of ventricular tachycardia with intra cardiac shocks: results in 33 patients," *Circulation* Vol. 75, p. 1037, 1997.

[57] L. S. Klein, H. T. Shih, K. Hackett, et al., "Radiofrequency catheter ablation of ventricular tachycardia in patients without structural heart disease," *Circulation* Vol. 85, p. 1666, 1992.

[58] P. Tchou, M. Jazayeri, S. Denker, et al., "Transcatheter electrical ablation of right bundle brand. A method of treating macroreentrant ventricular tachycardia attributed to bundle branch reentry," *Circulation,* Vol. 78, p. 246, 1988.

[59] J. Caceres, M. Jazayeri, J. McKinnie, et al., "Sustained bundle branch reentry as a mechanism of clinical tachycardia," *Circulation,* Vol. 79, p. 256, 1989.

[60] J. J. Langberg, J. Desai, N. Dullet, et al., "Treatment of macroreentrant ventricular tachycardia with radiofrequency ablation of the right bundle branch," *Am. J. Cardiol.* Vol. 63, p. 1010, 1989.

[61] M. E. Josephson, C. D. Gottleib, "Ventricular tachycardia associated with coronary artery disease," in D. P. Zipes and J. Jalife (Eds.), *Cardiac Electrophysiology: From Cell to Bedside*, Philadelphia: Saunders, 1990, p. 571.

[62] W. G. Stevenson, J. N. Weiss, J. Wiener, et al., "Slow conduction in the infarct scar. Relevance to the occurrence, detection, and ablation of ventricular reentry circuits resulting from myocardial infarction," *Am. Heart J.* Vol. 117, p. 452, 1989.

[63] F. Morady, M. Harvey, S. I. Kalbfleish, et al., "Radiofrequency catheter ablation of ventricular tachycardia in patients with coronary artery disease," *Circulation* Vol. 87, p. 363, 1993.

[64] T. Satoh, P. R. Stauffer, "Implantable helical coil microwave antenna for interstitial hyperthermia," *Internal. J. Hyperthermia,* Vol. 4, p. 497, 1988.

[65] P. Walinsky, A. Rosen, A. J. Greenspon, "Method and apparatus for high frequency catheter ablation," U.S. Pat. Vol. 4, p. 641, 649.

[66] R. M. Rosenbaum, A. J. Greenspon, S. Hsu, et al., "RF ablation for the treatment of ventricular tachycardia." *IEEE-MTT-S International Microwave Symposium Digest,* Vol. 2, p. 1155, 1993 .

[67] T. L. Wonnell, P. R. Stauffer, I. J. Langberg, "Evaluation of microwave and RF catheter ablation in a myocardial equivalent phantom model," *IEEE Trans. Biomed. Eng.* Vol. 39, p. 1086, 1992.

[68] J. G. Whayne, D. E. Haines, "Comparison of thermal profiles produced by new antenna designs for microwave catheter ablation (abstract)," *PACE* Vol. 15, p. 580, 1992.

[69] J. J. Langberg, T. L. Wonnell, M. Chin, et al., "Catheter ablation of the atrioventricular junction using a helical microwave antenna: A novel means of coupling energy to the endocardium," *PACE* Vol. 14, p. 2105, 1991.

[70] J. C. Lin, Y. J. Wang, R. J. Hariman, "Comparison of power deposition patterns produced by microwave and radio-frequency cardiac ablation catheters," *Electron. Lett.,* Vol. 30, pp. 922–923, 1994.

[71] J. C. Lin, "Biophysics of radio-frequency ablation," in S. K. S. Huan and D. J. Wilber (Eds.), *Radio-Frequency Catheter Ablation of Cardiac Arrhythmias: Basic Concepts and Clinical Applications,* 2nd ed., Armond, NY: Futura, 2000, pp. 13–24.

[72] S. Labonte, A. Blais, et al., "Monopole antennas for microwave catheter ablation," *IEEE Trans. Microwave Theory Tech.*, Vol. 44, No. 10, pp. 1832–1840, Oct. 1996.

[73] H. W. Gushing, "Electrosurgery as an aid to the removal of intracranial tumor with a preliminary note on a new surgical-current generator by W.T. Bovie," *Surg. Gyn. Obstet.*, Vol. 47, pp. 751–784, 1928.

[74] E. R. Cosman, B. J. Cosman, "Methods of making nervous system lesions," in *Medical Therapy of Movement Disorders Guide to Radio Frequency Lesion Generation in Neurosurgery,* Burlington, MA: Radionics Procedure Technique Series, 1974.

[75] W. W. Alberts, E. W. Wright, Jr., B. Feinstein, C. A. Gleason, "Sensory responses elicited by subcortical high frequency electrical stimulation in man," *J. Neurosurg.*, Vol. 36, pp. 80–82, 1972.

[76] J. F. Lehman (Ed.), *Therapeutic Heat and Cold*, Baltimore, MD: Williams & Wilkins, 1990.

[77] S. K. S. Huang (Ed.), *Radio-Frequency Catheter Ablation of Cardiac Arrhythmias*, Armond, NY: Futura, 1995.

[78] A. Rosen, H. Rosen, S. Edwards, "New frontiers for RF/microwaves in therapeutic medicine," in M. Golio (Editor-in-Chief), *RF and Microwave Handbook*, Boca Raton, FL: CRC Press, 2001, pp. 2.225–2.249.

[79] "RF ablation of bone mets brings relief to cancer pattients," http://www.diagnosticimaging.com/dinews/2002052102shtm. *Diagnostic Imaging Online*, May 21, 2002.

[80] A. Rabinovich, J. Fang, S. J. Scrivani, "Diagnosis and management of trigeminal neuralgia," *Columbia Dental Rev.*, Vol. 5, pp. 4–7, 2000.

[81] R. A. Weiss, M. Weiss, "Controlled radiofrequency endovenous occlusion using a unique radiofrequency catheter under duplex guidance to eliminate saphenous varicose vein reflux: A 2-year follow-up," *Dermatol. Surg.*, Vol. 28, pp. 38–42, 2000.

[82] A. Rosen, P. Walinsky, "Percutaneous transluminal microwave catheter angioplasty," U.S. Patent 4,643,186, Feb. 17, 1987.

[83] A. Rosen, et al., "Percutaneous transluminal microwave antioplasty catheter," *IEEE MTT-S Digest*, Vol. 2, p. 167, 1989.

[84] A. Rosen, et al., "Studies of microwave thermal balloon angioplasty in rabbits," *IEEE MTT-S Digest*, Vol. 2, p. 537, 1990.

[85] A. Rosen, et al., "Microwave thermal angioplasty in the normal and atherosclerotic rabbit model," *IEEE Microwave Guided Wave Lett.*, Vol. 1, No. 4, p. 73, Apr. 1991.

[86] P. Walinsky, A. Rosen, A. Martinez-Hernandez, et al., "Microwave balloon angioplasty," *J. Invasive Cardiol.*, Vol. 3, No. 3, p. 152, May–June 1991.

[87] A. Rosen, P. Walinsky, P. Herczfeld, "Microwaves in medical applications: Microwave balloon angioplasty," *MW '92 International Conference Proceedings,* Brighton, England, October 13–15.

[88] P. Walinsky, A. Rosen, A. Martinez-Hernandez, et al., "Microwave balloon angioplasty," in J. H. K. Vogel and S. B. King III (Eds.), *The Practice of Interventional Cardiology*, 2nd ed., Vol. 27, St. Louis, MO: Mosby Year Book, 1993, pp. 281–285.

[89] C. Landau, J. W. Currier, C. C. Haudenschild, et al., "Microwave balloon angioplasty effectively seals arterial dissections in an atherosclerotic rabbit mode," *J. Am. Coll. Cardiol.*, Vol. 23, p. 1700, 1994.

[90] S. S. Hsu, L. Hoh, R. M. Rosenbaum, A. Rosen, P. Walinsky, A. J. Greenspon, "A method for the in vitro testing of cardiac ablation catheters," *IEEE Trans. Microwave Theory Tech.*, Vol. 44, No. 10, pp. 1841–1847, Oct. 1996.

[91] A. Khebir, Z. Kaouk, P. Savard, "Modeling a microwave catheter antenna for cardiac ablation," *IEEE MTT-S Dig.*, Vol. 1, Orlando, May 16–20, 1995.

[92] C. K. Chou, G. W. Chen, A. W. Guy, et al., "Formulas for preparing phantom muscle tissue at various radiofrequencies," *Bioelectro. '84*, Vol. 5, 1984, pp. 436–441.

[93] R. D. Nevels, G. D. Arndt, G. W. Raffoul, J. R. Carl, A. Pacifico, "Microwave catheter design," *IEEE Trans. Biomed. Eng.*, Vol. 45, No. 7, July 1998, pp. 885–890.

[94] U. F. W. Franke, P. Rahmanian, J. Burkhardt, U. Stock, T. Wittwer, T. Wahlers, "Cardiac surgery and concomitant left atrial microwave ablation—A suitable strategy for chronic atrial fibrillation," *Thora Cardiovasc Surg 2002 Thema*, moderated poster session—new techniques/heart surgery, Feb. 20, 2002.

[95] Press release, ESC Virtual Press Office: "Microwave ablation is effective for the treatment of chronic atrial fibrillation," XXII Annual Congress of the ESC, Amsterdam, The Netherlands, Aug. 26–30, 2000.

[96] G. Triadafilopoulos, J. K. DiBaise, et al., "The Stretta procedure for the treatment of GERD: 6 and 12 month follow-up of the U.S. open label trial," *Gastrointest. Endosc.*, Vol. 55, No. 2, pp. 149–156, Feb. 2002.

[97] J. K. DiBaise, R. E. Brand, E. M. M. Quigley, "Endoluminal delivery of radiofrequency energy to the gastroesophageal junction in uncomplicated GERD: Efficacy and potential mechanism of action," *Am. J. Gastroenterol.*, Vol. 97, No. 4, pp. 833–842, 2002.

[98] G. Triadafilopoulos, J. K. DiBaise, et al., "Radiofrequency energy delivery to the gastroesophageal junction for the treatment of GERD," *Gastrointest. Endosc.*, Vol. 53, No. 4, pp. 407–415, Apr. 2001.

[99] W. O. Richards, S. Scholz, et al., "Initial experience with the stretta procedure for the treatment of gastroesophageal reflux disease," *Laparoendosc. Adv. Surg. Tech.*, Vol. 11, No. 5, pp. 267–273, Oct. 2001.

[100] G. Triadafilopoulos, D. S. Utley, "Temperature-controlled radiofrequency energy delivery for gastroesophageal reflux disease: The Stretta procedure," *J. Laparoendosc. Adv. Surg. Tech. A*, 2001. Vol. 11, No. 6, pp. 333–339, Dec.

[101] G. Triadafilopoulos, "Endoscopic therapies for gastroesophageal reflux disease," *Curr. Gastroenterol. Repts.*, Vol. 4, pp. 200–204, 2002.

[102] G. Triadafilopoulos, "Radiofrequency treatment for gastroesophageal reflux disease (the Stretta procedure)," *Up to Date On-line 10.1*, 2002, www.uptodate.com.

[103] D. S. Utley, M. Kim, et al., "Augmentation of lower esophageal sphincter pressure and gastric yield pressure after radiofrequency energy delivery to the gastroesophageal junction: A porcine model," *Gastrointest. Endosc.*, Vol. 52, No. 1, pp. 81–86, July 2000.

[104] I. B. Feldberg, N. J. Cronin, "A 9.2 GHz microwave applicator for the treatment of menorrhagia," *IEEE MTT-S Dig.*, Vol. 2, pp. 755–758, 1998.

[105] D. A. Hodgson, I. B. Feldberg, N. Sharp, et al., "Microwave endometrial ablation: Development, clinical trials and outcomes at three years," *Br. J. Obstet. Gynaecol.*, Vol. 106, pp. 684–694, 1999.

[106] N. C. Sharp, N. Cronin, I. Feldberg, et al., "Microwaves for menorrhagia: A new fast technique for endometrial ablation," *Lancet*, Vol. 346, pp. 1003–1004, 1995.

[107] M. P. Milligan, G. A. Etokowo, "Microwave endometrial ablation for menorrhagia," *J. Obstet. Gynaecol.*, Vol. 19, pp. 496–499, 1999.

[108] T. Fulop, I. Rakoczi, I. Barna, "NovaSure™ impedance controlled endometrial ablation system, long-term follow-up results," *2nd World Congress on Controversies in Obstetrics, Gynecology and Invertility*, Modizzi Editore S.p.A—MEDIMOND, 2001.

[109] J. Schepps, A. Rosen, "Microwave industry outlook—Wireless communications in healthcare," *IEEE Trans Microwave Theory Tech.*, Vol. 50, No. 3, pp. 1044–1045, Mar. 2002.

[110] R. Bansal (Ed.), *Handbook of Engineering Electromagnetics*, New York: Marcel Dekker, 2004.

[111] M. R. Tofighi, A. S. Daryoush, "Measurement techniques," in R. Bansal (Ed.), *Handbook of Engineering Elecromagnetics*, New York: Marcel Dekker, 2004.

[112] F. Sterzer, R. Paglione, F. Wozniak, et al., "A self-balancing microwave radiometer for non-invasively measuring the temperature of subcutaneous tissues during localized hyperthermia treatments of cancer," *IEEE MTT-S Int. Microwave Symp. Dig.*, Vol. 1, pp. 438–440, June 1982.

[113] R. W. Paglione, "Portable diagnostic radiometer," *RCA Rev.*, Vol. 47, Dec. 1986.

[114] F. Sterzer, "Microwave radiometers for non-invasive measurements of subsurface tissue temperatures," *J. Automedica*, Vol. 8, pp. 203-211, 1987.

[115] D. D. Mawhinney, T. Kresky, "Non-invasive physiological monitoring with microwaves," *Proc. National Aerospace and Electronics Conference*, Dayton, OH: NAECON, 1986.

[116] J. Lin, et al., "Microwave apex-cardiography," *IEEE Trans. MTT*, Vol. 27, No. 6, June 1979.

[117] S. S. Stuchly et al., "Monitoring of arterial wall movement by microwave doppler radar," *Proc. 1978 Symposium on Electromagnetic Fields in Biological Systems*, Ottawa, Canada, June 1978.

[118] I. Yamaura, "Measurements of 1.8–2.7 GHz microwave attenuation in the human torso, "*IEEE Trans. MTT*, Vol. 25, No. 8, pp. 707–710, Aug. 1977.

[119] C. C. Johnson, A. W. Guy, "Nonionizing electromagnetic wave effects in biological materials and systems," *Proc. IEEE*, Vol. 60, No. 6, pp. 692–718, June 1972.

[120] I. Yamaura, "Mapping of microwave power transmitted through the human thorax," *Proc. IEEE*, Vol. 67, No. 8, pp. 1170–1171, Aug. 1979.

[121] A. Rosen, W. P. Santamore, "Method of measuring blood perfusion," U.S. Patent 4,228,805, Oct. 21, 1980.

[122] F. A. Gibbs, *Proc. Soc. Exper. Biol. Med.*, Vol. 31, pp. 141–147, 1933.

[123] C. F. Schmidt et al., "The intrinsic regulation of the circulation in the hypothalamus of the cat," *Am J. Physiol.*, Vol. 108, pp. 241–263, 1934.

[124] J. Grayson, "Thermal conductivity of normal and infarcted heart muscle," *Nature*, Vol. 215, pp. 767–768, 1967.

[125] J. Grayson et al., "Internal calorimetry: Assessment of myocardial blood flour and heart production," *J. Appl. Phys.*, Vol. 30, No. 2, pp. 251–257, Feb. 1971.

[126] A. Rosen, P. Walinsky, "Microwave aided balloon angioplasty with lumen measurement," U.S. Patent 5,129,396, July 14, 1992.

[127] A. Rosen, P. Walinsky, "Microwave aided balloon angioplasty with guide filament," U.S. Patent 5,150,717, Sept. 29, 1992.

[128] A. Fastykovsky, "Effect of microwave therapy on the functional state of cardiovascular system in patients with a hypertensive disease," International Scientific Meeting on Microwaves in Medicine, Belgrade, Yugoslavia, April 8–11, 1991, *Digest of Papers*, pp. 100–105.

[129] Z. Djordjevic, M. Djordjevic, "Pacemaker protective undershirt," International Scientific Meeting on Microwaves in Medicine, Belgrade, Yugoslavia, April 8–11, 1991, *Digest of Papers*, pp. 230–232.

[130] C. C. Tamburello, L. Zanforlin, G. Tine, A. E. Tamburello, "Analysis of microwave effects on isolated hearts," *IEEE MTT Symp. Dig.*, pp. 805–808.

[131] V. V. Vorobyov, R. Khramov, "Hypothalamic effects of millimeter wave irradiation depend on location of exposed acupuncture zone of unanesthetized rabbits—statistical data included," *Am. J. Chinese Med.*, Vol. 30, No. 1, pp. 29–35, 2002.

[132] C. Tamburello, L. Zanforlin, G. Tine, A. E. Tamburello, "Analysis and modeling of microwave effects on isolated chick embrio hearts," in G. Franceschetti and R. Pierri (Eds.), *Italian Recent Advances in Applied Electromagnetics*, Liguori, Napoli, 1991, pp. 497–513.

[133] C. Tamburello, G. Tine, L. Zanforlin, "Interactions between pulse modulated microwaves and isolated chick embryo hearts," *Trans 1st EBEA Congress*, Bruxelles, Jan. 23–25, 1992, p. 352.

[134] J. G. Wu, W. Z. Huang, B. Y. Wu, "Effect of acupoint irradiation with Q-wave millimeter microwave on peripheral white blood cells in post-operational treatment with chemotherapy in stomach and colorectal cancer patients, "*Zhongguo Zhong Xi Yi Jie He Za Zhi*, Vol. 17, No. 5, pp. 286–288, May 1997.

[135] A. Rosen, W. Santamore, private communication.

[136] A. Rosen, H. Rosen, A. Fathy, private communication.

[137] B. W. Henderson, T. M. Busch, et al., "Photofrin photodynamic therapy can significantly deplete or preserve oxygenation in human basal cell carcinomas during treatment, depending on fluence rate," *Cancer Res.* Vol. 60, pp. 525–529, Feb. 1, 2000.

[138] T. J. Dougherty, C. J. Gomer, et al., "Photodynamic therapy," *J. Natl. Cancer Inst.*, Vol. 90, pp. 889–905, 1998.

[139] T. H. Foster, R. S. Murant, et al., "Oxygen consumption and diffusion effects in photodynamic therapy," *Radiat. Res.*, Vol. 126, pp. 296–303, 1991.

[140] T. Sitnik, B. W. Henderson, "The effect of fluence rate on tumor and normal tissue responses to photodynamic therapy *in vivo:* Effects of fluence rate," *Br. J. Cancer*, Vol. 77, pp. 1386–1394, 1998.

[141] F. Sterzer, "Localized heating of deep-seated tissues using microwave balloon catheters," in A. Rosen and H. Rosen (Eds.), *New Frontiers in Medical Technology*, New York: Wiley, 1995.

PROBLEMS

6.1. Utilizing a whip antenna at the end of a coaxial like, compare the microwave penetration at 2.45 GHz in the near field to that of the microwave radiation in the far field in

(a) muscle tissue

(b) fat.

6.2. At the RF frequency of 500 kHz (RF ablation), measure the muscle tissue ablation volume as a function of

(a) the electrode size

(b) the power level

(c) time.

6.3. At the RF frequency of 500 kHz (RF ablation), determine the best geometry of the electrode for maximum volume ablation size at minimum power level and minimum time.

6.4. Design your RF ablation experiments so as to mimic the real therapeutic modality taking blood flow into consideration.

6.5. Design and construct various microwave antennas at 915 MHz, 2.45 GHz, 10 GHz, and 60 GHz for possible use in

(a) the treatment of benign prostatic hyperplasia

(b) endometrial ablation

(c) microwave-assisted liposuction

(d) the treatment of any other condition which utilizes heat.

6.6. Determine the ablation mechanism in tissue due to microwaves and RF (500 kHz).

Index

Page references followed by f indicate figures, references followed by t indicate material in tables.

RF/Microwave Interaction with Biological Tissues, By André Vander Vorst, Arye Rosen, and
Youji Kotsuka
Copyright © 2006 by John Wiley & Sons, Inc.

Application-type wave absorber, 204
Applicators, 153–154
  for body cavity, 162f
  inductive heating, 166–174
  matching to a biological surface, 52–53
  microwave, 164
  microwave dielectric heating, 163–164
Applicator systems
  for breast hyperthermia, 170–174
  capacitive coupling, 157, 159
Argand diagram, 48
Arteries, lumen measurement of, 290–294
Atherosclerotic plaque, treatments for,
  303–304
Atoms, polarization of, 155f
Atrial fibrillation, 272f, 277
  chronic, 285
Atrioventricular nodule reentrant tachycardia
  (AVNRT), 271
Attenuation constant, 252
Autonomic nervous system, 64, 101
  functions of, 102
Auxiliary electrode, 169, 170
Axon, 65

Balloon angioplasty, coaxial cable in, 251–252,
  256–258
Balloon angioplasty catheter, microwave-
  aided, 290–294
Balloon catheter angioplasty, percutaneous
  transluminal, 304
Balloon catheters
  positioning of, 278f
  therapeutic temperatures using, 303
Balloon microwave catheter, in cancer
  treatment, 303
Barrett's esophagus (BE), 286
Basic exposure limitations, 53
Bead type thermistor sensor, 195
Beef blood, complex permittivity of, 79, 81f
Bei function, 165, 180
Bench-top tests, of biological solder, 297
Benign prostatic hyperplasia (BPH), 5, 267
Benzodiazepine receptors, 118–119
Ber function, 165, 180
Bessel differential equation, 179
Bessel functions, 179–180
  modified, 165
Beta dispersion, 75, 76, 77
Biodegradable thermoplastic hellow stent,
  304
Bioelectric effects/phenomena/processes,
  69
  natural, 64

Bioelectricity
  fundamentals of, 63–64
  importance of, 69
Bioheat equation, 98, 99
Biological cells, 65
  genotoxic effects on, 132
  period and hyperthermia sensitivity of, 185
Biological effects
  evaluating, 53
  frequency-dependent, 126
  of microwaves, 93–94
  of nonionizing radiation, 8
  of pulse-modulated radiation, 121
Biological liquids, measured data for, 77–80.
  See also Biological water
Biological materials
  as conductors, 40
  inhomogeneous, 84
Biological media, permittivity of, 75
Biological membranes, 127
Biological response, differences in, 93–94
Biological solder, 295, 296
  bench-top tests of, 297
  doping of, 297–299
  temperature measurement in, 299
Biological systems
  complex field distributions related to, 94
  energy availability in, 126–127
  RF/microwave effect on, 32
Biological tissues. *See also* Tissue(s)
  effect of RF/microwave radiation on, 63
  electric parameters of, 53
  interaction of EM fields with, 9
  penetration in, 39–44
  permittivity of, 49, 51
Biological water, 73–74
Biomembrane, inductive heating and, 167–168
Biosystems, properties of, 126. *See also*
  Biological systems
Blackbody, defined, 84
Blackbody radiation, 33–39, 38, 83–84
Blackbody spectral intensity (brightness), 35,
  36, 38. *See also* Total brightness
Blackbody temperature, 34
Blood–brain barrier (BBB), 102
  cerebral vascular system and, 105
  effects of microwave exposure on, 104–107,
  117
  effects of microwave fields on, 106
  opening of, 106–107
  permeability at high SAR, 106
  permeability of, 105–106
Blood flow rate, measuring within tissues,
  289–290